AF181082

H.-J. Gabius S. Gabius (Eds.)

Lectins and Cancer

With 79 Figures

Springer-Verlag
Berlin Heidelberg New York London
Paris Tokyo Hong Kong Barcelona
Budapest

Priv.-Doz. Dr. HANS JOACHIM GABIUS
Max-Planck-Institut
für experimentelle Medizin
Hermann-Rein-Straße 3
3400 Göttingen, FRG

Dr. SIGRUN GABIUS
Medizinische Universitätsklinik
Abteilung Hämatologie/Onkologie
Robert-Koch-Straße 40
3400 Göttingen, FRG

Cartoons by
PAUL GÜNTHER

ISBN-13:978-3-642-76741-8 e-ISBN-13:978-3-642-76739-5
DOI: 10.1007/978-3-642-76739-5

This work is subject to copyright. All rights are reserved, whether the whole or part of the material is concerned, specifically the rights of translation, reprinting, reuse of illustrations, recitation, broadcasting, reproduction on microfilms or in other ways, and storage in data banks. Duplication of this publication or parts thereof is only permitted under the provisions of the German Copyright Law of September 9, 1965, in its current version, and a copyright fee must always be paid. Violations fall under the prosecution act of the German Copyright Law.

© Springer-Verlag Berlin Heidelberg 1991
Softcover reprint of the hardcover 1st edition 1991

The use of registered names, trademarks, etc. in this publication does not imply, even in the absence of a specific statement, that such names are exempt from the relevant protective laws and regulations and therefore free for general use.

Typesetting: Camera ready by author

31/3145-543210 – Printed on acid-free paper

Preface

At first sight, the collection of the individual chapters of this book from apparently rather disparate disciplines may look like an assemblage of unconnected islands of knowledge in the field of lectinology. However, it was our deliberate intention to emphasize the inevitable necessity for interdisciplinary approaches to bridge gaps in order to turn the intriguing concept of the eminent physiological relevance of recognitive protein-carbohydrate interactions into clinically notable benefit. Since the growing attention to glycobiology and lectins is necessarily associated with an increasing difficulty to unexceptionally keep abreast of progress in areas that are not within the immediate individual focus of interest, the challenge to carefully think over techniques, paradigms and perspectives of particular sections of this broad discipline may well account for opening unexpected avenues of research and application. It is obvious that tools to glycohistochemically monitor the expression of determinants in this field need to be provided by chemical synthesis or biochemical purification, connecting chemistry, biochemistry and histochemistry. In addition to their value for in-depth histopathological or cell biological studies, aimed at uncovering correlations between the determined tumor parameters, influenced by differentiation or the microenvironment, and the biological behavior, the advantageous application of these biochemically defined substances for controlled immunomodulation or targeted delivery of pharmaceuticals with therapeutical impact has already revealed encouraging precedents of turning this concept into medical practice. Similarly, knowledge on the importance of glycobiological interactions in other defined systems with presently no evident relation to tumor biology may shape experimental approaches in this field. It is our firm conviction that patient and persistent efforts in various areas are required to achieve the desired solutions to the jig-saw puzzle-like problems of basic and applied tumor lectinology. Thus, the program of the INTERNATIONAL SYMPOSIUM ON LECTINS AND CANCER was purposefully designed to critically assess the status in certain aspects with supposed relevance, to convincingly point to meaningful perspectives and to creatively stimulate the imagination of future developments, unmistakably a crucial factor for the intently expected advances.

S. Gabius H.-J. Gabius

Contents

Molecular basis of lectin-carbohydrate interactions

N. Sharon

Department of Biophysics
The Weizmann Institute of Science
Rehovot 76100
Israel

Studies carried out during the last two decades, many of them with the aid of lectins, have firmly established that all cells come in a sugar coating, and that this coating usually changes in the course of normal processes of growth and development, as well as in pathological situations. Concurrently, it has been found that many cells express surface lectins, which may function as recognition molecules by combining with sugars on the surfaces of apposing cells (Sharon and Lis, 1989a, 1989b) (Table I). A recent example, which has created a great deal of excitement, is the identification of the cell adhesion molecules, named selectins or LEC-CAMs, that mediate adhesion of leukocytes to carbohydrates on endothelium and platelets during inflammation and blood clotting (Brandley *et al.*, 1990; Laskey, 1991; McEver, 1991; Rosen, 1990; Springer and Laskey, 1991).

The early findings that tumor cells frequently express, on their surfaces, carbohydrates and lectins different from those present on their normal counterparts, had raised hopes that these determinants could be used for diagnostic and therapeutic purposes (Gabius, 1988; Hakomori, 1989). In spite of much effort, this has not yet materialized. On the other hand, differences in lectin binding properties of distinct subpopulations of hematopoietic cells (Reisner *et al.*, 1978; Reisner and Sharon, 1980; Sharon, 1983) have provided a basis for the development of a technique for bone marrow purging, routinely employed in bone marrow transplantation of children born with severe combined immunodeficiency (also known as bubble children), and experimentally used for treatment of leukemic patients (Aversa *et al.*, 1990; Fischer *et al.*, 1990; O'Reilly *et al.*, 1990).

Table I
LECTIN-CARBOHYDRATE INTERACTIONS
IN BIOLOGICAL PROCESSES

Process	*Lectin in*
Blood coagulation	Platelets
Defense	Macrophages
Homing	Lymphocytes
Infection	Myxoviruses, bacteria, amoebae
Inflammation	Leukocytes, endothelial cells
Metastasis	Malignant cells
Symbiosis	Leguminous plants

Binding of carbohydrates to lectins can be considered as an example of the more general phenomenon of the specific reactions which occur between proteins and other types of small or large molecules (Sharon and Lis, 1990a). Thus, the transport functions of proteins, such as the carriage of hormones by specific plasma proteins, the binding of drugs to proteins, interaction between antigens and antibody or the formation of an enzyme-substrate complex, raise problems of a similar nature as those posed by the activity of lectins (Neuberger, 1984). There is little doubt, therefore, that a better understanding of lectin-carbohydrate interactions will shed light on the molecular basis of protein-ligand interactions in general. Although there is extensive literature on the interaction of lectins with carbohydrates (Sharon and Lis, 1989a,b; Reeke and Becker, 1988), relatively little is known about the molecular basis of these interactions.

To gain detailed information on the structure-function relationship of lectins, we have focused our attention on three such proteins, namely soybean agglutinin (SBA), peanut agglutinin (PNA), and *Erythrina corallodendron* lectin (ECorL), long under investigation in our laboratory. Their properties are summarized in Table II. They are members of the large family of legume lectins, all of which possess similar physicochemical properties and show extensive structural homologies, but have distinct carbohydrate specificities (Sharon and Lis, 1990b). Like other legume lectins, SBA, PNA and ECorL are readily obtainable from the seeds of the corresponding plants; they consist of 2 or 4 identical, or nearly identical, subunits, each of M.W. close to 30 kDa, possessing a single carbohydrate-binding site; they require tightly bound Ca^{2+} and Mn^{2+} (or another transition metal) for activity; they have a similar amino acid composition, in particular they are low in or completely devoid of sulfur-containing acids. Their structural similarity makes them an excellent system for comparative structure-function studies.

Table II
MOLECULAR PROPERTIES OF LEGUME LECTINS SPECIFIC FOR GALACTOSE AND/OR *N*-ACETYLGALACTOSAMINE

	SBA	*PNA*	*ECorL*
Molecular weight	120,000	110,000	56,000
Subunits	four	four	two
Carbohydrates	6%	0	4.5%
Metal ions	Mn^{2+}, Ca^{2+}	Mn^{2+}	Mn^{2+}, Ca^{2+}
Specificity			
Monosaccharide	GalNAc	Gal	Gal/GalNAc
Disaccharide	none	Galβ3GalNAc	Galβ4GlcNAc
Relative affinity			
GalNAc:Gal	40	<0.01	1.5-2.0
Disaccharide:Gal	1	50	40

As noted in Table II, SBA and ECorL are glycoproteins, whereas PNA is devoid of carbohydrate. Earlier work has led to the elucidation of the structure of the carbohydrate units of SBA (Dorland *et al.*, 1981) and ECorL (Ashford *et al.*, 1987) (Fig. 1). Both contain the pentasaccharide core, $Man_3(GlcNAc)_2$, present in all *N*-linked oligosaccharides, whether from animals, plants or microorganisms. In SBA, six additional mannose residues are attached to the core, forming a structure identical to that found, for example, in animal

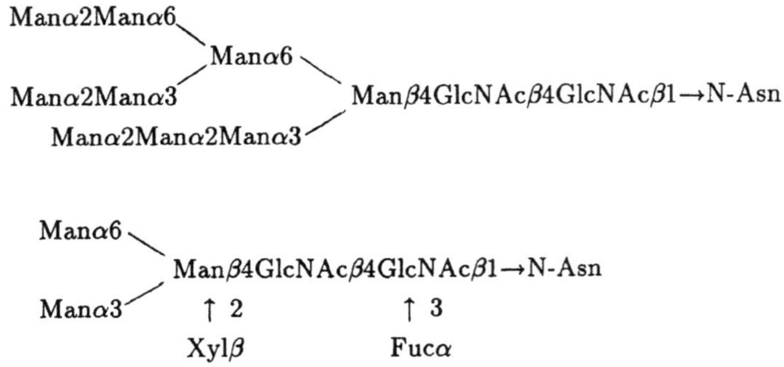

Figure 1. Carbohydrate units of SBA (top) and ECorL (bottom).

glycoproteins. In contrast, the carbohydrate unit of ECorL contains a xylose residue linked β1→2 to the β-linked mannose of the core, and an L-fucose linked α1→3 to the asparagine-linked *N*-acetylglucosamine. The heptasaccharide thus formed appears to be unique for plants, showing that these organisms have developed unusual pathways of processing *N*-linked oligosaccharides.

Although they are classified as Gal/GalNAc specific, SBA, ECorL and PNA differ markedly in their fine specificity and affinity for related mono- and oligosaccharides (Table II). Thus, SBA binds *N*-substituted (e.g. *N*-acetyl or *N*-dansyl) galactosamine much stronger than galactose, but has no preference for oligosaccharides containing these monosaccharides at their non-reducing end. Therefore, the combining site of SBA is most likely small and open (it can accommodate *N*-acetyl or *N*-dansyl groups at position 2 of the galactose ring). ECorL also binds galactose, *N*-acetylgalactosamine (the latter only slightly stronger than the former) and *N*-dansylgalactosamine (much stronger than galactose), suggesting that it too has an open binding site. However, it binds certain disaccharides, especially *N*-acetyllactosamine (Lis *et al.*, 1985) and other *N*-substituted derivatives of lactosamine (Kinzy *et al.*, 1988), considerably better than galactose, leading to the conclusion that the combining site of this lectin is an extended one. In contrast to SBA and ECorL, PNA does not bind *N*-acetylgalactosamine (at least not in its primary site) and combines with the disaccharide Galβ1→3GalNAc much stronger than with galactose.

Our current aims are:

1. To determine the three-dimensional structures of SBA, PNA and ECorL and of their complexes with sugars.
2. To understand the roles of various amino acids in the binding of sugars to the lectins.
3. To understand how lectins discriminate between different sugars (e.g., of galactose versus mannose).
4. To use the knowledge gained in order to modify the specificity of the lectins.

Since site-directed mutagenesis offers an excellent approach to studying the structure and function of proteins, down to the effect of a single amino acid, we started by cloning the cDNA's of the three lectins. This was achieved by conventional techniques of molecular genetics, first for ECorL by Rafael Arango and more recently also for SBA by Rivka Adar and for PNA by Esperanza Rodriguez, in our laboratory. Since the work on ECorL is more advanced, I shall confine my discussion to this lectin.

Examination of the hemagglutinating activity of extracts from seeds of *E. corallodendron* at various maturation stages revealed that the level of lectin increases markedly past mid-maturation (Arango *et al.*, 1990). Seeds at this stage of maturation, therefore, were used as a source of mRNA for the construction of an expression cDNA library in the vector λ Zap, which generates fusion proteins with an N-terminal portion of β-galactosidase. The library was screened with rabbit polyclonal anti-ECorL antiserum. Four immunopositive clones were isolated. Western blot analysis of cell extracts from one of the clones (p1EcL-B) showed a 36 kDa protein that reacted with the antiserum, as well as with a mouse monoclonal antibody raised against the lectin. DNA sequence analysis by the chain termination method revealed that clone pIEcL-C has an insert of 1017 bp with the entire coding sequence of ECorL, beginning with an initiation codon ATG at position 25 and ending with stop codon TAA at position 868. This fragment encodes a polypeptide of 281 amino acids consisting of a signal leader sequence of 26 amino acids and a mature protein of 255 amino acids.

The deduced amino acid sequence from this fragment is identical to the sequence of the first 244 amino acids of ECorL, as determined at the protein level (Adar *et al.*, 1989), except at 7 positions (Fig. 2). The presence, in the deduced sequence, of 11 additional amino acids not found in the chemically determined

cDNA VETISFSFSEFEPGNDNLTLQG**AA**L**ITQ**SGVLQLTKINQNGMPAWDST
Prot VETISFSFSEFEPGNDNLTLQG**DS**L**PET**SGVLQLTKINQNGMPAWDST

cDNA GRTLYAKPVHETRFSFSIEQPYTRPLPADGLVFFMGPTKSKPAQGYGY
Prot GRTLYAKPVHETRFSFSIEQPYTRPLPADGLVFFMGPTKSKPAQGYGY

cDNA LGIF**N**NSKQDNSYQTLGVEFDTFSN**Q**WDPPQVPHIGIDVNSIRSIKTQ
Prot LGIF**F**NSKQDNSYQTLGVEFDTFSN**P**WDPPQVPHIGIDVNSIRSIKTQ

cDNA PFQLDNGQVANVVIKYDASSKILHAVLVYPSSGAIYTIAVDVGLSGAT
Prot PFQLDNGQVANVVIKYDASSKILHAVLVYPSSGAIYTIAVDVGLSGAT

cDNA GAQRDAAETHDVYSWSFQASLPETNDAVIPTSNHNTFAI
Prot GAQRDAAETHDVYSWSFQASLPETNDAV

Figure 2. Comparison of the deduced amino acid sequence of ECorL cDNA (cDNA) to the sequence of the lectin as determined at the protein level (Prot). The amino acids that are different in the two sequences are in bold print. The leader sequence deduced from the cDNA is MATYKLCSVLALSLTLFLLILNKVNS.

one, may be due to incomplete chemical sequencing or, alternatively, due to a posttranslational modification at the C-terminus of the protein, as has been proposed for other legume lectins (Higgins *et al.*, 1983; Quinn and Etzler, 1989).

Comparison of the primary sequence of ECorL with those of other legume lectins reveals >10% invariant amino acids (Sharon and Lis, 1990b). These include essentially all the residues that ligate the tightly bound Ca^{2+} and Mg^{2+} required for binding of carbohydrates (Fig. 3). It is striking that the aspartic acid and asparagine, which form key polar contacts with the ligand in the high resolution X-ray structures of the sugar complexes of the mannose/glucose specific lectins that have been solved to date (Bourne *et al.*, 1990a,b; Derewenda *et al.*, 1989; Reeke and Becker, 1988), are also conserved in the other legume lectins, irrespective of their specificity (Fig. 4).

The complex of ECorL with lactose has been crystallized from methyl pentanediol; the crystals diffracted to 2.0 Å, and the structure of the ECorL-lactose complex was solved by molecular replacement, using the known structures of concanavalin A and pea lectin as a model (Shaanan *et al.*, 1991). A unique packing of molecules in the crystal allows the N-linked heptasaccharide of the lectin to settle into a single conformation stabilized by a number of interactions with the protein. Consequently, the whole carbohydrate chain can be seen with a clarity hitherto not observed in crystallographic studies of

```
Vicieae
        99                           120                        140
FBL     GGY-LGVF-NGKDYDKT-AQTVAVEFDTFYNAAW-DP-SNGKRHIGIDVN-TIKSI
LCL     GGY-LGVFYNGKEYDKT-SQTVAVEFDTFYNAAW-DP-SNKERHIGIDVN-SIKSV
LOL     GGY-LGVF-NSKDYDKT-SQTVAVEFDTFYNTAW-DP-SNGDRHIGIDVN-SIKSI
PSL     GGY-LGVF-NSAEYDKT-TQTVAVEFDTFYNAAW-DP-SNRDRHIGIDVN-SIKSV

Phaseoleae

DBL     GGY-LGVF-DSDVYNNS-AQTVAVEFDTLSNSGW-DP-SM--KHIGIDVN-SIKSI
ECorL   YGY-LGIFFNSKQDNSY--QTLGVEFDTFSNP-W-DP-PQ-VPHIGIDVN-SIRSI
LBL     GRL-LGLF-NKSENDINALT-VAVEFDTCHNLDW-D----KNS---IAVNLGIGSV
PHA-L   GGF-LGLFDGS-NSNFHT---VAVEFDTLYNKDW-DP---TERHIGIDVN-SIRSI
SBA     AGY-LGLF-N--ENESG-DQVVAVEFDTFRNS-W-DP-PN--PHIGINVN-SIRSI

Other

SL      GGY-LGIFKDAE-SNET---VVAVEFDTFSNR-W-DP-AN--SHIGINVN-SVKSK
LTA     GGFILGIFKDGS--GFN--QFVAVEFDSYHN-IW-DPKSLRSSHVGIDVN-SIMSL

Con A   GRL-LGLFPDAN AD-T---IVAVELDTYPNTDIGDP-SYP--HIGIDI-KSVRSK
         •                                                  •  •
        227         1            *  *  *  *       20     *
```

Figure 3. Amino acids of the metal binding sites of legume lectins. Vicieae: *Vicia faba* (FBL); *Lens culinaris* (LCL); *Lathyrus ochrus* (LOL); *Pisum sativum* (PSL). Phaseoleae: *Dolichos biflorus* (DBL); *E. corallodendron* (ECorL); *Phaseolus limensis* (LBL); *Phaseolus vulgaris* (PHA-L); *Glycine max.* (SBA). Other tribes: *Lotus tetragonolobus* (LTA); *Onobrychis viciifolia*, (SL); *Canavalia ensiformis* (Con A). The *Vicieae* lectins, SL and Con A are mannose specific; DBL, ECorL, LBL and SBA are Gal/GalNAc specific; LTA is L-fucose specific; and PHA-L binds only with oligosaccharides. The top numbering is for the FBL sequence; the bottom numbering is for Con A. The amino acids of Con A which are directly involved in the binding of metal are marked "*", and those which bind via a water molecule are marked "•".

glycoproteins. The conformation of the carbohydrate is within the range of accessible conformations in solution, as determined by NMR and energy calculations (Bouwstra *et al.*, 1990). The glycosylation site is located at a position which is part of the monomer-monomer interface in the other legume lectins of known three-dimensional structure. Disruption of the normal monomer-monomer assembly by the bulky carbohydrate chain, forces ECorL to form a dimer whose quaternary structure differs drastically from that of other members of the family.

In the complex of ECorL with lactose, hydrogen bonds occur between the carboxylate of Asp 89 and the axial 4-OH and equatorial 3-OH of the galactose,

```
Vicieae
         82                  100                        120          211
FBL      ADGFT-FFIAPVDT...GGYLGVF-NGKDYDKT-AQTVAVEFDTFYNA...GAEYA
LCL      ADGFT-FFIAPVDT...GGYLGVFYNGKEYDKT-SQTVAVEFDTFYNA...GAEFA
LOL      ADGFT-FFIAPVDT...GGYLGVF-NSKDYDKT-SQTVAVEFDTFYNT...GAEFA
PSL      ADGFT-FFIAPVDT...GGYLGVF-NSAEYDKT-TQTVAVEFDTFYNA...GAEYA

Phaseoleae
DBL      ADGIA-FALVPVGS...GGYLGVF-DSDVYNNS-AQTVAVEFDTLSNS...GLSEG
ECorL    ADGLV-FFMGPTKS...YGYLGIFFNSKQDNSY--QTLGVEFDTFSNP...GAQRD
LBL      DDPAWLFALVPVDS...GRLLGLF-NKSENDINALT-VAVEFDTCHNL...GA-Y-
PHA-L    ADGLA-FALVPVGS...GGFLGLFDGS-NSNFHT---VAVEFDTLYNK...GINKG
SBA      ADGLA-FFLAPIDT...AGYLGLF-N--ENESG-DQVVAVEFDTFRNS...G-LDI

Other
SL       GDGIT-FFLAPTDT...GGYLGIFKDAE-SNET---VVAVEFDTFSNR...G-DLV
LTA      TDGLV-FFLAPVGT...GGFLGIF-DGS-NGFN--QFVAVEFDSYHN-...G-NPE

Con A    ADGIA-FFISNIDS...GRLLGLFPDAN AD-T---IVAVELDTYPNT...G-LYK
          •                *                          •
         207              227        1                 15   98 **
```

Figure 4. Amino acids of the carbohydrate binding sites of legume lectins. For abbreviations and specificities, see legend to Figure 3. The amino acids of Con A which form contacts with the carbohydrate through their side chains are marked "•" and those which interact through their backbone by "*".

and between the amide of Asn 133 and the 3-OH (Fig. 5). Indeed, the same contacts with galactose have been observed in a complex of *Griffonia simplicifolia* lectin IV and the Le^b-OMe tetrasaccharide (Delbaere *et al.*, 1990).

As noted, both the asparagine and the aspartic acid present in the combining site of ECorL are also conserved in the mannose/glucose specific lectins (Fig. 4). Interestingly, in the complexes of these lectins with sugars, examined by high resolution X-ray crystallography, the aspartic acid forms hydrogen bonds with the equatorial 4-OH and the 6-OH of the ligand, whereas the amide of the asparagine is hydrogen bonded to the 4-OH (Fig. 5). The key role of hydrogen bonds with side chains of aspartic acid and asparagine in the binding of sugars to proteins has been emphasized by Quiocho (1986, 1989, 1990).

Consideration of the above findings leads to the conclusion that the ability of individual legume lectins to distinguish between mannose and galactose cannot be ascribed to the nature of the contact amino acids, but must be related to other steric and conformational features of the combining site.

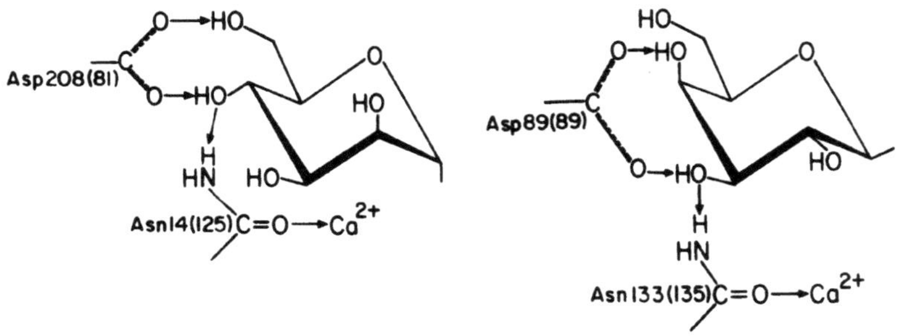

Figure 5. Comparison of key polar interactions involved in the binding of monosaccharides to legume lectins, based on high resolution X-ray crystallography. *Left*: Mannose bound to concanavalin A and *Lathyrus ochrus* lectin (numbers in parenthesis). *Right*: Galactose bound to ECorL and *Griffonia simplicifolia* lectin IV (numbers in parenthesis).

It is of interest to note that in wheat germ agglutinin, a non-legume lectin specific for *N*-acetylglucosamine and *N*-acetylneuraminic acid, neither aspartic acid nor asparagine participate in binding of the ligand. Instead, key polar interactions are formed, for example with hydroxyls of serine and tyrosine (Wright, 1990). Thus, as pointed out by Reeke and Becker (1988), nature seems to construct saccharide-binding sites on very different frameworks, as exemplified by the legume lectins with their large β structures and low cysteine content, on the one hand, and wheat germ agglutinin with its high cysteine content and virtual absence of regular secondary structures, on the other. In addition, chemical groups directly involved in binding are equally diverse and include not only side chain carboxylates and carboxamides, but also aliphatic and phenolic hydroxyls, as well as backbone amide groups. At this stage, it is therefore impossible to predict which contacts are involved in binding of sugars to lectins from other sources.

Acknowledgments

This paper was written during the tenure of a Greenberg Scholarship at the Oklahoma Medical Research Foundation, Oklahoma City, OK 73104, U.S.A. Work from the author's laboratory was supported in part by grants the United States-Israel Binational Science Foundation (#3727), the United States-Israel Binational Agricultural Research and Development Fund (#IS-1636-89), as well as by the Leo and Julia Forchheimer Center for Molecular Genetics and the Kimmelman Center for Macromolecular Assembly at the Weizmann Institute of Science. Thanks are due to Dr. Halina Lis and Dvorah Ochert for their help in writing this paper.

REFERENCES

Adar R, Richardson M, Lis H, Sharon N (1989) The amino acid sequence of *Erythrina corallodendron* lectin and its homology with other legume lectins. FEBS Lett 257:81-85

Arango R, Rozenblatt S, Sharon N (1990) Cloning and sequence analysis of the *Erythrina corallodendron* lectin cDNA. FEBS Lett 264:109-111

Ashford D, Dwek RA, Welply JK, Amtayakul S, Homans SW, Lis H, Taylor GN, Sharon N, Rademacher TW (1987) The $\beta 1 \rightarrow 2$-D-xylose and $\alpha 1 \rightarrow 3$-L-fucose substituted *N*-linked oligosaccharides from *Erythrina cristagalli* lectin: isolation, characterization and comparison with other legume lectins. Eur J Biochem 166:311-320

Aversa F, Reisner Y, Terenzi A, Carotti A, Pelicci P-G, Latini P, Martelli MF (1990) Lectin-separated BMT for leukemia patients after a conditioning regimen that includes thiotepa (TT). UCLA Symposium on "New Strategies in Bone Marrow Transplantation", Keystone, Colorado, January 20-27.

Bourne Y, Abergel C, Cambillau C, Frey M, Rouge P, Fontecilla-Camps.J-C (1990a) X-ray crystal structure determination and refinement at 1.9 A resolution of isolectin I from the seeds of *Lathyrus ochrus*. J Mol Biol 214:571-584

Bourne Y, Roussel A, Frey M, Rouge P, Fontecilla-Camps J-C, Cambillau C (1990b) Three-dimensional structures of complexes of *Lathyrus ochrus* isolectin I with glucose and mannose: fine specificity of the monosaccharide-binding site. Proteins: Struct Funct Gen 8:365-376

Bouwstra JB, Spoelstra EC, De Waard P, Leeflang BR, Kamerling JP, Vliegenthart JFG (1990) Conformational studies on the N-linked carbohydrate chain of bromelain. Eur J Biochem 190:113-122

Brandley BK, Swiedler SD, Robbins PW (1991) Carbohydrate ligands of the LEC cell adhesion molecules. Cell 63:861-863

Delbaere LTJ, Vandonselaar M, Prasad L, Quail JW, Pearlstone JR, Carpenter MR, Smillie LB, Nikrad PV, Spohr U, Lemieux RU (1990) Molecular recognition of a human blood group determinant by a plant lectin. Can J Chem 68:1116-1121

Derewenda Z, Yariv J, Helliwell JR, Kalb (Gilboa) AJ, Dodson EJ, Papiz MZ, Wan T, Campbell J (1989) The structure of the saccharide-binding site of concanavalin A. EMBO J 8:2189-2193

Dorland L, van Halbeek H, Vliegenthart JFG, Lis H, Sharon N (1981) Primary structure of the carbohydrate chain of soybean agglutinin; a reinvestigation by high-resolution ^{1}H-NMR spectroscopy. J Biol Chem 256:7708-7711

Fischer A, Landais P, Friedrich W, Morgan G, Gerritsen B, Fasth A, Porta F, Griscelli C, Goldman SF, Levinsky R, Vossen J (1990) European experience of bone-marrow transplantation for severe combined immunodeficiency. Lancet ii:850-854

Gabius HJ (1988) Tumor lectinology - at the intersection of carbohydrate chemistry, biochemistry, cell biology, and oncology. Angew Chem Intern Edit English 27:1267-1276

Hakomori S (1989) Aberrant glycosylation in tumors and tumor-associated carbohydrate antigens. Adv Canc Res 52:257-331

Higgins TJV, Chandler PM, Zurawski G, Button SC, Spencer D (1983) The biosynthesis and primary structure of pea seed lectin. J Biol Chem 258:9544-9549

Kinzy W, Lis H, Sharon N (1988) New derivatives of D-galactosamine and lactosamine and their application for studies of Gal/GalNAc specific lectins. 14th International Symposium on Carbohydrate Chemistry, Stockholm, August, Abst p 358.

Laskey LA (1991) Lectin cell adhesion molecules (LEC-CAMs): a new family of cell adhesion proteins involved with inflammation. J Cell Biochem 45:139-146

Lis H, Joubert FJ, Sharon N (1985) Isolation and properties of *N*-acetyllactosamine-specific lectins from nine *Erythrina* species. Phytochemistry 24:2803-2809

McEver RP (1991) Selectins: novel receptors that mediate leukocyte adhesion during inflammation. Thromb. Haemost. in press

Neuberger A (1984) What lectins and carbohydrates are made for? Biol Cell 51:113-114

O'Reilly RJ, Keever CA, Small TN, Brochstein J (1989) The use of HLA-non-identical T-cell-depleted marrow transplants for correction of severe combined immunodeficiency disease. Immunodef Rev 1:273-309

Quinn JM, Etzler ME (1989) *In vivo* biosynthetic studies of the *Dolichos biflorus* seed lectin. Plant Physiol 91:1382-1386

Quioccho FA (1986) Carbohydrate-binding proteins: tertiary structures and protein-sugar interactions. Ann. Rev. Biochem. 55:287-315

Quioccho FA (1989) Protein-carbohydrate interactions: basic molecular features. Pure Appl Chem 61:1293-1311

Quioccho FA (1990) Atomic structures of periplasmic binding proteins and the high affinity active transport systems in bacteria. Phil Trans Soc London B326:341-352

Reeke GN Jr, Becker JW (1988) Carbohydrate-binding sites of plant lectins. Curr Top Microbiol Immunol 139:35-58

Reisner Y, Itzicovitch L, Meshorer A, Sharon N (1978) Hemopoietic stem cell transplantation using mouse bone-marrow and spleen cells fractionated by lectins. Proc Natl Acad Sci USA 75:2933-2936

Reisner Y, Sharon N (1980) Cell fractionation by lectins. Trends Biochem Sci 5:29-31

Rosen S (1990) The LEC-CAMs: an emerging family of cell-cell adhesion receptors based upon carbohydrate recognition. Am J Respir Cell Mol Biol 3:397-402

Shaanan B, Lis H, Sharon N (1991) The three-dimensional structure at high resolution of *Erythrina corallodendron* lectin and its complex with lactose. Abst. XI International Symposium Glycoconjugates, Toronto, June.

Sharon N (1983) Lectin receptors as lymphocyte surface markers. Adv Immun 34:213-298

Sharon N, Lis H (1989a) Lectins as cell recognition molecules. Science 246:227-234

Sharon N, Lis H (1989b) Lectins, Chapman & Hall, London, 127pp

Sharon N, Lis H (1990a) Carbohydrate-protein interactions. Chem Brit 26:679-682

Sharon N, Lis H (1990b) The legume lectins - a large family of homologous proteins. FASEB J 4:3198-3208

Springer TA, Laskey LA (1991) Sticky sugars for selectins. Nature 349:196-197

Wright CS (1990) 2.2 Å resolution structure analysis of two refined *N*-acetylneuraminyl-lactose-wheat germ agglutinin isolectin complexes. J Mol Biol 215:635-651

N. Sharon

Invertebrate lectins: new aspects in retrospects

G. Uhlenbruck, F. G. Hanisch, and W. E. G. Müller*

Institute of Immunobiology
University of Cologne
Kerpener Straße 15
D-5000 Cologne 41
FRG

1.0 Introduction: Classification

Invertebrate lectins may be regarded as a special group of lectins, as there are several reasons for this: (a) their role in different defense mechanisms and in nutrition, (b) their specificity, for instance their anti-neuraminyl reactivity (and no mannose-specific lectins) and (c) their distribution in the haemolymph and in organ cells or defense-commited haemocytes.

From our point of view, three main groups of invertebrate lectins have so far been thoroughly investigated:

I. Snail lectins
II. Tridacna clam lectins and
III. Sponge lectins

As the first group has been described and been reviewed several times (Prokop and Uhlenbruck 1970, Schnitzler 1977, and Uhlenbruck et al. 1974), we want to focus the interest in

* Institute of Physiological Chemistry, University of Mainz, Duesbergweg, D-6500 Mainz, FRG

this paper on lectins from the Tridacna clam group and from various sponges.

1.1 Historical Background of Earlier Findings

Ten years before the Tridacna lectins have been discovered by us, the history, which led to this event, has been started by Prof. Otto Prokop in Berlin. He found and described the first well-characterized invertebrate lectins in the garden and in the land snail albumin gland, and also in the eggs of these snails (for review see Prokop and Uhlenbruck 1970). As a by-product of those investigations, the snail galactans (May and Weinbrenner 1938) have been re-discovered and a new group of protease-inhibitors were detected, because the Helix pomatia lectin seemed to be indigestible (Uhlenbruck et al. 1971).

It is interesting to note, that Prokop and his group in this connection also have described for the first time vertebrate lectins, when they investigated agglutinins in fish eggs too (Prokop and Uhlenbruck 1970). In analogy to the snail eggs, also in fish eggs a new group of proteinase-inhibitors has been discovered by us (Uhlenbruck et al. 1972).

1.2 The Tridacna Lectins: Aspects of Discovery

As already mentioned, the vertebrate lectins have been discovered in the course of investigations on fish eggs. The galactan polysaccharides however, have led much earlier to the discovery of vertebrate lectins, but in an indirect way, which clearly and evidently demonstrated the presence of a galactose-binding principle in the vertebrate liver. May and Weinbrenner (1938) injected large amounts of snail galactan into rabbits and found that in the way the amount of galactan disappeared from the blood, it accumulated selectively in the liver. In fact they made the same experiments, which Ashwell and his group made nearly 30 years later, but they could not interprete the phenomenon, which is due to the hepatic galactose-binding lectins, at that time and even did not follow it up, as they were mainly interested in the metabolism of galactan (called in those days "galactogen").

As the Helix pomatia snail galactan accumulated in our lab during the preparation of the snail`s lectin, we looked also for serological properties of this polysaccharide and found blood group H activity, because of the presence of L-galactose, detected with different catfish antisera (Baldo and Uhlenbruck 1973). The inclusion of the haemolymph of Tridacna maxima in our experiments was a pure chance, because we obtained and ordered it from a zoological catalogue in order to spend some money which was still left at the end of the year 1974. So we tested the crude haemolymph against our galactan preparation and found a very strong precipitation line, which was not due to any "anti-A" lectin contamination in the snail galactan sample (Uhlenbruck et al. 1975). On the contrary, the lectin was located in the electrophoretic analysis of the Tridacna haemolymph, moving to the anode, and was not identical with other "anti-galactan" lectins (Uhlenbruck et al. 1975).

1.3. Tridacna Clam Haemolymph as the Source of Tridacna Lectins

The source of the Tridacna lectins is the haemolymph of these clams, where it can account for up to 50% of the haemolymph proteins. The MW of these lectins ranges from 300 000-500 000

Dalton, the MW of of the subunits, obtained by -S-S- cleavage, is in the order of 22000 - 44000. Combining sites are calculated to be 10-12 in this acid protein, which is relatively labile against proteolytic degradation too (is easily split by matrix-bound proteases too. Main amino acids of the Tridacnin lectins are asparagin and glutaminic acid besides glycine. The carbohydrate content is about 5-10% (L-fucose 3%, D-galactose 1% and under 1%; D-glucose, D-mannose and both acetylated hexosamines (Baldo et al. 1978).

The haemolymph components of the various Tridacnid clams have been characterized by a two-dimensioned electrophoretic technique, which demonstrated typical "haemolymph fingerprints" (Kania et al. 1980).

1.4 The Purification of Lectins from Tridacnid Haemolymph

Several purification methods for the Tridacna haemolymph lectins have been successfully tried (Baldo et al. 1978), among those, which use affinity-chromatography with a poly-arabino-galactan column (Baldo and Uhlenbruck 1975a) or just acid activated Sepharose (Baldo and Uhlenbruck 1975b).

Tridacnin lectins can be purified by affinity chromatography on a lactose matrix, thus demonstrating that they bind equally well also to β-1-4-linked galactose as already been assumed by their reaction with different asiolo-glycoproteins on the N-acetyl-lactosamine type.

On the other hand, the reverse method, mainly the isolation of galactan polysaccharides and galactoglycoconjugates by a Tridacnin-Sepharose 4B affinity column, has been used by Gleeson et al (1979). As the lectin-binding is calcium-ion dependent, the glycoconjugates can be eluted in one step by washing the column with a calcium-free buffer.

1.5 Precipitating Properties of Tridacna Lectins

It has turned out already when the Tridacna lectins were discovered, that they react preferably with galacto-glycoconjugates, especially galactan polysaccharides. Therefore, they have also been named anti-galactans by us (Uhlenbruck et al 1975c, Uhlenbruck et al 1976a). Subsequently, because of their excellent precipitin properties, the Tridacna lectins have been successfully used in order to discriminate between the various galactans from different sources (Baldo and Uhlenbruck 1975, Uhlenbruck et al 1978b; Uhlenbruck et al 1976b, Uhlenbruck 1977b). In addition, they also "cross-react" with blood group substances (hog and human) with erythrocyte glycoproteins (bovine) and with bacterial polysaccharides (Pneumococcus type 14 polysaccharide) as has been found out very early (Uhlenbruck et al 1975a).

1.6 Precipitin Reactions as Tools

However, those lectins from invertebrates are not only of importance for the elucidation of plant arabinogalactans (Albersheim 1975, Gleeson 1977, Gleeson et al 1978, Gleeson and Clarke 1980; Fincher et al 1983), but also for the localization of arabinogalactan-protein structures in intracellular vesicles (Baldo et al 1978a). Of more practical interest is the isolation and characterization of allergens, for instance from house dust mite extracts by the help of Tridacna maxima lectin (Baldo and Uhlenbruck 1977), which can be measured then by a new radioallergosorbent test (Baldo et al 1976).

Besides comparative precipitin studies with various blood group substances, polysaccharides from invertebrates, plants and bacteria (Baldo and Uhlenbruck 1975d), the Tridacna lectins react preferably with serum glycoproteins, after they have been treated with neuraminidase (Uhlenbruck et al 1977c, Uhlenbruck et al 1977d, Uhlenbruck et al 1978c), a fact, which can also be seen in relationship to other lectins, for instance from plant origin (Uhlenbruck et al 1979; Chatterjee et al 1979).

1.7 Agglutinin Properties of Tridacna Lectins

When haemolymph samples from different Tridacnid clams were tested for their haemagglutination abilities, certain differences can be observed between the various lectins on the one side and different haemagglutination patterns on the other side can be seen with normal red cells and enzyme-treated erythrocytes from miscellaneous sources (Uhlenbruck et al 1978a).
Also certain bacteria, like neuraminidase treated Streptococci group B type II can be agglutinated by Tridacna lectins (Uhlenbruck et al 1985)

1.8 Mitogenic Abilities of Tridacna Lectins

Like all newly discovered lectins, the Tridacna lectins have been tested for mitogenic activity with respect to the response of human lymphocytes (Schumacher et al 1978). The Tridacna lectins were found to be as stimulating to human T-lymphocytes as it is known from the phythaemagglutinin PHA (Schumacher et al 1978, Uhlenbruck et al 1977e, f).

In addition we found, that Pneumococcus type 14 polysaccharide completely inhibited the mitogenic response, a fact, which also underlines the reactivity of Tridacna lectins with N-acetyl-lactosamine structures or lactose-derived molecules. The other Tridacna lectins show a similar, not less expressed mitogenic activity for instance Tridacna gigas.

1.9 The Receptor Problem of Tridacna Lectins

The mitogenic property of Tridacna lectins raises also the question: What is the receptor site of Tridacna lectin? It is generally accepted, deduced from the strong reaction with various galactans, that, as has been shown especially for the Tridacna maxima lectin, this anti-galactan lectin recognizes a digalactose structure, in which both sugars are linked β-1-6 glycosidically with each other (Eichmann et al 1976). One of the best macromolecular substances serving as an inhibitor, is pneumogalactan, which carries this carbohydrate structure in form of a side chain. However the reaction with several asialoglycoproteins, obviously due to N-acetyl-lactosamine residues, shows, that also this group can serve as "lectinotop".

2.0 Cross-Reacting Lectins and Antibodies

Cross-reacting lectins and antibodies have to be mentioned in this connection, because they also do react with the Tridacna maxima specific receptor, namely β-D-Gal(1-6)-D-Gal disaccharide structure. This has been convincingly demonstrated by Eichmann et al (1976). According to these fundamental experiments one can divide the β(1-6)-D-Gal anti-galactans in three groups:

I. The Tridacna lectins, which have been thoroughly discussed also in previous reviews (Uhlenbruck et al 1974, Uhlenbruck et al 1975d)

II. Sponge lectins, like the one from Axinella polypoides (Eichmann et al 1976, Bretting et al 1983) or the one from the sea anemone Cerianthus membranaceus (Baldo et al 1977), which all have the same or very similar "combining site".

III. The third group is represented by the immunoglobulin class of anti-galactans, which can be subdivided into three groups: myeloma proteins (Glaudemans 1975) monoclonal antibodies (Glaudemans 1987) and snake lectins (Uhlenbruck and Helmbold 1985)

2.1 The Biological Role of Tridacna Lectins

The biological role of Tridacna lectins has been extensively discussed in relationship to the so-called invertebrate immunity (Hildemann and Benedict 1975): For the Tridacna lectins we assumed, that they contribute less to a defense-mechanism, but more to the symbiosis with algae, which they culture and house for their nutrition (Uhlenbruck and Steinhausen 1977) by using the "old" algae for their metabolism. The lectins may serve for eliminating galactan-containing membrane constituents or for stimulating the growth of the algal symbionts. This would fit into the concept of algal symbiosis (Taylor, 1973; Muscatine and Greene, 1973) as well as into consideration about Molluscan immunobiology (Bayne, 1983).

3.0 New Aspects of the Geodia Lectin

The siliceous sponge Geodia cydonium produces in a distinct type of mucoid cells, the spherulous cells, large amounts of a lectin which reaches values of up to 14% of the extractable proteins in the extracellular milieu of the sponge (Müller et al 1983). The lectin has been attributed a particular function in the aggregation process of the sponge cells by being involved in the control of cell-cell contacts (Müller 1982). GCA has been reported to convert aggregation-deficient cells into aggregation-susceptible cells by interacting with the cell surface associated anti-aggregation receptor (Müller et al 1979a), a macromolecule which was found to weaken the binding between the aggregation factor and the aggregation receptor (Müller et al 1979b). A more direct involvement of the lectin in cell aggregation has been postulated on the basis of Ca^{2+} dependent lectin polymerization mediated by β-galactose expressing glycoconjugates (Conrad et al 1984) or by lectin self-recognition (Müller et al 1983).

The trimeric lectin is composed of two covalently linked subunits of Mr 13.8 and 13.0 kDa with isoforms distributed at pI 4.7 to 5.0, and one non-covalently associated subunit of Mr 12.2 kDa with pI 7.0 (Müller et al 1983, Hanisch et al 1984). The lectin molecule binds two moles of lactose per mole of native lectin in a Ca^{2+} independent manner (Müller et al 1983) and shows a rather broad binding specificity for β-galactosides of various linkage types in inhibition studies of haemagglutination and quantitative precipitation (Bretting 1981, Hanisch 1984). On the other hand, histochemical studies on human tissue sections have revealed that GCA binds to blood group A, B expressing cells more strongly than to blood group O specimen and that blood group dependent binding cannot be inhibited by even high concentrations of lactose (Vierbuchen et al 1991). This particular binding specificity, which is also Ca^{2+} independent, had previously not been observed in haemagglutination, where GCA unlike other A blood group specific lectins (DBA, HPA) did not clearly discriminate between A,B,O blood cells. On the other hand, binding studies in solid phase enzyme immunoassays have revealed that the lectin affinity for the blood group A trisaccharide GalNAcα1-3(Fucα1-2) Galβ exceeds that for lactose and other β-galactosides by appr. three orders of magnitude. Unlike the lectin from Helix pomatia GCA does not bind to the "A-like" core structure GalNAcα1- Ser/Thr (Tn-antigen) on desialylated ovine submaxillary mucin. Contrasting with the lectin from Dolichos biflorus which exhibits a decreasing binding specificity in the order A-trisaccharide > GalNAc > lactose, GCA prefers lactose vs. GalNAc by a factor of 50. The relative inhibitory potency of sugar inhibitors in GCA binding studies was found to be: A-trisaccharide (1.5×10^5) > B-trisaccharide (3×10^3) > lactose (50) > GalNAc (1).

It may be hypothesized that GCA exhibits two distinct binding specificities: a β-galactoside specificity localized on two of the lectin subunits and responsible for the blood group independent lectin binding and a blood group A (B) specificity attributable to the third lectin subunit. Blood group A specificity of GCA has no in-vivo correlate in sponge tissue, since the cells do not synthesize GalNAc containing glycoconjugates.

Acknowledgements

We wish to thank the Karl Jüngel-Stiftung for financial support and Ms C. Böttinger for skilfull help in preparing the manuscript.

References

Albersheim P (1975) The walls of growing plant cells. Scientific American, April 1975, pp 80-105

Baldo BA, Uhlenbruck G (1973) Cross-reactive human blood group H-active polysaccharide from Helix pomatia. I. Detection with catfish anti-H and eel sera. Immunology 25:649-661

Baldo BA, Uhlenbruck G (1975a) Tridacnin, a potential anti-galactan precipitin from the hemolymph of Tridacna maxima (Röding) In: Hildemann WH and Benedict AA (eds) Immunologic Phylogeny. New York Advances in Experimental Medicine and Biology. Plenum Press. New York London. 64:3-11

Baldo BA, Uhlenbruck G (1975b) Purification of Tridacnin, a novel anti-β-(1-6)digalactobiose precipitin from the haemolymph of Tridacna maxima (Röding). FEBS Lett 55:25-29

Baldo BA, Uhlenbruck G (1975c) Anti-galactan activity in Tridacna maxima (Röding) haemolymph. Immunology 29:1161-1170

Baldo BA, Uhlenbruck G (1975d) Quantitative precipitin studies on the specificity of an extract from Tridacna maxima (Röding). Carbohydr Res 40:143-151

Baldo BA, Turner KJ, Uhlenbruck G (1976) Isolation of a potent allergen from house dust mite by interaction with the lectin Tridacnin. Experientia 32:641-643

Baldo, BA, Uhlenbruck G (1977) Selective isolation of allergens. Clin Allergy 7:429-443

Baldo BA, Uhlenbruck G, Steinhausen G (1977) Invertebrate anti-galactans. A comparative study of agglutinins from the clam Tridacna maxima, the marine sponge Axinella polypoides and the anemone Cerianthus membranaceus. Comp Biochem Physiol 56A:343-351

Baldo BA, Neukom H, Stone BA, Uhlenbruck G (1978a) Reaction of some invertebrate and plant agglutinins and a mouse myeloma anti-galactan protein with an arabinogalactan from wheat. Aust J Biol Sci 31:149-160

Baldo BA, Sawyer WH, Stick RV, Uhlenbruck G (1978b) Purification and characterization of a galactan-reactive agglutinin from the clam Tridacna maxima (Röding) and a study of its combining site. Biochem J 175:467-477

Bayne CJ (1983) Molluscan immunobiology. In: The Mollusca, vol 5. Academic Press Inc., p 407

Bretting H, Phillips SG, Klumparf HJ, Kabat E (1981) A mitogenic lactose-binding lectin from the sponge Geodia cydonium. J Immunol 127:1653-1658

Bretting H, Jacobs G, Dannenberg F (1983) Structural similarities between Helix pomatia galactan and the carbohydrate moiety of hemocyanin. Comp Biochem Physiol 75B:269-276

Chatterjee BP, Vaith P, Chatterjee S, Karduck D, Uhlenbruck G (1979) Comparative studies of new marker lectins for alkali-labile and alkali-stable carbohydrate chains in glycoproteins. Int J Biochem 10:321-327

Conrad J, Uhlenbruck G, Zahn RK, Kurelec B, Jericevic B, Müller WEG (1984) The role of lectin I and of glycoconjugates in recognition of cells from the siliceous sponge Geodia cydonium. Biol Cell 51:287-294

Eichmann K, Uhlenbruck G, Baldo BA (1976) Similar combining specificities of invertebrate precipitins and mouse myeloma protein J 539 for β-(1-6)-galactans. Immunochemistry 13:1-6
Fincher GB, Stone BA, Clarke AE (1983) Arabinogalactan-proteins: Structure, biosynthesis, and function. Ann Rev Plant Physiol 34:47-70
Glaudemans CPJ, Zissis E, Jolley ME (1975) Binding studies of a mouse-myeloma immunoglobulin A having specificity for β-D-(1-6)-linked D-galactopyranosyl residues. Carbohydr Res 40:129-135
Glaudemans CPJ (1975) The interaction of homogeneous, murine myeloma immunoglobulins with polysaccharide antigens. Adv Carbohydr Chem Biochem 31:313-346
Glaudemans, CPJ (1987) Sevens structurally different murine monoclonal galactan-specific antibodies show identity in their galactosyl-binding subsite arrangements. Molecular Immunology 24:371-377
Gleeson PA (1977) Isolation of arabinogalactan proteins using insolubilized galactose-binding macromolecules. In: Arabinogalactan protein news. Proceedings of the Arabinogalactan Protein Club No 1, pp 30-33
Gleeson PA, Clarke AE, Jermyn MA, Knox RB (1978) Arabinogalactan proteins of the female reproductive tissues of Gladiolus gandavensis. Proc Aust Bio Soc II p 33
Gleeson PA, Jermyn MA, Clark AE (1979) Isolation of an arabinogalactan protein by lectin affinity chromatography on Tridacnin-Sepharose 4B. Anal Biochem 92:41-45
Gleeson PA, Clarke AE (1980) Antigenic determinants of a plant proteoglycan, the Gladiolus style arabinogalactan-protein. Biochem J 191:437-447
Hanisch FG, Saur A, Müller WEG, Conrad J, Uhlenbruck G (1984) Further characterization of a lectin and its in vivo receptor from Geodia cydonium. Biochim Biophys Acta 801:388-395
Hildemann WH, Benedict AA (eds) (1975) Immunologic Phylogeny. Plenum Press. New York London.
Kania J, Janssen E, Uhlenbruck G, Pearson R (1980) Comparative investigations on the protein components of the haemolymph fluid of the Tridacnid clam family using two-dimensional gel electrophoresis. Comp Biochem Physiol 66B:117-121
May F, Weinbrenner H (1938) Über die Ablagerung des Galaktogens im Warmblüterorganismus. Z Biol 99:199-216
Müller WEG, Kurelec B, Zahn RK, Müller I, Vaith P, Uhlenbruck G (1979a) Aggregation of sponge cells. Function of a lectin in its homologous biological system. J Biol Chem 254:7479-7481
Müller WEG, Zahn RK, Kurelec B, Müller I, Vaith P, Uhlenbruck G (1979b) Aggregation of sponge cells. Isolation and characterization of an inhibitor of aggregation receptor from the cell surface. Eur J Biochem 97:585-591
Müller WEG (1982) Cell membranes in sponges. Intern Rev Cytol 77:129-181
Müller WEG, Conrad J, Schröder C, Zahn RK, Kurelec B, Dreesbach K, Uhlenbruck G (1983) Characterization of the trimeric self-recognizing Geodia cydonium lectin I. Eur J Biochem 133:263-267

Muscatine L, Greene RW (1973) Chloroplasts and algae as symbionts in molluscs. Int Rev Cytol 36:137-169
Prokop O, Uhlenbruck G (1970) Protektine. In: Perlick E, Plenert W, Prokop O (eds)Fortschritte der Hämatologie. Vol. 1. Johann Ambrosius Barth. Leipzig. pp. 17-69.
Schnitzler S (1977) Hämagglutinine aus Wirbellosen unter besonderer Berücksichtigung des Anti-A_{HP}. In: Perlick E, Plenert W, Prokop O, Stobbe H. (eds) Blutgruppen, Band 4. Johann Ambrosius Barth Leipzig, pp. 34-186
Schumacher G, Uhlenbruck G, Chatterjee BP, Steinhausen G, Vaith P, van Mil A (1978) A new group of mitogenic lectins from invertebrate sources. Z Immun-Forsch 154:62-74
Taylor DL (1973) Algal symbionts of invertebrates. Ann Rev Microbiol 27:171-187
Uhlenbruck G, Sprenger I, Ishiyama I (1971) A new polyvalent proteinase-inhibitor occurring in the albumine gland of Helix pomatia. Z Klin Chem Klin Biochem 9: 361-362
Uhlenbruck G, Sprenger I, Hermann G, Franke H (1972) Fish eggs: Another new source of polyvalent proteinase-isoinhibitors. Z Naturforsch 27b, 322-323
Uhlenbruck G, Dahr W, Rothe A, Baldo BA (1974) Fakten und Folgerungen aus Forschungsergebnissen von Erythrozyten-Rezeptoren, heterophilen Agglutininen und Tumorzellmembranen. Westdeutscher Verlag. D-5670 Opladen (Forschungsbericht Nr. 2475)
Uhlenbruck G, Baldo BA, Steinhausen G (1975) Anti-carbohydrate precipitins and haemagglutinins in haemolymph from Tridacna maxima (Röding) Z Immunforsch 150:354-363
Uhlenbruck G, Steinhausen G, Gauwerky Ch, Baldo BA, Renwrantz L (1975) Über zwei weitere Präzipitine gegen das Galaktogen aus Helix pomatia. Biologisches Zentralblatt 94:205-210
Uhlenbruck G, Steinhausen G, Baldo BA, Kareem HA (1975) Antigalactans: New classes of anti-carbohydrate precipitins. Naturwiss 6:301
Uhlenbruck G, Steinhausen G, Baldo BA (1975) Galactane und Antigalactane. Verlag Joseph Stippak. D-5100 Aachen, Postfach 1262
Uhlenbruck G, Steinhausen G, Baldo BA (1976) Galactans and anti-galactans from invertebrates. Z Naturforsch 31c:205-206
Uhlenbruck G, Steinhausen G, Kareem HA (1976) Different glycosubstances and galactans in the albumin gland and eggs of Achatina fulica. Z Immun-Forsch 152:220-230
Uhlenbruck G, Steinhausen G (1977) Tridacnins: Symbiosis-profit or defense purpose? Dev Comp Immunol 1:183-192
Uhlenbruck G, Steinhausen G, Baldo BA (1977a) Different antigalactans in the haemolymph od Tridacna maxima and Tridacna gigas. Comp Biochem Physiol 56B:329-333
Uhlenbruck G, Steinhausen G, Palatnik M (1977b) Similarity of glycoproteo-galactans in the albumin glands from Achatina and Borus snail. Comp Biochem Physiol 57B:335-339
Uhlenbruck G, Steinhausen G, Schwick HG (1977c) A new marker for neuraminidase-treated human serum glycoproteins from the haemolymph of Tridacna maxima (Röding). J Clin Chem Clin Biochem 15:21-26

Uhlenbruck G, Haupt H, Reese I, Steinhausen G (1977d) Serumcholinesterase als Modell eines Glykoproteins. J Clin Chem Clin Biochem 15:561-564
Uhlenbruck G, Schumacher K, Steinhausen G, van Mil A (1977e) Tridacnin, a new mitogenic lectin from invertebrate sources. Z Immun-Forsch 153:265-267
Uhlenbruck G, Schumacher K, Steinhausen G, van Mil A (1977f) Über eine neue Gruppe von mitogen wirkenden Lektinen aus Invertebraten. In: Schuhmacher K, Grosser KD (eds) Aktuelle Probleme der Inneren Medizin. FK Schattauer Verlag, Stuttgart. pp 187-193
Uhlenbruck G, Karduck D, Pearson R (1978a) Different Tridacnins in different Tridacnid clams: a comparative study. Comp Biochem Physiol 63b:125-129
Uhlenbruck G, Steinhausen G, Geserick G, Prokop O (1978b) Further comparative studies of glycosubstances and proteins from different snail albumin glands. Comp Biochem Physiol 59b:285-288
Uhlenbruck G, Baldo BA, Steinhausen G, Schwick HG, Chatterjee BP, Horejsi V, Krajhanzl A, Kocourek J (1978c) Additional precipitation reactions of lectins with human serum glycoproteins. J Clin Chem Clin Biochem 16:19-23
Uhlenbruck G, Vaith P, Karduck D, Haupt H, Müller WEG (1979) Anti-galactan Lectins. In: H. Peeters (ed) Protides of Biological Fluids. Pergamon Press. Oxford New York. pp 595-598
Uhlenbruck G, Helmbold W (1985) Schlangengift als Quelle von B- und T-Lymphozyten-Mitogenen. Dtsch Ärztebl 82:3176-3177
Uhlenbruck G, Lütticken R, Böhmer G, Janssen E, Pulverer G (1985) Group B streptococus type II antisra have anti-galactan specificities. Zbl Bakt Hyg A 259:179-187
Vierbuchen M et al (1991) in press, J Histochem

Narcissus and Gerardia lectins: Tools for the development of a vaccine against AIDS and a new ELISA to quantify HIV-gp 120

W. E.G. Müller, J. M. S. Forrest*, S.-H. Chang, B. E. Weiler, H. Schäcke, E. Matthes+ and M. Bachmann

Institut für Physiologische Chemie,
Abteilung "Angewandte Molekularbiologie",
Universität, Duesbergweg 6,
D-6500 Mainz, Germany

Abstract

The lectins, isolated from Narcissus pseudonarcissus [NPL] and Gerardia savaglia [Gerardia-lectin], are specific for terminal Man(α1-3)Man units. They react very specifically with the external glycoprotein gp120 of the AIDS-virus, the human immunodeficiency virus type 1 (HIV-1). They prevent in vitro the infection of cells with HIV. These lectins were tools for identifying antigens [mannans from Candida albicans and from Saccharomyces cerevisiae, comprising Man(α1-3)Man and Man(α1-2)Man linked residues] which were successfully used to raise in rabbits neutralizing antibodies against HIV infection in vitro. The NPL was used to develop a novel enzyme-linked immunoassay [ELISA] system for quantifying HIV-gp120 within the range of 0.6 to 20,000 ng/ml.

Introduction

Antibodies recognize the carbohydrate side chains in glycoproteins fairly frequently and thus can cross-react with similar oligosaccharide sequences of other glycoproteins. One way to overcome this problem is to screen for lectins which recognize slight differences in the structures of the

---

* Scottish Crop Research Institute, Invergowrie, Dundee, U.K.
+ Zentralinstitut für Molekularbiologie, D-1115 Berlin-Buch, Germany

oligosaccharide side chains. One successful application of this strategy is outlined in this article. It is shown that two lectins interact exclusively with the carbohydrate side chains of the glycoprotein (gp) 120 of the human immunodeficiency virus type 1 (HIV-)1 and thus prevent binding of the virus to the cell-surface glycoprotein CD4, which is expressed on a subset of T-cells and macrophages. Our knowledge of the specificity of these lectins was utilized in two ways. First it was used to develop model antigens with a similar carbohydrate structure to the lectin-receptor on the virus (the gp120); they were used to raise "lectin-like" antibodies. These antibodies displayed neutralizing activity against an HIV infection in vitro. In a second approach the lectins were used to develop a novel ELISA test system.

Problems with the development of a vaccine against HIV infection

The CD4 receptor, which is expressed on a subset of T-cells and macrophages, serves as the receptor for HIV-1 (Klatzmann et al., 1984; Dalgleish et al., 1984; Sattentau and Weiss, 1988). Both monoclonal antibodies to CD4 and soluble, recombinant CD4 were found to block infection of cells by HIV-1 (Smith et al., 1987; Fisher et al., 1988; Hussey et al., 1988; Clapham et al., 1988 and Traunecker et al., 1988). In strategies to develop a vaccine against HIV infection, gp120 was found to have some neutralizing epitopes; e.g. the amino acids 288-467 leads to the development of neutralizing antibodies in chimpanzees (Goudsmit et al., 1988). However, only in a few experiments it was possible to produce a protective response by immunizing chimpanzees with recombinant gp120 (Hu et al., 1987 and Berman et al., 1988). Moreover, it was found that neutralizing antibodies made in animals against recombinant gp120 block in in vitro tests the infectivity of only a few isolates (Berman et al., 1988). In another approach soluble CD4 or CD4 immunoadhesins (Capon et al.,1989) have been shown to block HIV-1 infectivity in vitro.

Gp120, which consists of more than 20 potential N-glycosylation sites, is heavily glycosylated. One-half of the molecular weight of gp120 is made up of oligomannosidic oligosaccharides (Allan et al., 1985; Geyer et al., 1988; Mizuochi et al., 1988; Richardson et al., 1988). The trimming reactions which convert the $Glc_3Man_9GlcNAc_2$ to high mannose-type oligosaccharides on gp120/160 display some peculiarities. For example, some of the carbohydrate moieties remain in the $(Man)_9GlcNAc_2$ form and are not trimmed further by mannosidase IA/B. In addition more than 50% of the glycans of gp120 do not contain N-acetylneuraminic acid residues at their termini (Geyer et al., 1988). This is in contrast to most mannose-containing oligosaccharides of cellular glycoproteins which are predominantly sialylated at their termini (Baenziger and Green 1988 and Montreuil, 1984). The extent of glycosylation of gp120 seems to play an important role in virus-induced cell fusion as native gp120, but not unglycosylated recombinant gp120 protein backbone is involved in envelope-mediated syncytium formation (Putney et al., 1986; Walker et al., 1987; Gruters et al., 1987 and Montefiori et al., 1988). This observation is supported by studies reported in this review which show that mannose-binding lectins, e.g. Concanavalin A or Gerardia-lectin (Müller et al, 1988a) prevent virus infectivity and syncytium formation.

Table 1. Cytoprotective effects of Narcissus-lectin [NPL] and Gerardia-lectin [GL] on growth of different HIV-1 and HIV-2-infected cells. U927 monocytes and the human T cell lines MT-2 and CEM were infected with HIV-1_{IIIB} or different HIV-2 strains. The 50% cytoprotective concentration (IC_{50}) represents that concentration at which growth of HIV-infected cells reached 50% of the growth rate of the uninfected cells during an incubation period of 7 d; the 50% cytotoxic concentration (TC_{50}) represents the concentration which reduces the growth rate of infected cells to 50%; the antiviral index (AI) is calculated from the ratio TC_{50}:IC_{50}. n. d.: not determined. (Taken from Weiler et al., 1990; Müller et al., 1988b)

Virus/cells system	Lectin	IC_{50} (µg/ml)	TC_{50} (µg/ml)	AI
MT-2/HIV-1	NPL	4.97	> 100	> 25
	GL	0.35	> 50	>143
CEM/HIV-1	NPL	2.18	> 100	> 46
	GL	0.27	> 50	>185
U937/HIV-1	NPL	7.31	> 100	> 14
	GL	1.26	> 50	> 40
CEMx174/HIV-2_{ST}	NPL	6.58	> 100	> 15
	GL	n.d.	-	-
U937/HIV-2_{MS}	NPL	8.57	> 100	> 12
	GL	n.d.	-	-

Inhibition of HIV infection by lectins

In a basic study it was established that 65% of the intracellular HIV gp120 molecules contained partially processed 8 or 9 mannose residues in their glycan portion (Geyer et al., 1988). The high-mannose type N-linked glycans consist of four outermost α-1,2-linked mannose residues and two inner α-1,3- as well as two innermost α-1-6-linked mannose residues.

Table 2. Neutralization of HIV-1 induced Jurkat cell fusion by Narcissus-lectin [NPL] and Gerardia-lectin [GL]. $HIV\text{-}1_{IIIB}$-infected H9 cells were mixed with uninfected Jurkat cells either in the presence or absence of lectin (0 - 3 μg/ml); where indicated a monosaccharide (500 μM) was added. Syncytium formation was scored 4 and 24 h later [-, no syncytia; 1+, 2+, 3+ and 4+ increasing size of syncytia] (Taken from Weiler et al., 1990; Müller et al., 1988a)

Lectin	Concentration (μg/ml)	Monosaccaride (μM)	Syncytium formation 4 h	24 h
None	0	-	4+	4+
NPL	0.3	-	3+	3+
	1.0	-	1+	1+
	3.0	-	-	-
	3.0	D-mannose	4+	4+
	3.0	D-glucose	-	-
	3.0	D-fructose	-	-
	3.0	D-galactose	-	-
GL	0.3	-	2+	2+
	1.0	-	1+	1+
	3.0	-	-	-
	3.0	D-mannose	3+	3+

Lifeson et al. (1986) and Robinson et al. (1987) reported that the mannose-specific lectin concanavalin-A completely inactivated HIV. However, this lectin binds not only to the virus but also to the cell surface as well. More interestingly, two other lectins, one isolated from the coral Gerardia savaglia [Gerardia-lectin] (Kljajic et al., 1987) and the other from Narcissus pseudonarcissus [NPL] (VanDamme et al., 1988) displayed a strong inhibitory action on HIV infection in vitro by binding specifically to the carbohydrate part of gp120 (Müller et al., 1988a; Weiler et al., 1990); the anti-HIV effect was measured both in HIV-1 and HIV-2 systems. The antiviral effect of the two lectins was established not only on the basis of cytoprotection assays but also by measuring the inhibition of HIV production. As summarized in Table 1, the antiviral index (AI) of the two lectins is > 12 to > 185. These data were confirmed recently by Balzarini et al. (1991). Moreover, addition of the two lectins during co-incubation of Jurkat cells with H9 cells producing HIV-1 strongly inhibited syncytium formation mediated by the virus (Table 2). At concentrations of 3 μg/ml no syncytium formation of Jurkat cells could be measured. This is further proof that the lectins inhibit infection of cells by HIV-1. Addition of D-mannose to the cultures during the incubation periods of the lectins, abolished - as expected - the cytoprotective effect displayed by them; D-glucose, D-fructose and D-galactose had no influence.

In contrast to concanavalin-A, a lectin which is highly specific for α-(1,2)-linked mannose oligosaccharides and does not react with α-(1,3)-mannan (Liener et al., 1986), both the Gerardia-lectin and NPL display pronounced specificity for α-(1,3)-mannosyl residues (Kljajic et al., 1987; VanDamme et al., 1988). Hence we assume that these lectins bind preferentially to non-sialylated $Man_5GlcNAc_2$ oligosaccharide chains on gp120.

Search for model antigens to raise neutralizing antibodies

The mannan preparation (Zymosan A) from bakers' yeast (<u>Saccharomyces cerevisiae</u>) is known to have an α-1,6-linked mannose main chain and in the 2-position α-substituted α-1,2- and α-1,3-mannose residues (Gorin 1958). Applying the precipitation reaction described by So and Goldstein (1967) 10 μg of this mannan was reacted with 10 μg of the three mannose-specific lectins; under these conditions concanavalin A precipitated 7.5 μg of mannan, NPL 8.3 μg, and Gerardia-lectin 8.6 μg during a 1 h incubation period (37° C). As summarized in Table 3 the lectin-yeast mannan interactions were differentially inhibited by the mannan test preparations used. NPL and Gerardia-lectin function was efficiently inhibited by Man(α1-3)Man units and to a smaller extent by Man(α1-2)Man units; the reverse specificity has been already reported for concanavalin A (Liener et al., 1986). The mannan preparation from <u>Candida albicans</u> (serotype A) which has the following structure (Fukazawa et al., 1980):

M(α1-3)[M(α1-2)M]$_4$-M ┐
                                    [M-(1-6)M]$_2$
[M(α1-2)]$_5$-M ────────┘

strongly inhibited the mannan-NPL- and mannan-Gerardia-lectin precipitation; however, it is a poorer inhibitor of the interaction with concanavalin (50% inhibition at > 700 μg/ml).

Based on these findings we selected the two described mannan preparations as model antigens for raising neutralizing antibodies against gp120. First of all, we used the Zymosan A preparation which consists of 70-75% polysaccharide [comprising α-1,2- and α-1,3-mannose residues] and 13-17% protein (DiCarlo and Fiore, 1958); secondly, we chose the C. albicans mannan (serotype A) preparation [also α-1,2- and α-1,3-linked terminal mannose residues], which was covalently linked to bovine serum albumin [BSA] (Müller et al., 1991). In addition we used the mannan from C. albicans (serotype B) also coupled to BSA (Müller et al., 1991). This mannan has the following structure:

[M-(α1-2)]$_3$-M-(α1-3)-M-(α1-2)-M ┐
                                              [M-(1-6)M]$_2$
[M-(α1-3)]$_2$-[M-(α1-2)]$_2$-M ─────┘

<u>Table 3.</u> Inhibition of lectin-yeast mannan precipitation by different manno-oligosaccharides. The following lectins were used; concanavalin A [Con A], Gerardia-lectin [GL] and Narcissus-lectin [NPL]. M: mannose- and GN: GlcNAc unit. The 50 % inhibition values obtained are given in mM; the values for the C. albicans mannan are given in μg/ml.

Oligosaccharide	50% inhibition Con A	 GL	 NPL
M(α1-2)M	4.9 mM	6.3 mM	7.4 mM
M(α1-3)M	> 15 mM	0.7 mM	1.9 mM
M(α1-6)M	9.3 mM	4.9 mM	5.3 mM
M(α1-6)┐ M(α1-6)┐ M(α1-3)┘ M(β1-4)GN(β1-4)GN-Asn M(α1-3)┘	5.9 mM	0.8 mM	0.5 mM
M(α1-3)M(α1-2)M┐ C. albicans M(α1-2)M(α1-2)M┘	>700 μg	245 μg	157 μg

Neutralizing antibodies against the model antigens

Antibodies against Zymosan [termed: anti-Man-Z] and against C. albicans mannan both of serotype A and serotype B [termed: anti-Man-A and B] were raised in rabbits (Müller et al. 1989, 1990 a-c and 1991). The cultures were infected with the virus then cultivated for 4 to 12 d. As a measure of virus production the reverse transcriptase [RT] level was determined in the culture medium. As shown in Table 4 an anti-Man-Z concentration of 10 µg/ml reduced virus production by 99% during the incubation period of 4 d; the same inhibitory effect was measured with anti-Man-A. The neutralizing activity of the anti-Man-B preparation was less pronounced; at a concentration of 5 µg/ml virus release was inhibited by 74% and at a concentration of 10 µg/ml by 85%. After a prolonged period of incubation the extent of virus production increased again.

Table 4. Effect of anti-Man-antibodies on HIV production by H9 cells. The antibodies were used for the experiments in the HIV/H9 cell system at the indicated concentrations. To test for virus release, cells were removed by centrifugation and the supernatant was assayed for reverse transcriptase (RT) activity (Müller et al., 1988b) at day 4, 8 and 12.

Antibody concentration (µg/ml)		RT activity (cpm x 10^{-5}) days after infection 4 d	8 d	12 d
anti-Man-Z:	0	2.936	2.843	2.739
	5	0.857	1.963	2.593
	10	0.024	0.043	0.049
anti-Man-A:	0	2.936	2.843	2.739
	5	0.037	0.145	0.758
	10	0.015	0.029	0.041
anti-Man-B:	0	3.042	2.892	2.671
	5	0.793	1.854	2.490
	10	0.518	1.827	2.254

Western blotting studies revealed that the anti-Man antibodies reacted exclusively with HIV-gp120/gp160. Applying the technique of immunofluorescence (using FITC- or RITC-labelled anti-Man antibodies) we could not detect any interaction between the antibodies and untreated H9 cells. However, the anti-Man antibody preparations reacted with the cell surface of HIV-infected living H9 cells. Applying the technique of confocal laser scanning microscopy it was possible to demonstrate that the anti-Man antibody preparations reacted with the gp120 on the surface of HIV-infected living cells. The studies using the anti-Man antibodies (from rabbits) and subsequently the anti-gp120 antibodies (from sheep) revealed that both anti-Man-Z, anti-Man-A and anti-Man-B antibodies effectively mask the gp120 molecules on the cell surface (Müller et al., 1991).

Our findings which were confirmed recently (Hansen et al., 1990) support the strategy of using viral glycans (or their mimics) as candidates for antiviral immunotherapy or vaccine development. Animal studies are available which show that the mannan preparation, used in the present study, can be administered intraperitoneally (Bouwens and Wisse, 1988) and intravenously (Kai et al., 1988). Hence, it can be assumed that also in humans antibody reactions against that mannan antigen can occur which might neutralize HIV-1 and in consequence, prevent HIV-1 infection.

Lectin-based enzyme-linked immunoassay [Lectin-ELISA]

The second application of our findings, that NPL displays a specific anti-HIV activity, culminated in the development of a sensitive ELISA system for the quantification of gp120 in serum/media (Weiler et al., 1991); Fig. 1.

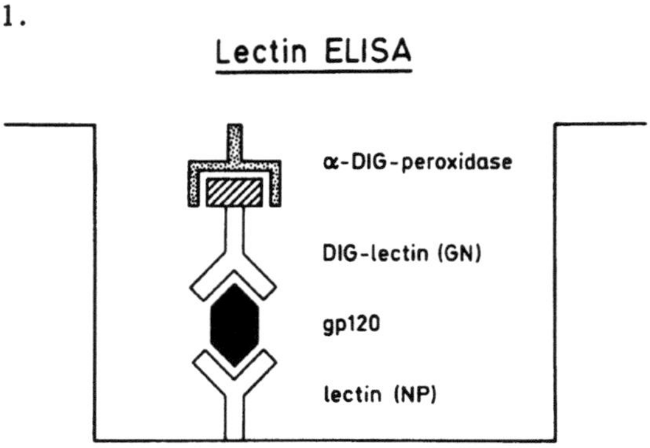

Fig. 1. Scheme of Lectin-ELISA. The NP-lectin was bound to the microtiter well; NPL displayed the ability to bind gp120. In the next step *Galanthus nivalis* (GN) lectin, covalently linked with digoxigenin (DIG), was bound to the formed gp120/NPL complex. This final complex was quantified by the α-DIG peroxidase reaction.

Microtiter plates were coated with NPL. Then 50 µl of a sample, containing gp120, was added (8 h; 4° C). After washing, digoxigenin-labelled lectin from Galanthus nivalis, also a mannose-specific lectin (Shibuya et al., 1988) was added and incubated. After washing, anti-digoxigenin/peroxidase was added and the immunocomplexes were detected and quantified by adding the substrate o-phenylene diamine (containing hydrogen peroxide); after 4 min the reaction was stopped and the optical density determined at 490 nm.

The ELISA test was used to determine the amount of gp120 in the supernatant of HIV-1 infected CEM cells at various periods of time after virus infection. As summarized in Table 5, 3 days p. i. RT activity could be measured in the culture supernatant indicating the release of virus particles into the medium. The appearance as well as the amount of the HIV-p24 antigen parallels that of RT. However, under the same in vitro conditions gp120 was detectable only 2 d p. i. and hence before the appearance of RT. This finding indicates that gp120 may be shed into the medium prior to the appearance of complete virus there.

Table 5. Lectin-based ELISA for determination of HIV-gp120. Determination of (i) virus production (measured on the basis of reverse transcription [RT] activity in the culture medium, (ii) the amount of HIV-1 p24 in the culture medium and (iii) the amount of gp120 in the culture medium at different periods of time after HIV-1 infection. The concentrations of gp120 were determined by the Lectin-ELISA.

Days p.i	RT activity (cpm x 10^{-5}/ml)	p24 (ng/ml)	gp120 (ng/ml)
0	< 100	< 1	< 0.6
1	< 100	< 1	< 0.6
2	< 100	<'1	2.1
3	350	1.2	12.5
4	12,400	43.7	262.9
5	41,750	142.5	918.7
6	53,150	189.1	1,058.4

The ELISA was also used as a screening system for compounds which block the binding of gp120. As an example, Sulphoevernan which binds to gp120 in the presence of Ca^{2+} (Weiler et al., 1990) inhibited at a concentration of 3 ng/assay binding of gp120 to NPL (Weiler et al., 1991). In a third application the ELISA was successfully used to detect antibodies recognizing carbohydrate moieties of gp120.

This ELISA system may also be suitable for measurement of the gp120 level in patients; perhaps it can be used as one prognostic parameter for estimating the progression of the disease.

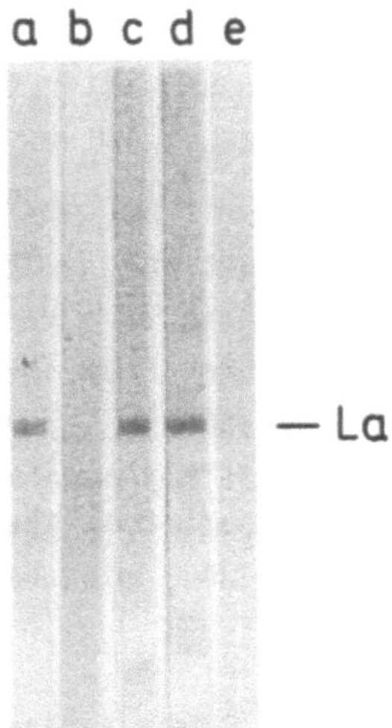

Fig. 2. Characterization of the nuclear La protein as a glycoprotein with lectin-like properties. Homogeneously purified La protein isolated from rat liver nuclei was blotted against the digoxigenin-labelled lectin from Galanthus nivalis (GNA) (lane a). After treatment with endoglycosidase H, the GNA binding affinity disappeared (lane b). Interestingly, if blotted La protein was incubated with a biotinylated La protein sample followed by an incubation with streptavidin-peroxidase complex it became evident that La protein has self-recognizing properties (lane c). Similarly, La protein had binding capability to other mannosylated proteins such as the gp-120 of HIV (lane d). In this case blotted La protein was in the first step incubated with gp-120, in a second step with an anti-gp-120 antibody developed in goat, followed by an incubation with an anti-goat peroxidase conjugate. Nitrocellulose strip containing blotted La protein was treated with La-absorbed GNA (lane e).

Interaction between gp120 and the nuclear lectin-like protein La

One of the targets of autoantibodies present in sera of patients with certain autoimmune diseases such as Lupus erythematosus is the so called "La protein". Recently we characterized this protein as a transcription termination factor of RNA polymerase III (Bachmann et al., 1990). Interestingly, the nuclear form of this protein seems to be a glycoprotein. La protein is stained during the proteoglycan reaction, which bases on a periodate oxidation and dioxygenation process, positively (not shown). Moreover, the nuclear La protein reacts well with the mannose specific lectin from Galanthus nivalis (GNA) (Fig. 2, lane a). This binding dissappears if La protein is treated with endoglycosidase H, which indicates that nuclear La protein belongs to the high mannose type glycoproteins (Fig. 2, lane b). The mannosylated La protein recognizes its own molecular species (Fig. 2, lane c) and also other mannosylated proteins, for example HIV-gp-120 (Fig. 2, lane d). Therefore, La protein represents a self binding nuclear lectin-like protein. The sugar residues reside in the 29 kD fragment of the 43 kD-La protein, which is formed during a proteolytic degradation process of La protein (not shown). Analyzing the cDNA sequence of La protein, published independently by several groups, one potential N-glycosylation site exists in this 29-kD region.

Acknowledgements
This work was supported by grants from the Bundesgesundheitsamt (AI 02 II-032-87), the Bundesministerium für Forschung und Technologie (under the coordination of the Kernforschungsanlage Jülich; no. 0319207A8) and the Yugoslav-German cooperation programme, coordinated by GKSS (Internationales Büro), Geesthacht.

References

Allan JS, Coligan JE, Barin F, McLane MF, Sodrowski JG, Rosen CA, Haseltine WA, Lee TH, Essex M (1985) Major glycoprotein antigens that induce antibodies in AIDS patients are encoded by HTLV-III. Science 228: 1091-1094

Baenziger JU, Green ED (1988) Pituitary glycoprotein hormone oligosaccharides: structure, synthesis and function of the asparagine-linked oligosaccharides on lutropin, follitropin and thyrotropin. Biochem Biophys Acta 947: 287-306

Bachmann M, Pfeifer K, Schröder HC, Müller WEG (1990) Characterization of the autoantigen La as a nucleic acid dependent ATPase/dATPase with melting properties. Cell 60: 85-63

Balzarini J, Schols D, Neyts J, VanDamme E, Peumans W, De Clercq E (1991) α-(1-3)- and α-(1-6)-D-mannose-specific plant lectins are markedly inhibitory to human immunodeficiency virus and cytomegalovirus infections in vitro. Antimicr Agents Chemother: in press

Berman PW, Groopman JE, Gregory T, Clapham PR, Weiss RA, Ferriani R, Riddle L, Shimasaki C, Lucas C, Lasky LA (1988) Human immunodeficiency virus type 1 challenge of chimpanzees immunized with recombinant envelope glycoprotein gp120. Proc Natl Acad Sci USA 85: 5200-5204

Bouwens L, Wisse E (1988) Tissue localization and kinetics of pit cells or large granule lymphocytes in the liver of rats treated with biological response modifiers. Hepatology 8: 46-52

Capon DJ, Chamow SM, Mordenti J, Marsters SA, Gregory T, Mitsuya H, Byrn RA, Lucas C, Wurm FM, Groopman JE, Smith DH (1989) Designing CD4 immunoadhesins for AIDS therapy. Nature 337: 525-532
Clapham PR, Weber JN, Whitby D, McIntosh K, Dalgleish AG, Maddon PJ, Deen KC, Sweet RW, Weiss RA (1988) Soluble CD4 blocks the infectivity of diverse strains of HIV and SIV for T cells and monocytes but not for brain and muscle cells. Nature 337: 368-370
Dalgleish AG, Beverley PCL, Clapham PR, Crawford DH, Greaves MP, Weiss RA (1984) The CD4 antigen is an essential component of the AIDS retrovirus. Nature 312: 763-767
DiCarlo FJ, Fiore JV (1958) On the compositon of Zymosan. Science 127: 756-757
Fisher RA, Bertonis JM, Meier W, Johnson VA, Costopoulos DS, Liu T, Tizard R, Walker BD, Hirsch,MS, Schooley RT, Flavell RA (1988) HIV infection is blocked in vitro by recombinant soluble CD4. Nature 331: 76-78
Fukazawa Y, Nishikawa A, Suzuki M, Shinoda T (1980) Immunological basis of the serologic specificity of the yeast: immunochemical determinants of several antigenic factors of yeasts. In: Preusser H (ed) Medical mycology, Zbl Bakt Suppl 8. Gustav Fischer, Stuttgart, pp 127-136
Geyer H, Holschbach C, Hunsmann G, Schneider J (1988) Carbohydrates of human immunodeficiency virus. J Biol Chem 262: 11760-11767
Gorin PAJ (1973) Rationalization of carbon-13 magnetic resonance spectra of yeast mannans and structurally related oligosaccharides. Can J Chem 51: 2375-2383
Goudsmit J, Debouck C, Meloen RH, Smit L, Bakker M, Asher DH, Wolff AV, Gibbs CJ, Gajdusek DC (1988) Human immunodeficiency virus type 1 neutralization epitope with conserved architecture elicits early type specific antibodies in experimentally infected chimpanzees. Proc Natl Acad Sci USA 85: 4478-4482
Gruters, RA Neefjes JJ, Tersmette M, Matthijs de Goede REY, Tulp A, Huisman HG, Miedema F, Ploegh HL (1987) Interference with HIV-induced syncytium formation and viral infectivity by inhibitors of trimming glycosidase. Nature 330: 74-77
Hansen JES, Clausen H, Nielsen C, Teglbjaerg LS, Hansen LL, Nielsen CM, Dabelsteen E, Mathiesen L, Hakomori SI, Nielson JO (1990) Inhibition of human immunodeficiency virus (HIV) infection in vitro by anticarbohydrate monoclonal antibodies. J Virol 64: 2833-2840
Hu SL, Fultz P, McClure HM, Eichberg JW, Thomas EK, Zarling J, Singhal MC, Kosowski SG, Swenson RB, Anderson DC (1987) Effect of neutralization with a vaccinia-HIV env recombinant on HIV infection of chimpanzees. Nature 328: 721-723
Hussey RE, Richardson NE, Kowalski M, Brown NR, Chang HC, Siliciano RF, Dorfman T, Walker B, Sodroski J, Reinherz EL (1988) A soluble CD4 protein selectively inhibits HIV replication and syncytium formation. Nature 331: 78-81
Kai C, Szigeti R, Aman P, Klein E (1988) Activation of B lymphocytes with human serum-treated zymosan. Cell Immunol 116: 467-474
Klatzmann D, Champagne E, Chamaret S, Gruest J, Guetard D, Hercend T, Gluckman JC, Montagnier L (1984) T-lymphocyte T4 molecule behaves as the receptor for human retrovirus LAV. Nature 312: 767-768
Kljajic Z, Schröder HC, Rottmann M, Cuperlovic M, Movesian M, Uhlenbruck G, Gasic M, Zahn RK, Müller WEG (1987) A D-mannose-specific lectin from Gerardia savaglia that inhibits nucleocytoplasmic transport of mRNA. Eur J Biochem 169: 97-104

Liener IE, Sharon N, Goldstein IJ (eds) (1986) The Lectins. Academic Press Orlando
Lifeson J, Coutre S, Huang E, Engelman E (1986) Role of envelope glycoprotein carbohydrate in human immunodeficiency virus (HIV) infectivity and virus-induced cell fusion. J Exp Med 164: 2101-2106
Mizuochi T, Spellman MW, Larkin M, Solomon J, Basa LJ, Feizi T (1988) Carbohydrate structures of the human-immunodeficiency-virus (HIV) recombinant envelope glycoprotein gp120 produced in Chinese-hamster ovary cells. Biochem J 254: 599-603
Montefiori DC, Robinson WE, Mitchell WM (1988) Role of protein N-glycosylation in pathogenesis of human immunodeficiency virus type 1. Proc Natl Acad Sci USA 85: 9248-9252
Montreuil J (1984) Spatial conformation of glycans and glycoproteins. Biol Cell 51: 115-132
Müller WEG, Renneisen K, Kreuter MH, Schröder HC and Winkler I (1988a) The D-mannose-specific lectin from Gerardia savaglia blocks binding of human immunodeficiency virus type I to H9 cells and human lymphocytes in vitro. J Acquir Immun Defic Syndr 1: 453-458
Müller WEG, Sarin PS, Sun D, Rossol S, Voth R, Rottmann M, Hess G, Meyer zum Büschenfelde KH, Schröder HC (1988b) Dual biological activity of apurinic acid on human lymphocytes. Antiviral Res: 191-204
Müller WEG, Reuter P, Schaffrath M, Schröder HC and Winkler I (1989) Antibodies to mannan inhibit infection of lymphocytes by HIV-1 in vitro. AIDS-Forschung (AIFO) 4: 465-466
Müller WEG, Reuter P, Maidhof A, Schröder HC, Uhlenbruck G, and Winkler I (1990a) Mannan: An antigen for raising neutralizing polyclonal antibodies to the AIDS virus. In: Schauzu M (ed) Progress in AIDS Research in the Federal Republic of Germany (bga-Schriften 1/1990). MMV Medizin-Verlag München, pp 205-209
Müller WEG, Schröder HC, Reuter P, Maidhof A, Dindorf W, Uhlenbruck G, und Winkler I, (1990b) Neutralisierende, polyclonale Antikörper gegen das Humane Immundefizienz-Virus Typ 1 (HIV-1), die die Kohlenhydrat-Struktur des gp120 erkennen. AIFO 5: 78-82
Müller WEG, Schröder HC, Reuter P, Maidhof A, Uhlenbruck G, and Winkler I (1990c) Polyclonal antibodies to mannan from yeast also recognize the carbohydrate structure of gp120 of the AIDS virus: an approach to raise neutralizing antibodies against HIV infection in vitro. AIDS 4: 159-162
Müller WEG, Bachmann M, Weiler BE, Schröder HC, Uhlenbruck G, Shinoda T, Shimizu H, and Ushijima H, (1991) Antibodies against defined carbohydrate structures of Candida albicans protect H9 cells against infection with human immunodeficiency virus-1 in vitro. J Acquir Immun Def Syndr, in press
Putney SD, Matthews TJ, Robey WG, Lynn DL, Robert-Guroff M, Mueller WT, Langlois AJ, Ghrayeb J, Petteway SR, Weinhold KJ, Fischinger PJ, Wong-Staal F, Gallo RC, Bolognesi DP (1986) HTLV-III/LAV-neutralizing antibodies to an E. coli-produced fragment of the virus envelope. Science 234: 1392-1395
Richardson NE, Brown NR, Hussey RE, Vaid A, Matthews TJ, Bolognesi DP, Reinherz EL (1988) Binding site for human immunodeficiency virus coat protein gp120 is located in the NH_2-terminal region of T4 (CD4) and requires the intact variable-region-like domain. Proc Natl Acad Sci USA 85: 6102-6106
Robinson WE, Montefiori DC, Mitchell WM (1987) Evidence that mannosyl residues are involved in human immunodeficiency virus type 1 (HIV-1) pathogenesis. AIDS Res and Human Retrov 3: 265-281

Sattentau QI, Weiss RA (1988) The CD4 antigen: physiological ligand and HIV receptor. Cell 52: 631-633
Shibuya N, Goldstein IJ, VanDamme ELM, Peumans WJ (1988) Binding properties of a mannose-specific lectin from the snowdrop (Galanthus nivalis) bulb. J Biol Chem 263: 728-743
Smith DH, Byrn RA, Marsters SA (1987) Blocking of HIV-1 infectivity by a soluble, secreted form of the CD4 antigen. Science 328: 1704-1707
So LL, Goldstein IJ (1967) Protein-carbohydrate interaction; IV. J Biol Chem 242: 1617-1622
Traunecker A, Luke W, Karjalainen K (1988) Soluble CD4 molecules neutralize human immunodeficiency virus type 1. Nature 331: 84-86
VanDamme EJ, Allen AK, Pneumans WJ (1988) Related mannose-specific lectins from different species of the family Amaryllidaceae. Physiol Plantar 73: 52-57
Walker BD, Kowalski M, Goh WC, Kozarsky K, Krieger M, Rosen C, Rohrschneider L, Haseltine WA, Sodroski J (1987) Inhibition of human immunodeficiency virus syncytium formation and virus replication by castanospermine. Proc Natl Acad Sci USA 84: 8120-8124
Weiler BE, Schröder HC, Stefanovich V, Stewart D, Forrest JMS, Allen LB, Bowden J, Kreuter MH, Voth R, and Müller WEG (1990) Sulfoevernan, a polyanionic polysaccharide, and the Narcissus lectin potently inhibit HIV infection by binding to viral envelope protein. J Gen Virol 71: 1957-1963
Weiler BE, Schäcke H., Bachmann M, Mills J, Gilbert M, Brigido L, Matthes E, Forrest J, and Müller WEG (1991) Human immunodeficiency virus: novel enzyme-linked immunoassays for the quantitation of envelope glycoprotein 120. Application in diagnosis and drug screening. J Virol Methods, in press

Lectin histochemistry of human thyroid tumors

I. Damjanov and M. Sobrinho-Simoes

Thomas Jefferson University
Jefferson Medical College
Philadelphia, PA 19107, USA

Abstract. The binding of fluorescein isothiocyanate (FITC) labeled lectins to thyroid cells in histologic sections was analyzed and the data obtained on tumors of several histologic types were compared. The affinity and pattern of lectin binding to tumor cells differed from the binding to normal cells. Most tumors showed considerable cellular heterogeneity which could be visualized readily by lectin histochemistry. No constant or predictable lectin binding patterns typical of malignant transformation could be delineated. The variation in the expression of lectin binding glycoconjugates in tumor cells in histologic sections limits the use of labeled lectins in diagnostic histopathology. Nevertheless FITC labeled lectins may serve as valuable reagents for probing the heterogeneity of tumors, identification and/or isolation of tumor associated and specific glycoconjugates and for studying tumor cell biology.

Introduction

Lectin isolated from plants or from lower animals on the basis of their carbohydrate binding specificity may be readily derivatized and used as FITC or horseradish peroxidase labeled cytochemical or histochemical reagents (for review see Damjanov, 1987). We have used lectin histochemistry to study the expression of complex glycoconjugates in various biologic processes such as the formation of trophoblast in early mouse embryos (Wu et al., 1983), spermatogenesis (Lee and Damjanov, 1984) or the uterine changes that occur during pregnancy (Lee and Damjanov, 1985). Lectin histochemistry was also used to define differences in the cellular composition of normal adult organs such as the endocervix (Yen et al., 1986) or the gallbladder (Karayannopoulou and Damjanov, 1987). The changing patterns of expression of glycoconjugates as found in pathologic conditions like metaplasia (Karayannopoulou and Damjanov, 1988) or neoplasia (Lee et al., 1985) have also been explored in several organ systems.

Commercial availability of derivatized plant and animal lectins has made it possible to explore their use in oncologic histopathology. However, despite numerous efforts, the overall results have been less than encouraging (Walker, 1989) and the diagnostic application of lectin histochemistry has not produced yet notable results, nor has it led with a few notable exceptions, to routine usage of this technique in diagnostic pathology. Nevertheless the use of lectins for probing cellular glycoconjugates has been expanding, and therefore it seems appropriate to discuss the advantages and disadvantages of this approach in histopathology, and illustrate salient points with data obtained in the study of thyroid tumors.

Materials and Methods

Formaldehyde-fixed, paraffin-embedded tissues were retrieved from the contributing pathology departments. The blocks were sectioned routinely and slides were studied by lectin histochemistry as reported previously (Lee and Damjanov, 1984; Yen et al., 1988). The lectins used in the study are listed in Table 1 together with the inhibiting sugars. Some of the material discussed here has been published before (Sobrinho-Simoes and Damjanov, 1986; Sobrinho-Simoes et al., 1990).

Table 1. - Lectins Used to Study Thyroid Tumors

Lectin	Source	Inhibiting Sugar
ConA	Canavalia ensiformis	Methyl α-mannopyranoside
LCA	Lens culinaris	Methyl α-mannopyranoside
WGA	Triticum vulgare (wheat germ)	D-GlcNAc
LAA	Laburnum alpinum	D-GlcNAc
STA	Solanum tuberosum	D-GlcNAc
GSA-II	Griffonia simplicifolia	D-GlcNAc
DBA	Dolichos biflorus	D-GalNAc
SBA	Glycine maximus	D-GalNAc
VVA	Vicia villosa	D-GalNAc
HAA	Helix aspersa	D-GalNAc
HPA	Helix pomatia	D-GalNAc
MPA	Maclura pomifera	D-GalNAc
PHA-E	Phaseolus vulgaris (E_4)	D-GalNAc
PHA-L	P vulgaris (L_4)	D-GalNAc
PNA	Arachis hypogaea (peanut)	D-Gal
GSA-I	G simplicifolia	D-Gal
RCA-I	Ricinus communis	D-Gal
RCA-II	R communis	D-Gal
UEA-I	Ulex europaeus	L-Fucose
LTA	Lotus tetragonolobus	L-Fucose

Results and Discussion

The normal thyroid gland reacts extensively with several lectins. However, none of the lectins could be used as a unique marker for any of the histologic components of this endocrine gland (Sobrinho-Simoes and Damjanov, 1986). Malignant growth was usually associated with altered lectin binding (Fig. 1). This change could involve a loss of lectin binding sites (Fig. 1a), or a change in the pattern of lectin binding (Fig. 1b,c). Apical surface staining of papillary carcinoma with Con A (Fig. 1b) or diffuse cytoplasmic staining of poorly differentiated follicular carcinoma (Fig. 1c) with LCA illustrate the changes associated with malignancy. These changes reflect in part the newly acquired patterns of growth, such as papillary or

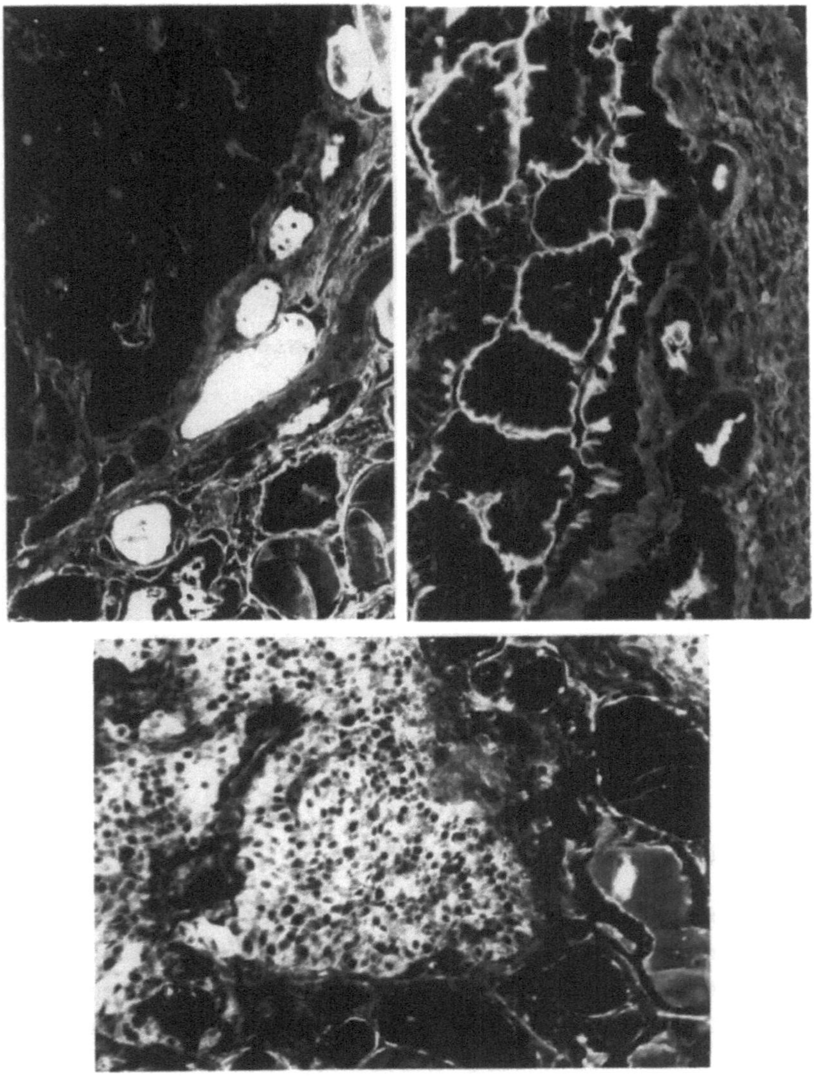

Fig. 1. Tumors may show a different lectin binding pattern than the normal thyroid.
a. The normal thyroid (lower part) shows strong reactivity with LCA, whereas the follicular carcinoma (left upper corner) is unreactive. x 160.
b. Papillary carcinoma (left part) shows apical staining of tumor cells with FITC-Con A. A few follicles with Con A reactive colloid are seen in to the right. x 160.
c. Con A reacts diffusely with the cytoplasm of this solid carcinoma infiltrating between the thyroid follicles. The colloid and the follicular cells are partially reactive with Con A. x 160.

solid arrangement of cells, but could also be related to restructuring at the cellular level. Papillary tumors seem to be composed of polarized cells, whereas the cells forming solid tumors do not show any preferential orientation. The favorable prognosis of papillary tumors may be in part due to the well differentiated nature of tumor cells and their limited propensity for invasion.

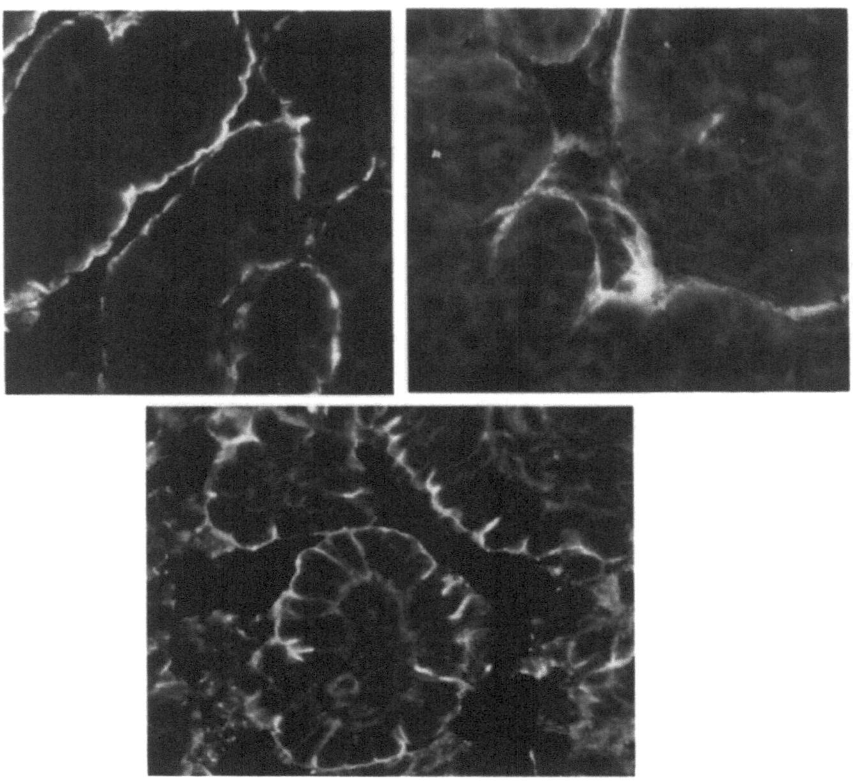

Fig.2. Apical lectin binding of papillary carcinoma cells.
a. FITC - STA binds strongly to essentially all tumor cells. x 240.
b. FIT-DBA binds strongly to some tumor cells and is unreactive with others. x240.
c. FTIC-HPA reacts with the apical cell surface and the lateral cell borders. x 240.

Apical cell membrane of papillary tumor reacted with many more FITC lectins than the normal thyroid cells. These include, among others, Con A, STA, MPA, HPA, SBA, RCA-I and UEA-I (Fig. 2), indicating that the apical surface of papillary tumor cells is rich on glycoconjugates expressing N-acetyl-D-glucosamine, N-acetyl-D-galactosamine and alpha-L-fucose residues respectively. The apical surface of these tumor cells is covered with microvilli, which are not found on normal thyroid cells. The appearance of microvilli seems to be associated with newly acquired biochemical cell membrane properties, which probably reflect new functions assigned to this cell surface. The nature of these functional changes and the significance of cell surface specialization remain unknown. It should, however, be noted that apical surface of papillary tumors show considerable lectin binding heterogeneity with regard to some lectins, such as DBA (Fig. 2b). The reactivity of other lectins, such as HPA (Fig. 2c) is not limited to the apical cell surface but extends to the lateral cell membranes as well. Hence, one can conclude that the apical cell membrane differs from other cell surfaces but at the same time has some features in common with at least the contiguous lateral cell membranes. Furthermore some lectins, such as WGA reacted with the basolateral aspects of cell membranes in a punctuate manner suggestive of reactivity with cell-matrix or cell-cell junctions (Fig. 3). On the basis of all this one could conclude that the lectins reveal complex tumor cell surface heterogeneity, reflecting their heterogenous functions and specialization.

Several lectins were found to be reactive with tumors cell cytoplasm rather than with the cell membranes, although some reacted with both. Lectins that react with cytoplasm of tumor cells often reveal internal structural disorganization (Fig. 4). In better differentiated tumors the same lectin may, however, react exclusively with the Golgi apparatus. These cells display marked uniformity and all show comparable staining (Fig. 5). Likewise some lectins like MPA react with the cytoplasmic granules

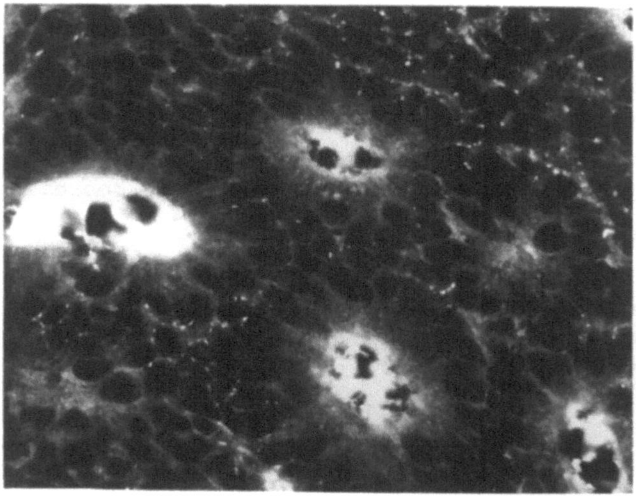

Fig. 3. FITC-WGA reacts with the colloid in the follicular carcinoma but also in a punctate manner with baso-lateral cell membranes. x 240.

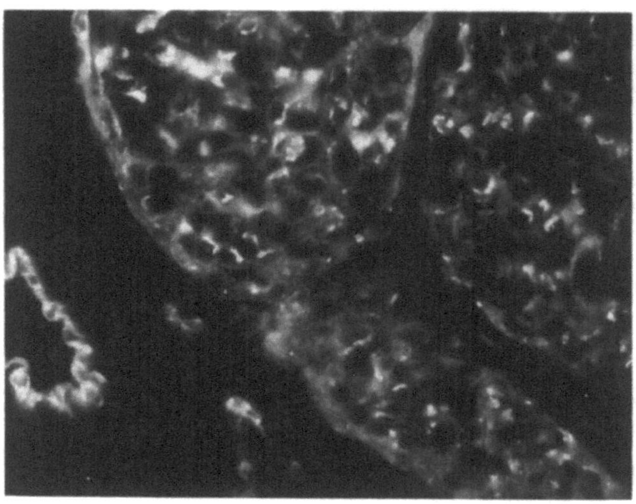

Fig.4. Irregular staining of a solid poorly differentiated follicular carcinoma with FITC-HPA. x 240.

probably corresponding to the secretory material in the rough endoplasmic reticulum and the Golgi apparatus (Fig. 6). Finally there are lectins, like STA, MPA and WGA (Fig. 3) which react also with the extracellular colloid. The components of the colloid synthesized in the cytoplasm may be unreactive but becomes lectin reactive upon secretion into the intercellular spaces.

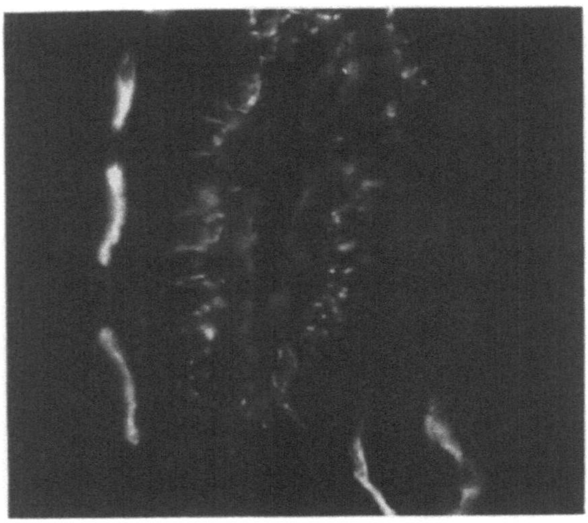

Fig.5. FITC-HPA reacts with the Golgi apparatus of a well differentiated follicular carcinoma. The tumor was from a patient of A blood group and therefore the blood vessels are also reactive. x 240.

Most notable is the fact that the colloid produced by tumor cells may differ from that produced by normal thyroid cells. The colloid produced by tumors of different histologic types such as follicular and medullary carcinoma may have common lectin binding properties.

Lectins reacting with the cytoplasm of tumor cells reveal remarkable heterogeneity within tumor cell populations (Fig. 7). In contrast to cell surface binding lectins, which illustrate the heterogeneity of papillary and follicular tumors, the lectins binding to the cytoplasmic

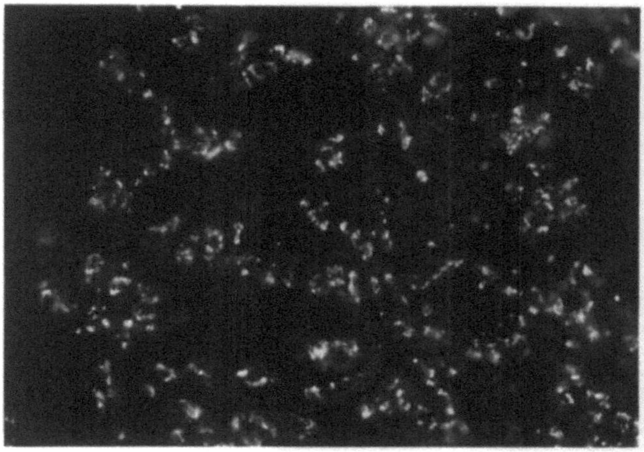

Fig.6. FITC-MPA reacts in a coarsely granular pattern with cell cytoplasm of a well differentiated follicular carcinoma. The granules correspond to the secretory material in the endoplasmic reticulum and the Golgi apparatus. x 160.

glycoconjugates reveal, in a more pronounced manner, the heterogeneity of solid tumors.

Conclusions

Lectin histochemistry of solid tumors has so far proven to be of little practical value in diagnostic pathology. Nevertheless, this approach provides a remarkable new dimension to histopathology and could serve as a bridge between connecting morphology and biochemical oncology. The morphologic patterns revealed in tumor tissue sections by lectin histochemistry evade for the time being unequivocal explanations and lack correlations with functional or cell biological parameters. Nevertheless, we believe that the leads from lectin histochemical studies of tumors could serve as a starting point for biochemical isolation of tumor associated glycoconjugates and elucidation of the role that the glycoconjugates play in malignancy.

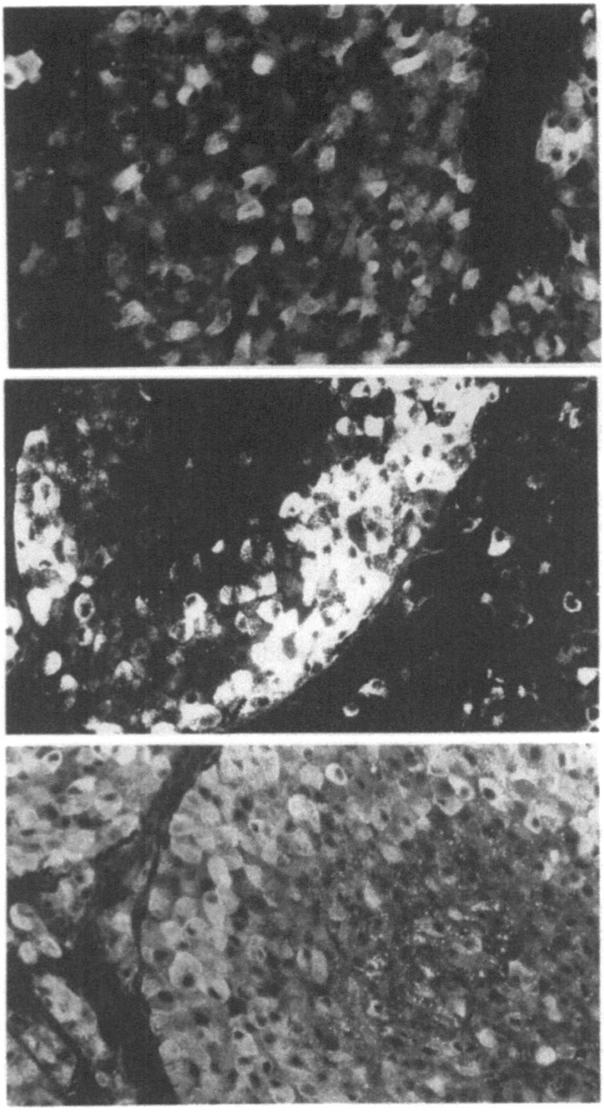

Fig.7. Three solid poorly differentiated follicular carcinomas reacted with FITC-MPA show different forms of tumor cell heterogeneity. x 160.

References

Damjanov I (1987) Lectin cytochemistry and histochemistry. Lab Invest 57:5-20

Karayannopoulou G, Damjanov I (1987) Lectin binding sites in the human gallbladder and cystic duct. Histochemistry 88:75-83

Karayannopoulou G, Damjanov I (1988) Lectin histochemistry of goblet cells in metaplastic epithelium of human gallbladder. Modern Pathol 1:132-134

Lee M-C, Damjanov I (1984) Anatomic distribution of lectin binding sites in mouse testis and epididymis. Differentiation 27: 74-81

Lee M-C, Damjanov I (1985) Pregnancy-related changes in the human endometrium revealed by lectin histochemistry. Histochemistry 82:275-280

Lee M-C, Talerman A, Oosterhuis JW, Damjanov I (1985) Lectin histochemistry of classical and spermatocytic seminoma. Arch Pathol Lab Med 109: 938-942

Sobrinho-Simoes M, Damjanov I (1986) Lectin histochemistry of papillary and follicular carcinoma of the thyroid gland. Arch Pathol Lab Med 110:722-729

Sobrinho-Simoes M, Sambade C, Nesland JM, Holm R, Damjanov I (1990) Lectin histochemistry and ultrastructure of medullary carcinoma of the thyroid gland. Arch Pathol Lab Med 114:369-375

Walker, RA (1989) The use of lectins in histopathology. Pathol Res Pract 185:826-835

Wu T-C, Wan Y-J, Damjanov I (1983) Fluorescein conjugated Bandeiraea simplicifolia lectin as a marker of endodermal, yolk sac and trophoblastic differentiation in the mouse embryo. Differentiation 24:55-59

Yen Y, Lee M-C, Salzmann M, Damjanov I (1986) Lectin binding sites on human endocervix: A comparison with secretory and proliferative endometrium. Anat Rec 215:262-266

Yen Y, Schmiemann C, Damjanov I (1988) Lectin histochemistry of uterine adenocarcinoma. Arch Pathol Lab Med 112:791-793

Neoglycoconjugates: Fundamentals and recent progresses

Y. C. Lee and R. T. Lee

Biology Department
Johns Hopkins University
3400 North Charles Street
Baltimore, Maryland
21218 U. S. A.

1. Introduction

Studies of lectins have been greatly benefitted by the use of synthetic carbohydrate conjugates. Proteins modifies with carbohydrate derivatives can be used for precipitin reactions or binding studies. Solid matrices containing carbohydrates of definitive structure can be used for isolation of lectins or cell adhesion studies. We have coined a term "neoglycoproteins" to describe the proteins artificially modified with carbohydrates of known structure. The usage of prefix "neo" was later extended to lipids modified with carbohydrates ("neoglycolipids"), and currently "neoglycoconjugates" is used to describe artificially constructed carbohydrate conjugates, including derivatives of proteins, lipids, and many other types of compounds, such as solid matrices ("glyco-matrices").

Several reviews have already appeared on the subject of "neoglycoproteins" and "neoglycoconjugates" (Stowell & Lee; 1980; Aplin & Wriston, 1981; Lee & Lee, 1982; & Magnusson, 1986). In this article, only a brief account will be given on what have been reviewed, and more emphasis will be placed on the newer development. Some workers prefer the use of "pseudo" in lieu of "neo" as a prefix to this group of compounds. However, the term "pseudo-sugar" is in wide use to indicate those compounds in which the ring oxygen is replaced with carbon -- i.e., "Carba-sugars" (Suami & Ogawa, 1990). Thus the use of "pseudo" may cause confusion and is not recommended.

2. Modification of proteins with carbohydrate derivatives

Table I. Various types of conjugation between carbohydrates and proteins (see Stowell & Lee, 1980; Aplin & Wriston; 1981; Lee & Lee, 1982; & Magnusson, 1986).

Protein side chains (amino acids)	Functional group of carbohydrate der. (reaction type)
amino (Lys, N-terminals)	carboxyl (amidation) carbonyl (amination) imidate (amidine) isothiocyanate (thioureidation)
phenol (Tyr)	diazo (diazo coupling)
thiol (Cys)	maleimido (Michael addition)
carboxyl (Asp, Glu, C-terminals)	amino (amidation)

<u>a. Side chains of proteins useful for modification</u> --- Proteins represent a group of compounds quite suitable for modification with carbohydrate derivatives, because of their solubility in aqueous solution and their variety in the side chains that can be used for modification. In addition, they are very often bioactive (enzymes, immunoglobulins, hormones, etc.), and their modification with carbohydrates provides additional functions (e.g., targeting, adhesion, stability). Carbohydrates can be even attached to natural glycoproteins, if need be. In such an instance, the endogenous carbohydrates can be removed or

at least partially destroyed so that they do not participate in the intended use of newly formed neoglycoproteins. In such a case, the use of thioglycosides is advantageous (Stowell & Lee, 1980, Magnusson, 1986).

In **Table** I are listed protein side chains and the linkage types useful for neoglycoproteins. As can be seen, the most frequently used functional group for the modification of proteins is ε-amino group. This is because the ε-amino groups are usually quite accessible, and also because the chemistry of modification of amino groups is well developed in conjunction with peptide synthesis.

b. Improved methods for reductive amination --- Reductive amination (Gray, 1974, Stowell & Lee, 1980, Magnusson, 1986) is one of the gentlest methods for attaching reducing oligosaccharides to proteins or other materials containing amino-groups. It is widely used because of the operational simplicity. The advantage of this method is that the reaction is carried out at near neutrality and at room temperature. The modified amino groups remain positively charged. With ready availability of "endo"-ß-hexosaminidase or glycopeptide amidases (Maley *et al.*, 1989), the preparation of reducing oligosaccharides from N-glycosides of glycoproteins is quite feasible. The original application of reductive amination for preparation of neoglycoprotein was with sodium cyanoborohydride (Gray, 1974), but pyridine borane or other borane complexes (Cabacungan *et al.*, 1982) may be the attractive alternative. The disadvanatage of this method is that the rate of reaction is rather slow, frequently requiring several days (Gray, 1974). The slow rate of the reaction is presumably caused by the low concentration of the acyclic form of reducing sugar. Thus, one way to overcome this shortcoming is to use 1-thioglycosides possessing ω-aldehydo-group in the aglycon to facilitate rapid reductive amination (Lee & Lee, 1980). A similar principle was used for a heterobifunctional reagent to be described below (5, c). Another recommended remedy is to use borate buffer in the reaction mixture, which was reported to accelerate the rate of reaction in some cases (Roy *et al.*, 1984).

Reductive amination necessarily converts the reducing terminus into an acyclic form. If sacrificing the ring structure of the reducing terminus is acceptable, it is possible to improve both the reaction rate and efficiency of conjugation of the reductive amination by first performing selective periodate oxidation of the reducing terminus to create free aldehydo group(s). This approach is especially useful for preparation of neoglycoconjugates from O-glycosides which are usually obtained by ß-elimination followed by reduction of the released terminus. Exploitation of this approach gave birth to a new class of neoglycolipids (Tang & Feizi, 1987, Stoll *et al.* , 1990), which proved to be useful for thin layer chromatographic separation of oligosaccharide components (as "neoglycolipids"), and even for mass-spectroscopic analysis of the separated components (Lawson *et al.*, 1990).

c. Conjugation of polysaccharides to proteins --- A successful application of the above-mentioned selective periodate oxidation of reduced polysaccharide is reported for the modification of hyaluronic acid with 1,6-diaminohexane (Raja *et al.* , 1984). Another interesting way of incorporating an aldehydic function to polysaccharides is to react a polysaccharide with chloroacetaldehyde dimethyl acetal. After deacetalization, the modified polysaccharide is ready to be conjugated to proteins by reductive amination (Bøgwald *et al.*, 1986). This method, however, cannot specifically direct the point of modification on the polysaccharides.

Certain glycosaminoglycans (e.g., heparin) possess free or potentially free amino groups, and after treating with nitrous acid, 2,5-anhydro-D-mannose residue is generated with concomitant glycosidic bond cleavage. Small oligosaccharides containing this active aldehydo-group at the reducing terminus is quite suitable for conjugation to proteins by reductive amination.

3. Glycosides useful for preparation of neoglycoconjugates

<u>a. ω-Unsaturated glycosides</u> --- Allyl glycosides have been used as co-monomer in preparation of polyacrylamide gel for affinity gel electrophoresis (Hořejší & Kocourek, 1974). Similar gel layers prepared with longer aglycons were tested for rat hepatocyte culture (Kobayashi & Akaike, 1990). Thioglycosides with a terminal double bond were conveniently synthesized by addition of the SH-group of 1-thio-sugars to <u>bis</u>-methylenediacrylamide (Lee *et al.*, 1979). n-Pentenyl glycosides, originally developed for activation of anomeric carbon, and utilized successfully in synthesis of some complex oligosaccharides can also be used as co-monomer as illustrated in Fig. 1 (Mootoo *et al.*, 1989).

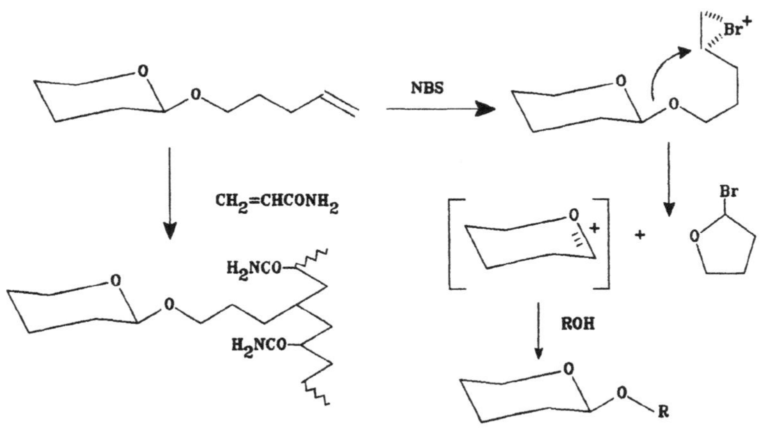

Fig. 1. n-Pentenyl glycosides for glycosylation and polymerization.

Glycosides with shorter and longer aglycon with ω-unsaturation, though not useful for glycosylation, can also be used for polymerization. More recently, Nishimura et al. applied the same principle to glycosides with terminal-unsaturated aglycon of C3, C4, C5, and C11 length (Nishimura *et al.* , 1990, Akaike *et al.*, 1989).

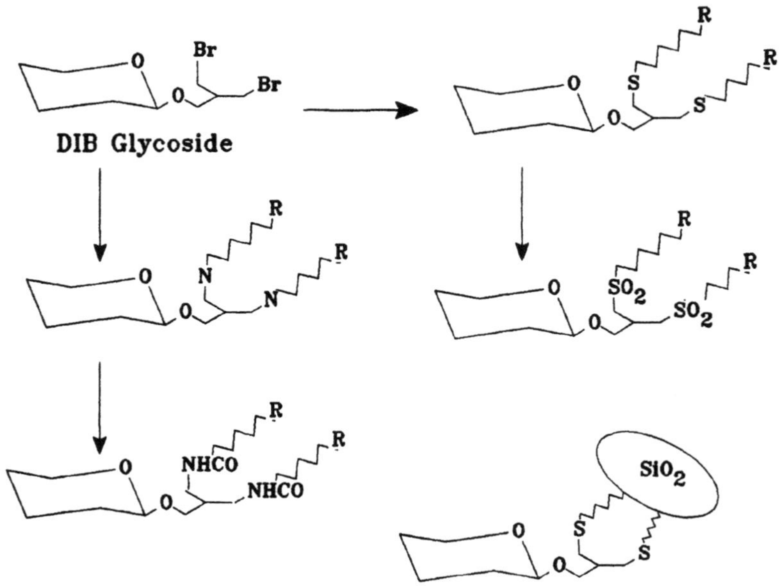

Fig. 2. DIB Glycosides for neoglycoconjugates.

__b. Bromoalkyl glycosides__ --- An interesting versatile approach has been proposed by Magnusson et al. for preparation of a wide variety of neoglycoconjugates (Magnusson, 1986, Magnusson *et al.* , 1990). In the most recently described general account of this approach, dibromoisobutyl (DIB) glycoside is used as the common intermediate (Fig. 2). The DIB glycoside can be reacted with thiols to yield __bis__-sulfide, which can be oxidized into sulfones to increase hydrophilicity. Alternatively, the bromo groups can be transformed into amino groups, which in turn can be acylated with an appropriate carboxyl group. The bromide can also be converted to alcohols, which allows other modifications.

One of the major applications of this approach is to prepare double-pronged neoglycolipids which resemble closely natural glycolipids. The two-pronged aglycons can also be used for immobilization of the glycosides onto silica gel surface. The DIB group can be converted to allylic bromide by the action of tetrabutylammonium fluoride. This provides an avenue for a single-chain aglycon.

4. Conjugation of glycosylamine derivatives

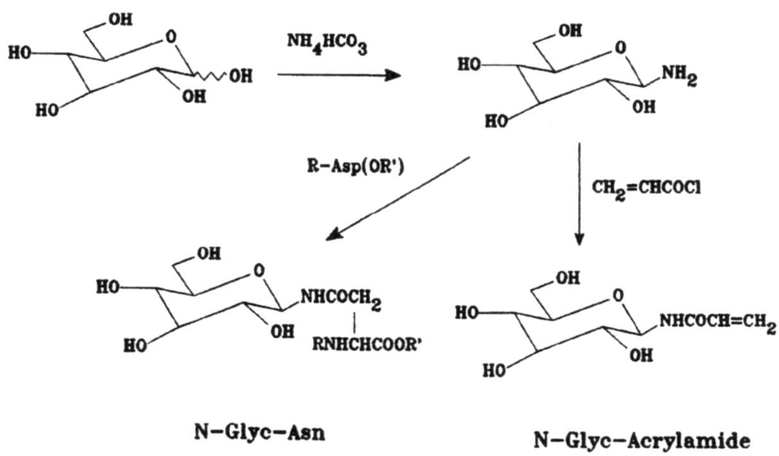

Fig. 3. Conjugation of glycosylamines

Since natural glycoproteins contain N-glycosides of Asn, attempts have been made over the past two decades to chemically synthesize "natural type" glycopeptides. Chemical syntheses of glycopeptides containing N-glycoside have been reviewed (Garg & Jeanloz, 1985). The most common method for preparation of an Asn-linked oligosaccharide is to prepare glycosyl halide, which is converted to glycosyl azide. Hydrogenolysis of the azidogroup yields glycosylamine which is coupled to ß-COOH of suitably protected aspartic acid derivative. In a more convenient method (Likhosherstov *et al.*, 1990; Chernyak *et al.* , 1990), reducing oligosaccharides are kept in saturated ammonium bicarbonate for several days, and after removal of excess ammonium bicarbonate, the glycosylamine is purified with a column of cation exchange resin. The glycosylamine can be acryloylated for polymerization or acylated with Asp-derivatives to form Asn-oligosaccharides (**Fig.** 3).

5. Conjugation *via* glycamine derivatives

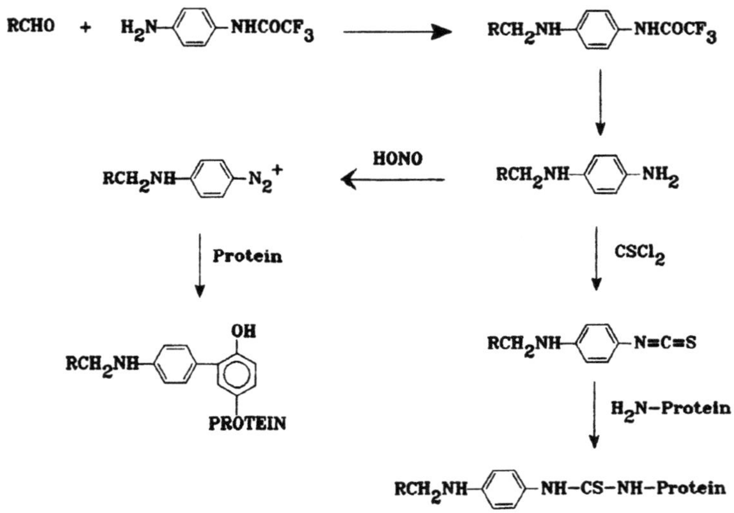

Fig. 4. Reductive amination with 4-trifluoroacetamidoaniline.

<u>a. 4-Trifluoroacetamido-aniline</u> --- Utilizing the fact that anilino-amino group is more effective in reductive amination, Kallin et al. (Kallin *et al.*, 1986) introduced the 4-trifluoroacetamido-aniline as a derivatization agent (**Fig. 4**). After the reducing sugar is attached to amino-group via reductive amination, the TFA-group is removed to expose the new amino group which can be diazotized or can be converted to isothiocyanate for conjugation to the proteins. The efficiency

of derivatization with this compound is quite high, though introduction of aromatic ring may be unwelcome.

<u>b. Heterobifunctional reagents</u> --- Frequently, glycopeptides rather than reducing oligosaccharides are more readily available. Glycopeptides possesses at least one amino group and one carboxyl group at the terminals, which can be utilized for conjugation to proteins. If there are other side chains, such as phenolic group, present, they may be utilized also.

A recently developed heterobifunctional reagent reacts with amino groups of glycopeptides to provide i) a spacer arm and ii) an aldehydo-group. The latter group then can be used for conjugation to proteins *via* amino groups.

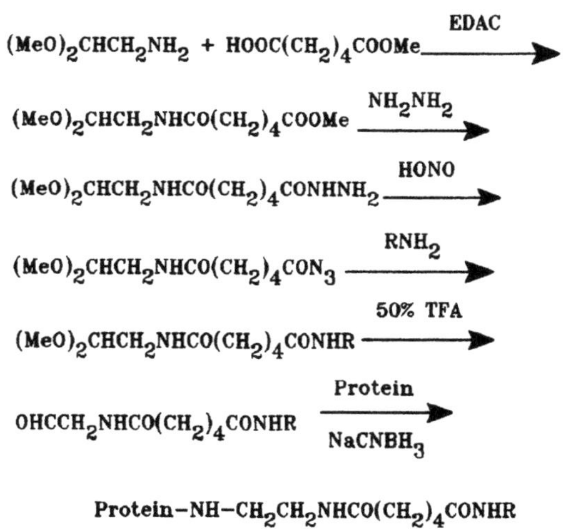

Fig. 5. Heterobifunctional reagent for glycopeptide coupling.

6. Conversion of natural glycolipids into neoglycoproteins

In some cases, procedural advantages may be gained (e.g., solubility, ligand density) by converting glycolipids into neoglycoproteins. A good example for this is the conversion of gangliosides into neoglycoproteins. Lysogangliosides can be selectively re-N-acetylated on the amino sugars, leaving only the amino group on sphingosine available.

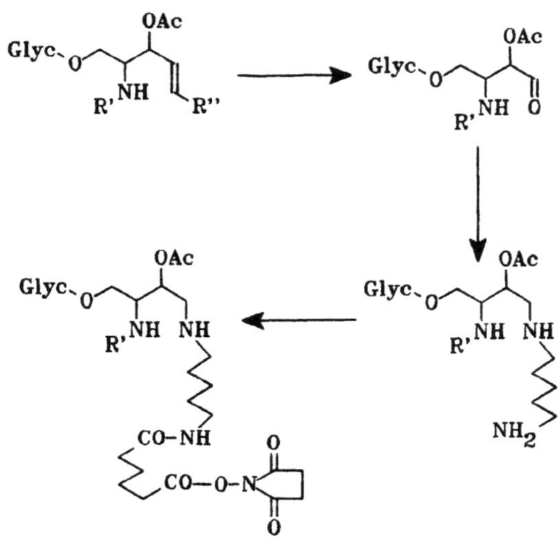

Fig. 6. Intermediates for preparation of neogangliproteins.

Schnaar and coworkers (Tiemeyer *et al.* , 1989, Yasuda *et al.*, 1987) modified lysogangliosides with bis(sulfosuccinimidyl)suberate or similar reagents to couple to bovine serum albumin. Alternatively, oxidation of the double bond in the sphinganine leads to oligosaccharide derivative containing a terminal-aldehydo group that can be used for coupling to proteins via an extender arm (1,6-diaminohexane) (Tiemeyer *et al.* , 1989) (Fig. 6). These neoglycoproteins (sometimes called "neogangliproteins") were used to show that a membrane receptor for ganglioside is associated with central nervous system myelin (Tiemeyer *et al.*, 1990). Fishman et al (Pacuszka & Fishman, 1990) used neo-gangliproteins to prove that the interaction of cholera toxin with GM_1 requires not only the oligosaccharide chain of GM_1, but the lipid moiety is also essential.

7. Use of enzymes in neoglycoproteins

Use of enzymes can be very fruitful in construction of neoglycoproteins. They can be used either to directly link carbohydrate derivatives to the side chains of proteins or to modify the preexisting sugar residues.

a. Addition of glycosyl units to preformed neoglycoconjugates --- Enzymatic attachment of glycoside units to peptide side chains, except for the case of transglutaminase to be described below, remains to be difficult. However, addition of glycosyl units to preformed neoglycoproteins with mono- or disaccharide units are feasible with glycosyl transferases. Addition of glycosyl unit with glycosyltransferase is very specific with respect to both anomeric configuration and the position of attachment. In a classic example of such an application, Paulson et al. (Paulson *et al.*, 1984) used α-2,3- and α-2,6-specific sialyl transferase to resialylate red blood cells had been previously desialylated enzymatically. These "synthetic glycoconjugates" containing specific sialyl linkages allowed them to probe viral binding specificities. Berman et al. (Berman *et al.*; 1985, Berman *et al.*, 1986) incorporated ^{13}C-labeled Gal onto GlcNAc with a galactosyl transferase for binding studies by CMR.

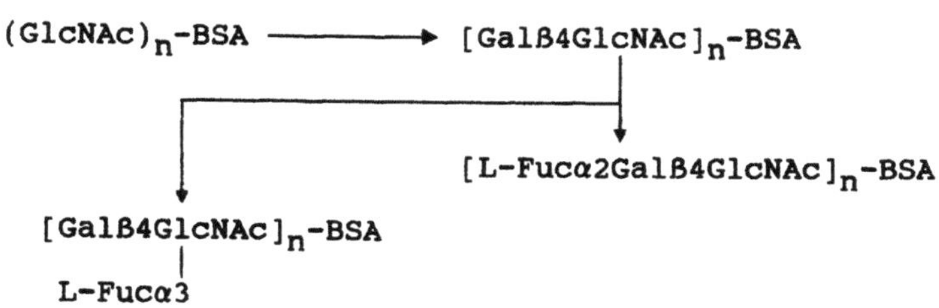

Fig. 7. Enzymatic glycosylation of neoglycoproteins

Hill and coworkers used a Gal-transferase and two different L-Fuc-transferases, respectively, to sequentially add Gal and L-Fuc to GlcNAc-AI-BSA to produce L-fucose-containing neoglycoproteins for their studies of fucose-binding proteins (Lehrman & Hill, 1986, Lehrman *et al.*, 1986a; Lehrman *et al.*, 1986b) (**Fig.** 7). This type of combined chemical-enzymatic approach (Sabesan & Paulson, 1986) is expected to increase in future.

b. Use of transglutaminase --- An ingenious method of attaching glycopeptides by the action of transglutaminase was devised (Yan & Wold, 1984). The γ-COOH of Glu in the protein (in which all the amino groups have been masked) and α-amino group of glycopeptides are conjugated by this reaction. In the case of modification of ß-casein with ovalbumin Asn(oligosaccharides), 4 Glu-sites were readily utilized in this enzymatic reaction.

c. Glycoconjugates of non-covalent attachment. --- The well known avidin-biotin interaction is so strong that it can almost be considered to be a covalent bond. A unique application of this interaction was invented (Chen & Wold, 1986, Shao & Wold, 1987) by binding biotinylated glycopeptides or biotinylated 6-aminoalkyl glycosides to avidin or streptavidin. The use of the avidin-(biotinyl glycopeptide) allows investigation of the orientation of sugar moieties from the known geometry of combining sites of the avidin. This type of complex finds another application in studies of the influence of peptide backbone on the assembly of oligosaccharide chains. (Chen *et al.*, 1986; Yet *et al.*, 1988; Shao & Wold, 1987; Shao *et al.*, 1987).

A similar approach based on the enzyme-coenzyme complexing was also investigated. For example, aspartate aminotransferase complexed with phosphopyridoxylated Asn-$GlcNAc_2Man_5$ (Chen & Wold, 1984) has been used in studies of binding by alveolar macrophages (Ohsumi *et al.*, 1988).

8. Glyco-matrices

Insoluble matrices containing mono- or oligo-saccharides as well as glycoproteins have been used to study cell adhesion (Schnaar, 1984; Largent *et al.*, 1984). Hepatocytes grown on lactose-carrying polystyrene (PVLA) have been shown to adhere much more efficiently than those grown on comparable layers without terminal galactosyl residues (Kobayashi *et al.*, 1986; Kobayashi *et al.*, 1985; Kobayashi & Akaike, 1990). PVLA was also found to be useful in separating rat liver parenchymal cells from non-parenchymal cells (Akaike *et al.*, 1989). As mentioned above, omega-unsaturated glycosides can be incorporated into polyacrylamide (Nishimura *et al.*, 1990).

9. Neoglycoconjugate applications

a. Neoglycoenzymes --- When enzymes are modified with carbohydrate residues (neoglycoenzymes), they can be used for glycohistochemistry or lectin detection in solid-phase assays. For example, bacterial ß-galactosidase modified with p-aminophenyl α-D-mannopyranoside via amide bond was useful in determination of Con. A immobilized on plastic microtiter plates. Lactose-modified ß-galactosidase was effective in histochemical detection of galactoside-specific lectin (Gabius *et al.*, 1989). Proper choice of the carbohydrate prosthetic groups would enable this approach to gain higher specificity. As in other neoglycoproteins, the density of the carbohydrate prosthetic group may also aid in causing tighter binding, and thus increase sensitivity.

b. Lengths of spacer arms --- Glycosyl groups are most usually attached to backbones via a spacer arm. The length of the spacer arm can be quite important in manifestation of biological activities. When polyacrylamide containing N-hydroxysuccinimide esters were prepared with insertion of 1 or 2 6-aminohexanoic acid (Pless *et al.*, 1983), hepatocytes appear to bind with greater efficiency than the gels without a spacer arm (R. Schnaar, personal communication). Gabius et al. (Gabius, 1990) reported that during isolation of certain lectins, the lengths of the spacer arms caused different affinity for different isolectins.

REFERENCES

Akaike, T, Kobayashi, A, Kobayashi, K, Sumitomo, H (1989) Separation of parenchymal liver cells using a lactose-substituted styrene polymer substratum. *J Bioactive and Compatible Polymers* **4**:51-56

Aplin, JD, Wriston, JC (1981) Preparation, properties, and applications of carbohydrates of proteins and lipids. *Critical Rev in Biochem* **10**:259-306

Berman, E, Brown, J, Lis, H, Sharon, N (1985) Binding of [^{13}-C]galactose-labeled N-acetyllactosamine to Erythrina cristagalli agglutinin as studied by ^{13}C-NMR. *Eur J Biochem* **152**:447-451

Berman, E, Lis, H, James, TL (1986) Binding of [^{13}-C]-galactose-enriched hen ovalbumin to Erythrina cristagalli agglutinin as studied by 13C-NMR spectroscopy. *Eur J Biochem* **161**:589-594

Bøgwald, J, Seljelid, RHJ (1986) Coupling of polysaccharides activated by means of chloroacetaldehyde dimethyl acetal to amines or proteins by reductive amination. *Carbohydr Res* **148**:101-107

Cabacungan, JC, Ahmed, AI, Feeney, RE (1982) Amine boranes as alternative reducing agents for reductive alkylation of proteins. *Anal Biochem* **124**:272-278

Chen, VJ, Wold, F (1984) Neoglycoproteins: Preparation of non-covalent glycoproteins through high-affinity protein-(glycosyl) ligand complexes. *Biochemistry* **23**:3306-3311

Chen, VJ, Wold, F (1986) Neoglycoproteins: Preparation and properties of complexes of biotinylated asparagine-oligosaccharides with avidin and streptavidin. *Biochemistry* **25**:939-944

Chen, VJ, Yan, S-CB, Wold, F (1986) Neoglycoprotein models for the study of glycoprotein processing. *Microbiol* **1986**: 297-302

Chernyak, AY, Weintraub, A, Norberg, T, Kallin, E (1990) Preparation of oligosaccharide-polyacrylamide conjugates and their use as antigens in enzyme immunoassay (EIA). *Glycoconjugate J* **7**:111-120

Gabius, H-J (1990) Influence of type of linkage and spacer on the interaction of ß-galactoside-binding proteins with immobilized affinity ligands. *Anal Biochem* **189**:91-94

Gabius, S, Hellmann, K-P, Hellmann, T, Brinck, U, Gabius, H-J (1989) Neoglycoenzymes: A versatile tool for lectin detection in solid-phase assays and glycohistochemistry. *Anal Biochem* **182**:447-451

Garg, HG, Jeanloz, RW (1985) Synthetic N- and O-glycosyl derivatives of L-asparagine, L-serine, and L-threonine. *Adv Carbohydr Chem Biochem* **43**:135-201

Gray, GR (1974) The direct coupling oligosaccharides to proteins and derivatized gels. *Arch Biochem Biophys* **163**:426-428

Horesji, V, Kocourek, J (1974) Affinity electrophoresis: Separation of phytohemagagluatinins on O-glycosyl polyacrylamide gels. *Methods Enzymology* **34**:178-181

Kallin, E, Lönn, H, Norberg, T (1986) New derivatization and separation procedure for reducing oligosaccharides. *Glycoconjugate J* **3**:311-319

Kobayashi, A, Akaike, T, Kobayashi, K, Sumitomo, H (1986) Synthesis and function of polystyrene derivatives having pendent oligosaccharides. *Makromol Chem Rapid Commun* **7**: 645-650, abstract.

Kobayashi, K, Akaike, T (1990) Hepatocyte adhesion using carbohydrate polymer as substrate material and expression of sugar chain high density effect. *Trends in Glycosciences and Glycotechnology* **2**:26-33

Kobayashi, K, Sumitomo, H, Ina, Y (1985) Enhanced adhesion and survival efficiency of liver cells in culture dishes coated with a lactose-carrying styrene homopolymer. *Polymer J* **17**: 567-575

Largent, BL, Walton, KM, Hoppe, CA, Lee, YC, Schnaar, RL (1984) Carbohydrate-specific adhesion of alveolar macrophages to mannose-derivatized surfaces. *J Biol Chem* **259**:1764-1769

Lawson, AM, Chai, W, Cashmore, GC, Stoll, MS, Hounsell, EF, Feizi, T (1990) High-sensitivity structural analyses of oligosaccharide probes (neoglycolipids) by liquid-secondary-ion mass spectrometry. *Carbohydr Res* **200**:47-57

Lee, RT, Cascio, S, Lee, YC (1979) A simple method for the preparation of polyacrylamide gels containing thioglycoside ligands. *Carbohydr Res* **95**:260-269

Lee, RT, Lee, YC (1980) Preparation and some biochemical properties of neoglycoproteins produced by reductive amination of thioglycosides containing an ω-aldehydo-aglycon. *Biochemistry* **19**:156-163

Lee, YC & Lee, RT (1982) *in The Glycoconjugates, Vol IV, Academic Press, N.Y.*, pp.57-83.

Lehrman, MA, Haltiwanger, RS, Hill, RL (1986a) The binding of fucose-containing glycoproteins by hepatic lectins. The binding specificity of the rat liver fucose lectin. *J Biol Chem* **261**:7426-7432

Lehrman, MA, Hill, RL (1986) The binding of fucose-containing glycoproteins by hepatic lectins. Purification of a fucose-binding lectin from rat liver. *J Biol Chem* **261**:7419-7425

Lehrman, R, Pizzo, SV, Imber, MJ, Hill, RL (1986b) The binding of fucose-containing glycoproteins by hepatic lectins. Re-examination of the clearance from blood and the binding to membrane receptors and pure lectins. *J Biol Chem* **261**:7412-7418

Likhosherstov, LM, Piskarev, VE, Derevitsakaya, VA, Kochetkov, NK (1990) Synthesis of N-glycoproteins with a natural type of carbohydrate-peptide bond. (Russian). *Bioorg Chem* **16**: 124-126

Magnusson, G, Ahlfors, S, Dahmen, J, Jansson, K, Nilsson, U, Noori, G, Stenvall, K, Tjornebo, A (1990) Prespacer glycosides in glycoconjugate chemistry. Dibromoisobutyl glycosides for the synthesis of neoglycolipids, neoglycoproteins, neoglycoparticles, and soluble glycosides. *J Org Chem* 55:3932-3946

Maley, F, Trimble, RB, Taraentino, AL, Plummer, JTH (1989) Characterization of glycoproteins and their associated oligosaccharides through the use of endoglycosidases. *Anal Biochem* **180**:185-204

Mootoo, DR, Konradsson, P, Fraser-Reid, B (1989) n-Pentenyl glycosides facilitate a stereoselective synthesis of the pentasaccharide core of the protein membrane anchor found in Trypanosoma brucei. *J Amer Chem Soc* **111**:8540-8542

Nishimura, S, Matsuoka, K, Kurita, K (1990) Synthetic glycoconjugates: Simple and potential glycoprotein models containing pendant N-acetyl-D-glucosamine and N,N'-diacetylchitobiose. *Macromolecules* 23:4182-4184

Ohsumi, Y, Chen, VJ, Yan, S-CB, Wold, F, Lee, YC (1988) Interaction between new neoglycoproteins and the D-Man/ L-Fuc receptor of rabbit alveolar macrophages. *Glycoconjugate J* 5:99-106

Pacuszka, T, Fishman, PH (1990) Generation of cell surface neoganglioproteins. GM1-neoganglioproteins are nonfunctional receptors for cholera toxin. *J Biol Chem* **265**: 7673-7678

Paulson, JC, Rogers, GN, Carroll, SM, Higa, HH, Pritchett, T, Milks, G, Sabesan, S (1984) Selection of influenza virus variants based on sialyloligosaccharide receptor specificity. *Pure & Appl Chem* **56**:797-805

Pless, DD, Lee, YC, Roseman, S, Schnaar, RL (1983) Specific cell adhesion to immobilized glycoproteins demonstrated using new reagents for protein and glycoprotein immobilization. *J Biol Chem* **258**:2340-2349

Raja, RH, LeBoeuf, RD, Stone, GW, Weigel, PH (1984) Preparation of alkylamine and 125I-radiolabeled derivatives of hyaluronic acid uniquely modified at the reducing end. *Anal Biochem* **139**:168-177

Roy, R, Katzenellenbogen, E, Jennings, HJ (1984) Improved procedures for the conjugation of oligosaccharides to protein by reductive amination. *Can J Chem* **62**:270-275

Sabesan, S, Paulson, JC (1986) Combined chemical and enzymatic synthesis of sialyloligosaccharides and characterization by 500-MHz ^{1}H and ^{13}C NMR spectroscopy. *Biochemistry* **108**: 2068-2080

Schnaar, R (1984) Immobilized glycoconjugates for cell recognition studies. *Anal Biochem* **143**:1-13

Shao, M-C, Chin, CCQ, Caprioli, RM, Wold, F (1987) The regulation of glycan processing in glycoproteins. The effect of avidin on individual steps in the processing of biotinylated glycan derivatives. *J Biol Chem* **262**:2973-2979

Shao, M-C, Wold, F (1987) The use of avidin-biotinylglycan as the model for in vitro glycoprotein processing. *J Biol Chem* **262**:2968-2972

Stoll, MS, Hounsell, EF, Lawson, AM, Chai, W, Feizi, T (1990)

Microscale sequencing of O-linked oligosaccharides using mild periodate oxidation of alditols, coupling to phospholipid and TLC-MS analysis of the resulting neoglycolipids. *Eur J Biochem* **189**:499-507

Stowell, CP, Lee, YC (1980) Neoglycoproteins. The preparation and application of synthetic glycoproteins. *Adv in Carbohydr Chem and Biochem* **37**:225-281

Suami, T, Ogawa, S (1990) Chemistry of carba-sugars (pseudosugars) and their derivatives. *Adv Carbohydr Chem & Biochem* **48**:22-90

Tang, PW, Feizi, T (1987) Neoglycolipid micro-immunoassays applied to the oligosaccharides of human milk galactosyltransferase detect blood-group related antigens on both O- and N-linked chains. *Carbohydr Res* **161**:133-143

Tiemeyer, M, Swank-Hill, P, Schnaar, RL (1990) A membrane receptor for gangliosides is associated with central nervous system myelin. *J Biol Chem* **265**:11990-11999

Tiemeyer, M, Yasuda, Y, Schnaar, RL (1989) Ganglioside receptors on rat brain membranes. *J Biol Chem* **264**:1671-1681

Yan, S-CB, Wold, F (1984) Neoglycoproteins: In vitro introduction of glycosyl units at glutamines in β-casein using transglutaminase. *Biochemistry* **23**:3759-3765

Yasuda, Y, Tiemeyer, M, Blackburn, CC, Schnaar, RL (1987) *in "New Trends in Ganglioside Research: Neurochemical and Neuroregenerative Aspects" (Ledeen R, Hogan, EL, Tettamanti G, Yates, AJ, & Yu, H, eds), Springer-Verlag*, pp.229-243.

Yet, M-G, Shao, M-C, Wold, F (1988) Effects of the protein matrix on glycan processing in glycoproteins. *FASEB J* **2**:22-31

Neoglycoproteins and lectins in human lung cancer: A survey

K. Kayser and H.-J. Gabius

Department of Pathology, Thoraxklinik, D-6900 Heidelberg, and Department of Chemistry, Max-Planck-Institute for Experimental Medicine, D-3400 Göttingen

Introduction

Although the "cancer problem" is now understood to be closely associated with abnormalities of the genetic function or dysfunction of uncontrolled proliferating cells, it is still elusive how these genetic abnormalities cause the complex alterations of the malignant phenotype. Mammalian life is only possible in a quasistationary cellular system with strong rules controlling cellular functions, death and replacement. Therefore, a set of information has to be exchanged between different cells and cellular compartments. One natural source for information storage and exchange is encoded in the structure of sugar sequences. Plant lectins have proven valuable to assess aberations in these structures upon malignant transformation (Damjanov,1987; Walker,1989). Notably, monoclonal antibodies to certain carbohydrate structures enabled to detect their differentiation-dependent expression in the lung (Itai et al.,1980). It is also instructive to refer to the important role of defined carbohydrates in viral infection (Rogers et al.,1986). Alterations in the pattern of expression of receptors for the carbohydrate part of glycoconjugates have similarly been related to the state of tumor differentiation (Gabius 1986; 1990; Lotan et al.,1989a,1989b), to metastatic potential of tumors (Gabius et al.,1987; 1990; Gabius and Engelhardt, 1988; Lotan and Raz,1988; Raz et al.,1990), or to early preneoplastic changes (Huber et al.,1986;Harris et al.,1987). Evidently, changes in expression of cell surface carbohydrate structures have been described to affect tumor cell invasion in vitro (Schallier et al.,1988). Studies on murine model systems have already indicated that endogenous lectins are present on lung tumor cells, their expression being markedly affected by the microenvironment (Kieda and Monsigny,1986; Glaves et al.,1989; Vidal-Vanaclocha et al.,1990) Glycohistochemical approaches for visualization of this receptor class have been initiated with human tissue specimens (Kayser an Gabius,1988; Kayser et al.,1989a;1989b).These techniques allow a exact overview of differencences of expression in a whole tumor cross section.

Although general aggreement about the potential biological significance of sugar receptors in mammalian cells is reached, little is known about their actual roles in tumor development an growth. The broad biochemical spectrum of sugar structures in glycoconjugates and their possible interactions with receptors lead to a complex diversity of this level. A further problem can be seen in the fact that information transfer and the generation or dissolution of information is usually closely associated with information storage, i.e. the exhibition of certain biochemical information signals is not only influenced by the momentary conditions, but also related to the history of the cells involve This fact may explain the observation reported by several author that the binding capacities of lectins and neoglycoproteins to normal tissue can well vary to a large amount and are not always constant (Kawai et al.,1988; Itai el al.,1990; Kayser et al.,1991). The influence of the microenvironment definitely deserves attention in this respect (Glaves et al.,1989; Vidal-Vanaclocha et al.,1990). In the following, we try to give an overview about the binding capacities of a panel of "simple" neoglycoproteins and lectins to normal and altered human lung parenchyma and to certain malignant lung tumors.

Non-tumorous Lung Parenchyma

The binding capacities of normal lung parenchyma and bronchial epithelial cells to lectins are usually weak and not always consistent (Kawai et al.,1988). Peanut lectin which recognizes Gal-ß-(1-3)-GalNAc, the lectins of horse gram, which recognizes alpha-GalNAc, Castor bean which recognizes ß-D-Gal, and gorse which recognizes alpha-L-fucose have been reported to bind in onl

5%-50% of the cases analyzed. Areas of metaplasia and dysplasia usually show an increased number of positively reacting cells (Alvarez-Fernandez and Carretero-Albinana,1990), but again show a high variability in the staining intensity and percentage.
The application of neoglycoproteins to normal and altered lung parenchyma has been reported to give the following results: Neoglycoproteins which specifically recognize receptors for lactose (Lac), melibiose (Mel), mannose (Man), fucose (Fuc) and fucoidan (Fucoid) have been found not to bind in all or in the broad majority of cases of normal tissue under the applied experimental conditions, i.e. non-altered lung parenchyma (Kayser et al.,1991). The result holds true for pneumocytes and bronchial epithelial cells. A positive staining of the pneumocytes was only seen for maltose-carrying neoglycoprotein (Mal) in a small percentage (20%) of cases. The staining behaviour is, however, completely different in lung parenchyma which was affected by chronic infectious diseases such as tuberculosis, chronic inflammatory diseases such as sarcoidosis or in lung parenchyma exposed to extrinsic agents such as radiation or cytostatic therapy. Neoglycoproteins, reacting with receptors that are specific for Fuc, Mal and Man disclose characteristic changes in those cases. Lung parenchyma infected with tuberculosis shows binding capacities specific for Fuc, which are twice as frequent as lung parenchyma altered by sarcoidosis. The opposite was found for Mal (Kayser et al.,1991). Lung parenchyma exposed to cytostatic therapy or radiation exhibits increased percentage of specific binding capacities for Mal, Fuc and Man. These affinities are still present even years after the termination of the cytostatic regimes. Histomorphologically, detectable damages of the lung parenchyma are usually not seen. If present, the binding capacities are more intensive and stronger.
In summary, the reported data of binding capacities of normal and altered lung parenchyma to lectins and endogenous lectins indicate, that:
1. Both sugar antigens and sugar receptors are usually not demonstrable or inconsistently present in epithelial cells of normal, i.e. healthy and "untached" lung parenchyma, under the experimental conditions used.

2. Infectious diseases such as tuberculosis "switch on" certain sugar-binding capacities, especially those related to Fuc.
3. Chronic inflammatory diseases such as sarcoidosis show a similar reaction pattern of endogenous lectins compared to that seen in tuberculosis. However, specific differences exist, for example a stronger binding capacity to Mal and a weaker binding capacity to Fuc.
4. Toxic damages of the lung parenchyma such as radiation or damages related to cytostatic drug regimes induce alterations o the sugar-binding capacities. These alterations are related to those seen in infectious and inflammatory disorders of the lung The binding capacities, however, are usually not as intensive a frequent. They are closely related to the lung damage that is histomorphologically observeable.
5. There is evidence that the alterations of endogenous recepto for neoglycoproteins refer to infectious or inflammatory damage of the epithelial cells, being no longer present at the time of investigation. They can be used as a "constant state" (memory) reflecting the "history" of the cells.
6. Alterations of the binding capacities of human lung parenchym cells for neoglycoproteins precede histomorphologically detecta damages of the lung parenchyma.

Primary Lung Cancer

Data of lung carcinomas with respect to the binding capacities f plant lectins generally show that the lung tumors lack a specifi profile for each of the major histological cell types (Alvarez-Fernandez and Carretero-Albinana,1990). These reports, however, have to be discussed in detail: Epidermoid carcinomas, adenocarcinomas and small cell anaplastic carcinomas have been reported to stain positively to lectins specific for alpha-D-mannose (Canavalia ensiformis, ConA), alpha-L-fucose (Ulex europaeus 1), D-galactose (Bandeirea simplicifolia) and D-GalNAc (Dolichos biflorus) in 50% - 100% of the analyzed cases (Alvarez Fernandez and Carretero-Albinana,1990). Staining with different types of lectins with the same nominal specificity have been reported to be not invariably positive or even negative; i.e. on sugar specific lectin form a certain group may bind to tumor cel in contrast to another lectin with the same nominal sugar specificity which is not reactive. The results clearly substantiate that the detailed specificities of such related lectins are not identical. With respect to histomorphological application it is interesting to note that small cell anaplastic carcinomas express lectin receptors in a smaller percentage of cases and weaker intensity compared to non-small cell carcinoma (Shiba et al.,1984; Alvarez-Fernandez and Carretero-Albinana,1990). There is evidence that the pattern of lectin affinities is to some extent related to the grade of differentiation of the carcinomas. The more differentiated the carcinomas are, the stronger and more frequent are the binding capacities. Major differences of lectin binding capacities between human tissue and rodents have not been found (Shiba et al.,1984).

Table 1: Binding capacities of human lung carcinomas (N = 179) to various neoglycoproteins (in percent)

Neoglycoprotein	Tumor Cell Type			
	Epidermoid	Adeno	Large cell	Small cell
Mal	65	88	56	10
Cel	56	70	70	20
Fuc	42	81	25	5
Man	74	81	56	4
GlcNAc	78	75	50	42
Sial	76	48	50	29
Glnc[a)]	80	58	89	33
Rham	47	71	31	21
Lac	41	52	*	41
Mel	20	31	*	0
Hep	42	35	19	8

*) not determined [a)]glucuronic acid

The reported data show, in general, that binding of lectins with the same nominal sugar specificity (for example alpha D-mannose by Canavalia ensiformis or by Lens culinaris) is not related to the cell type of human lung cancer or normal lung parenchyma (Shiba et al.,1984; Kawai et al.,1988; Alvarez-Fernandez and Carretero-Albinana, 1990; Itai et al.,1990). However, if a positive staining was seen, it was related to the degree of tumor differentiation. The staining behaviour of primary human lung carcinoma to a panel of neoglycoproteins is demonstrated in Table 1. Probes that are specific for receptors for Mal, Fuc, Man, Cel and Glnc bind to cells of the majority of non-small cell lung carcinomas. The percentage of positive cases was reported to be 60%-80% (Kayser al.,1989a; 1989b). In contrast, small cell anaplastic carcinomas were found to react positively only in 5%-20% of the analyzed cases. The general statement that the expression of lectin-bindi capacities is related to the degree of cellular differentiation holds true for the histochemical detection of endogenous lectins in primary lung cancer. In addition, remarkable differences were seen between small cell lung carcinomas and non-small cell lung carcinomas: Small cell lung carcinomas show weak binding capacities to Mal, Fuc and Man. These neoglycoproteins can be us to aid in the differential diagnosis of small cell lung cancer versus non-small cell lung cancer with a similar sensitivity and specificity as the commonly applied neuroendocrine markers (neuron-specific enolase, synaptophysin, bombesin) (Kayser et al.,1988; 1989a; 1989b; 1990). Again, no predection for the expression of certain characteristic sugar receptors in relation to the cell type or immunohistochemical properties can be given. The distinction of metastatic adenocarcinoma into the pleura fro primary epithelial mesothelioma of the pleura still remains a diagnostic problem. Monoclonal antibodies specific for carcinoembryonic antigen (CEA), Leu M1 (CD15), and HLA125 provi a tool that is helpful in this respects (Kwee et a.,1982; Sheiba et al.,1986). However, the reported specificities and sensitivities are far below an acceptable 5% error probability (Kwee et al.,1982; Kayser et al.,1987; Kawai et al.,1988b). The expression of the endogenous lectin, specific for GlcNAc, is of similar specificity and sensitivity compared to the mentioned monoclonal antibodies. The reported specificity is 90%-100% for metastatic adenocarcinoma. Its sensitivity is 70%-80% (Kayser et al.,1989). These data could be confirmed in a larger prospective study. The other applied neoglycoproteins failed to be of major importance for this specific diagnostic problem.

The data of the application of exogenous lectins and neoglycoproteins indicate that:

1. The application of lectins is not of pronounced importance in the diagnosis of human lung cancer. Lectins with the same nominal specificity may give completely different results. Therefore, the interpretation of the results is difficult.

2. Heterogeneity could be observed in the staining behaviour of lectins in human lung cancer.
3. Endogenous lectins usually show binding capacities which are related to the degree of tumor differentiation. The more differentiated the tumors are, the stronger and more frequent are the binding capacities. This statement holds true for rather "simple" ligands as parts of neoglycoproteins, especially for Mal, Fuc, Man.
4. The application of neoglycoproteins may be used as diagnostic aid for distinguishing non-small cell carcinoma from small cell anaplastic carcinoma of the lung. The application of Mal, Fuc, and especially Man is a useful extension to the neuroendocrine markers due to their not absolutely reliable staining behaviour (neuroendocrine markers stain positively for small cell anaplastic carcinomas in about 60% and negatively for non-small cell carcinoma in about 90%; Mal, Fuc, Man will not yield staining for small cell anaplastic carcinomas in about 95% and reach positively for non-small cell anaplastic carcinomas in about 70%).
5. Similarly, the neoglycoprotein with conjugated GlcNAc exhibits a remarkable potential to serve as diagnostic aid for separating epithelial mesotheliomas from metastatic adenocarcinomas into the pleura with sensitivity and specificity comparable to that of CEA, Leu M1 and HLA125.

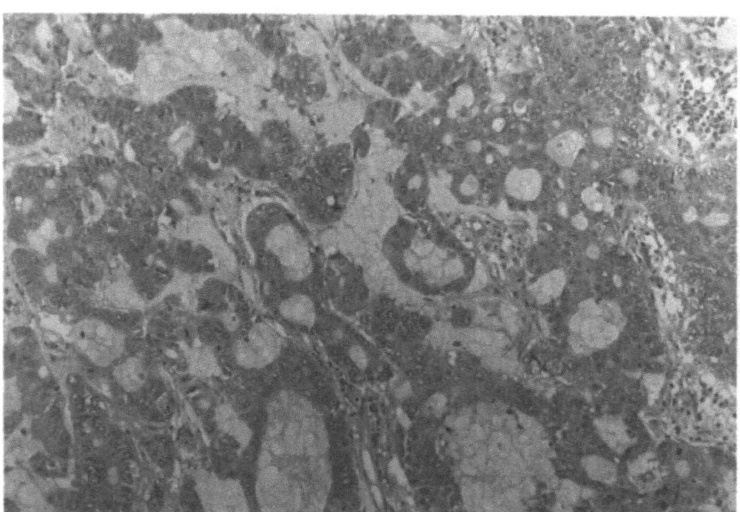

Fig. 1: Moderately differentiated Adenocarcinoma with MAL-specific binding capacities, ABC, x180.

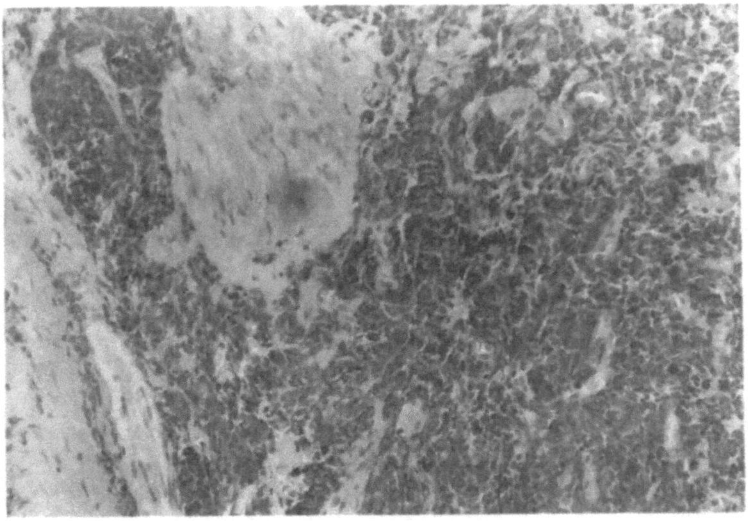

Fig.2 : Small cell anaplastic carcinoma with positive binding capacities to GlcNac, ABC, x 180

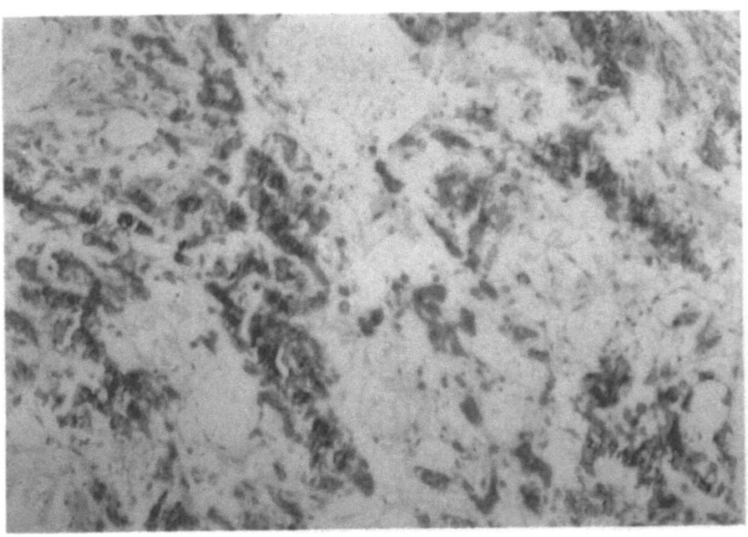

Fig. 3: Metastatic chondrosarcoma into the lung with positive binding capacities to Man, ABC, x 180.

Human Lung Metastases

Data of lectin binding to intrapulmonary metastases in murine models reveal that metastases into the lung exhibited a phenotype with positivity to PNA, SBA, UEA (Kahn and Baumal,1985; Kahn et al.,1988). It is, however, generally acknowledged that intratumoral heterogeneity and influences during tumor outgrowth will require attention for unequivocal delineation of metastases-associated parameters. Studies of neoglycoprotein-binding to sections of intrapulmonary metastases reveal a high frequency of receptors for galNAc and Rham in intrapulmonary metastases. Within the heterogeneous set of intrapulmonary metastases the variability of binding capacities for Rham was low and within the statistical range of significance. A high variability was seen for the binding capacities, that are specific for sialic acid (Sial) and Cel. The binding capacities to other neoglycoproteins (Hep, Man, Fuc, Mal) showed some inconsistent differences between epithelial and mesenchymal malignancies (Gabius and Kayser,1989). The data are shown in Table 2.

Table 2: Binding capacities of intrapulmonary metastases (N = 72) to various neoglycoproteins (in percent)

Neoglycoprotein	Sarcomas	Germ cell tumors	Carcinomas
Mal	51	55	80
Cel	37	55	45
Fuc	37	18	60
Man	57	64	87
GalNAc	60	55	73
Sial	60	55	73
Rham	77	73	95
Hep	46	36	73

The data may be interpreted as follows:

1. The high frequency of positive binding to Rham-specific probes suggests that either Rham-specific receptors are related to an "environment" effect of the lung parenchyma or to some metastatic potential of the malignancy.
2. The binding capacities to Sial and Cel-specific neoglycoproteins are probably related to a certain "homing" effect of the analyzed metastases.

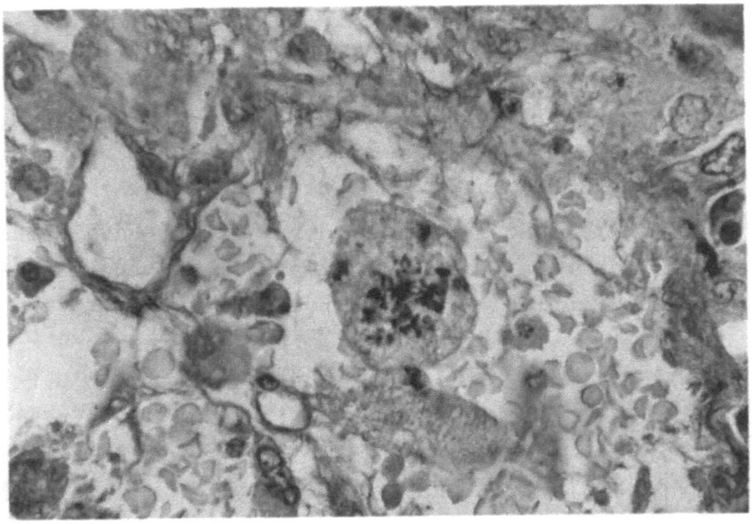

Fig. 4: Metastatic osteosarcoma into the lung with positive binding capacities to Rham, ABC, x 240.

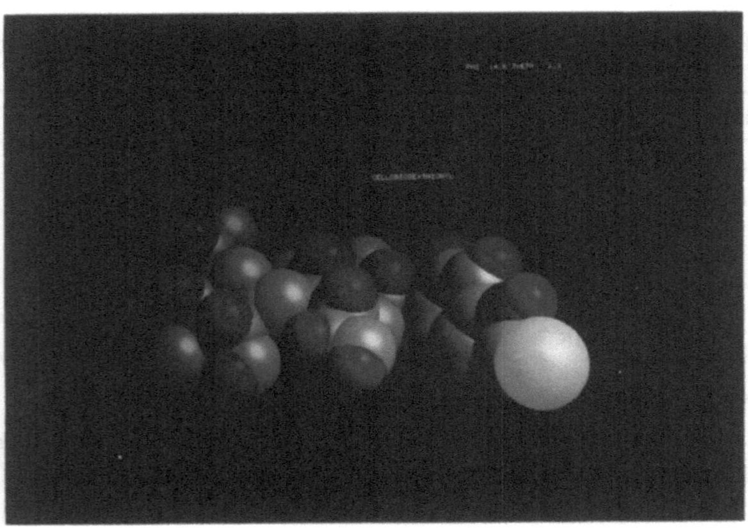

Fig. 5: Electron shell cofiguration of the cellobiose-thionyl molecule used for glycohistochemical detection of Cel-specific binding capacities (With courtesy of Dr. Bischoff, Institute of Chemistry, University of Heidelberg).

3. No major differences of expression of endogenous lectins between intra-pulmonary metastases and primary lung carcinomas can at present unequivocally defined.

Conclusion

The histochemical application of neoglycoproteins to visualize endogenous lectins is of practical importance in the histomorphological analysis of human lung cancer and offers reasonable approach for tumor diagnosis and the recognition of "pre-histomorphological" damage of the lung parenchyma. It may be considered as useful supplement in addition to immunohistochemical techniques and to the analysis of genetic defects (chromosomal aberations, etc). Besides the easy handling, the tools to detect endogenous lectins have the advantage that the biochemical configuration of the ligand is defined in molecular terms. The electron shell configuration of the applied probes can be visualized by computergraphics, as shown in Fig. 5. These tools may offer new aspects in the analysis of functional states in normal and altered human cells.

References

Alvarez-Fernandez E, Carretero-Albinana L (1990) Lectin histochemistry of normal bronchopulmonary tissues and common forms of bronchogenic carcinoma. Arch Pathol Lab Med 114:475-481

Damjanov I (1987) Lectin cytochemistry and histochemistry. Lab Invest 57:5-20

Gabius HJ, Engelhardt R, Rehm S, Deerberg F, Cramer F (1986) Differences in the pattern of endogenous lectins from spontaneous rat mammary tumors. Tumor Biol 7:71-81

Gabius HJ, Bandlow G, Schirrmacher V, Nagel GA, Vehmeyer K (1987) Differential expression of endogenous sugar-binding proteins (lectins) in murine tumor model system with metastatic capacity. Int J Cancer 39:643-648

Gabius HJ, Engelhardt R (1988) Sugar receptors of different types in human metastases to lung and liver. Tum Biol 9:21-36

Gabius HJ, Kayser K (1989) Elucidation of similarities of sugar receptor (lectin) expression of human lung metastases from histogenetically different types of primary tumors. Anticancer 9:1599-1604

Gabius S, Schirrmacher V, Franz, H, Joshi SS, Gabius HJ (1990) Analysis of cell surface sugar receptor expression by neoglycoenzyme binding and adhesion to plastic-immobilized neoglycoproteins for related weakly and strongly metastatic cel lines of murine tumor model systems. Int J Cancer 46:500-507

Glaves D, Gabius HJ, Weiss L (1989) Site-associated expression endogenous tumor lectins. Int J Cancer 44:506-511

Harris L, Preat V, Farber E (1987) Patterns of ligand binding t normal, regenerating, preneoplastic, and neoplastic rat hepatocytes. Cancer Res 47:3954-3958

Huber BE, Glowinski IB, Thorgeirsson SS (1986) Transcriptional post-transcriptional regulation of the asialoglycoprotein recep on normal and neoplastic rat liver. J Biol Chem 216:12400-12407

Itai S, Nishikata J, Takahashi N, Tanaka O, Matsubara Y, Hasega S, Yanai N, Takaoka K, Arii S, Tobe T, Kannagi R (1990) Differentiation-dependent expression of l and sialyl l antigens the developing lung of human embryos and in lung cancers. Cance Res 50:7603-7611

Kahn HJ, Baumal R (1985) Difference in lectin binding in tissue sections of human and murine malignant tumors and their metastases. Am J Pathol 119:420-429

Kahn HJ, Brodt P, Baumal R (1988) Lectin binding by liver and l metastasizing variants of the murine Lewis lung carcinoma. Am J Pathol 132:180-185

Kawai T, Greenberg SD, Titus JL (1988a) Lectin histochemistry o normal lung and pulmonary adenocarcinoma. Modern Path 1:485-492

Kawai T, Greenberg SD, Truong LD, Mattioli CA, Klima M, Titus J (1988b) Differences in lectin binding of malignant pleural mesothelioma and adenocarcinoma of the lung. Am J Pathol 130:40 410

Kayser K, Bosslet K, Kiefer B (1987) Differential diagnosis of mesothelioma and metastatic adenocarcinoma of pleura by means o indirect immunoperoxydase technique combined with syntactic structure analysis. TumDiagn Therapie 8:28-35

Kayser K, Gabius HJ: Histomorphological characterization of crabohydrate-binding proteins in human lung cancer. In: Gabius H Nagel GA (eds) Lectins and glycoconjugates in oncology. Springe Berlin, pp 87-96, 1988

Kayser K, Schmid W, Ebert W, Wiedenmann B (1988) Expression of neuroendocrine markers (neuronspecific enolase, synaptophysin and bombesin) in carcinoma of the lung. Path Res Pract 183:412-417

Kayser K, Heil M, Gabius HJ (1989a) Is the profile of binding of a panel of neoglycoproteins useful as a diagnostic marker in human lung cancer? Path Res Pract 184:621-629

Kayser K, Gabius HJ, Ciesiolka T, Ebert W, Bach S (1989b) Histopathologic evaluation of application of labeled neoglycoproteins in primary bronchus carcinoma. Hum Pathol 20:352-360

Kayser K, Gabius HJ, Köhler A, Runtsch T (1990) Binding of neuroendocrine markers and biotinylated sex hormones and the survival in human lung cancer. Lung Cancer 6:171-183

Kayser K, Gabius HJ, Carl S, Bubenzer J (1991) Alterations of human lung parenchyma after cytostatic therapy. APMIS 99:121-128

Kieda C, Monsigny M (1986) Involvement of membrane sugar receptors and membrane glycoconjugates in the adhesion of 3LL cell subpopulations to cultured pulmonary cells. Invasion Metastasis 6:347-366

Kwee Ws, Veldhuizen RW, Golding RP, Mullink H, Stan J, Donner R, Boon MJ (1982) Histologic distinction between malignant mesothelioma, benign pleural lesion and carcinoma metastasis. Virch Arch A 397:287-299

Lotan R, Raz A (1988) Endogenous lectins as mediators of tumor cell adhesion. J Cell Biochem 37:107-117

Lotan R, Carralero D, Lotan D, Raz A (1989a) Biochemical and immunochemical characterization of K-1735P melanoma galactoside-binding lectins and their modulation by differentiation inducers. Cancer Res 49:1261-1268

Lotan R, Lotan D, Carralero DM (1989b) Modulation of galactoside-binding lectins in tumor cells by differentiation-inducing agents. Cancer Lett 48:115-122

Raz A, Zhu D, Hogan V, Shah N, Raz T, Karkash R, Pazerini G, Carmi P (1990) Evidence for the role of 34-KDa galactoside-binding lectin in transformation and metastasis. Int J Cancer 46:871-877

Rogers GN, Herrler G, Paulson JC, Klenk H-D (1986) Influenza C virus uses 9-0-acetyl-N-acetylneuraminic acid as a high affinity receptor determinant for attachment to cells. J Biol Chem 261:5947-5951

Schallier DCC, Bolscher JGM, Smets LA: Alterations in cell surf carbohydrates affecting invasion in vitro. In: Gabius HJ, Nagel (eds) Lectins and glycoconjugates in oncology. Springer, Berlin pp 87-96, 1988

Sheibani K, Battifora H, Burke JS (1986) Antigenic phenotype of malignant mesotheliomas and pulmonary adenocarcinomas: an immunohistologic analysis demonstrating the value of Leu M1 antigen. Am J Pathol 123:212-219

Shiba M, Ohiwa T, Klein-Szanto AJP (1984) Letin-binding sites i preneoplastic and neoplastic lesions of human and rodent respiratory tracts. JNCI 72:43-51

Vidal-Vanaclocha F, Barbera-Guillem E, Weiss L, Glaves D, Gabiu HJ (1990) Quantitation of endogenous lectin expression in 3LL tumors, growing subcutaneously and in the kidneys of mice. Int Cancer 46:908-912

Walker RA (1989) The use of lectins in histopathlogy. Path Res Pract 185:826-835

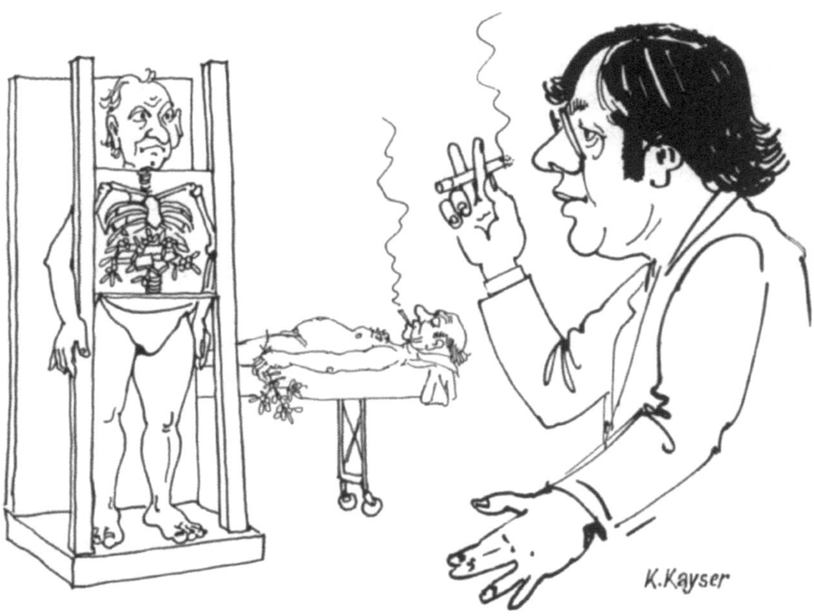

How to localize β-galactoside-specific endogenous lectins and their ligands in tissue sections: A comparison of the application of different histochemical tools in breast cancer

A. Schauer, S. Gabius[1], U. Brinck, and H.-J. Gabius[2]

University of Göttingen
Medical School
Department of Pathology
Robert-Koch-Str. 40
3400 Göttingen
FRG

Introduction

Endogenous lectins and respective tissue glycoconjugates constitute a molecular recognitive system also present in tumor cells (Gabius, 1988a). Disturbances of this system in malignancy can play a key role to modify the social behaviour not only of tumor cells which encompasses various steps within tumor growth and spread (Barondes, 1981; Uhlenbruck et al., 1986; Gabius, 1988a, b, 1991; Lotan and Raz, 1988; Monsigny et al., 1988; Sharon and Lis, 1989). This assumption has provided the rational to at first disclose presence of endogenous lectins in breast tumors and then to demonstrate that neoglycoproteins will allow to localize their specific receptors in this tumor class (Gabius et al., 1986, 1988a; Schauer et al., 1988).

Having conclusively shown in our initial reports on this topic that further studies are unquestionably warranted, we are now interested in comprehensively evaluating the potential of the glycohistochemical tools that are appropriate to detect tumor-

[1] University of Göttingen, Medical School, Department of Hematology/Oncology, Robert-Koch-Str. 40, 3400 Göttingen, FRG
[2] Max-Planck-Institute for Experimental Medicine, Department of Chemistry, Hermann-Rein-Str. 3, 3400 Göttingen, FRG

associated changes in the expression of endogenous lectins and their ligands for refinements of diagnostic procedures.

Our objectives consist of evaluation of the impact of the structural design of the carbohydrate part of neoglycoproteins on the efficiency of localization of lectins, comparison of glyco- and immunohistochemical detection of lectins and application of the results in a cell biological model for therapy. Moreover, another neoligandoprotein is exploited in search of its receptor, namely a neohormoprotein, to illustrate the value of this concept.

Histochemical tools with changes in the carbohydrate density and the linker region: Synthesis and standardized application

Neoglycoproteins containing lactose or D-galactose moieties as ligands for lectin localization were synthesized by diazo coupling of the appropriate derivative of p-aminophenyl lactoside, prepared from p-nitrophenyl lactoside after its catalytic reduction in the presence of palladium on charcoal (lac-BSA (diazo)), by coupling of p-isothiocyanato lactoside, prepared from p-aminophenyl lactoside by reaction with thiophosgene (lac-BSA(thio)), by reductive amination of lactose in the presence of sodium cyanoborohydride (lac-BSA(red amin)) and by carbodiimide-mediated coupling of the monosaccharide after its reaction to the aliphatic spacer (2,3-epoxypropane)-4-oxybutyric acid (gal-BSA(epi)), as outlined in detail previously (McBroom et al., 1972; Gordon et al., 1987; Gabius et al., 1988a, b, 1989a, b, 1990).

Glycohistochemical processing of paraffin-embedded and formalin-fixed surgical biopsy specimens of 20 invasive ductal carcinomas including 3 axillary lymph node metastases was performed by incubation with 100 µg/ml labelled neoglycoprotein for 4 hours and staining with 3-amino-9-ethyl-carbazole following a standard protocol (Gabius et al., 1988a, b).

Effect of coupling mode of the sugar to the carrier on the histochemical reactions

Average binding of each of the four different types of neoglycoprotein to tumor cells of the different cases of invasive ductal carcinoma is given in Table 1. Coupling of the diazo derivative of p-aminophenyl lactoside led in most cases to the relatively highest extent of staining in terms of number of stained cells and staining intensity. When the product of coupling of the diazonium derivative of p-aminophenyl lactoside is comparatively analyzed in relation to the other three types of neoglycoprotein, it proved to generate the strongest reaction. Neoglycoproteins, prepared by the attachment of the carbohydrate ligand by the p-isothiocyanato derivative or via an aliphatic linker after its reaction with the C_6-hydroxyl group of the sugar moiety were rather equally well effective. Reductive amination with concomitant ring opening at the reducing end of the disaccharide lactose, however, resulted in a neoglycoprotein, yielding the lowest extent of staining. The biotin content of the probes varied from 9-14 groups per carrier molecule.

Table 1: Synopsis of binding of neoglycoproteins to tumor cells of 20 invasive ductal carcinomas.

	---------- lac[a]	----------	----------	gal
	BSA_{10}[b]	BSA_{17}	BSA_{30}	BSA_{22}
Cytoplasm	3.3/2.2	1.0/0.6	2.4/1.5	2.8/1.6
Nuclei	2.5/1.6	1.3/1.0	2.8/1.9	2.1/1.3

The percentage of positive cells is grouped into the categories: 0 = 0 %, 1 = 0-5 %, 2 = 5-20 %, 3 = 20-50 % and 4 = 50-100 % and is given in front of the diagonal streak (.../).
The intensity of the staining reaction is grouped into the categories: 0 = no staining, 1 = weak, but significant staining, 2 = medium staining, 3 = strong staining, 4 = very strong staining and is given behind the diagonal streak (/...).

[a] Abbreviations for sugars: lactose, galactose.

[b] Abbreviations for synthesis of neoglycoproteins: diazo coupling (BSA_{10}), reductive amination (BSA_{17}), thionylation (BSA_{30}) and EDC-mediated coupling after reaction with an aliphatic linker (BSA_{22}).

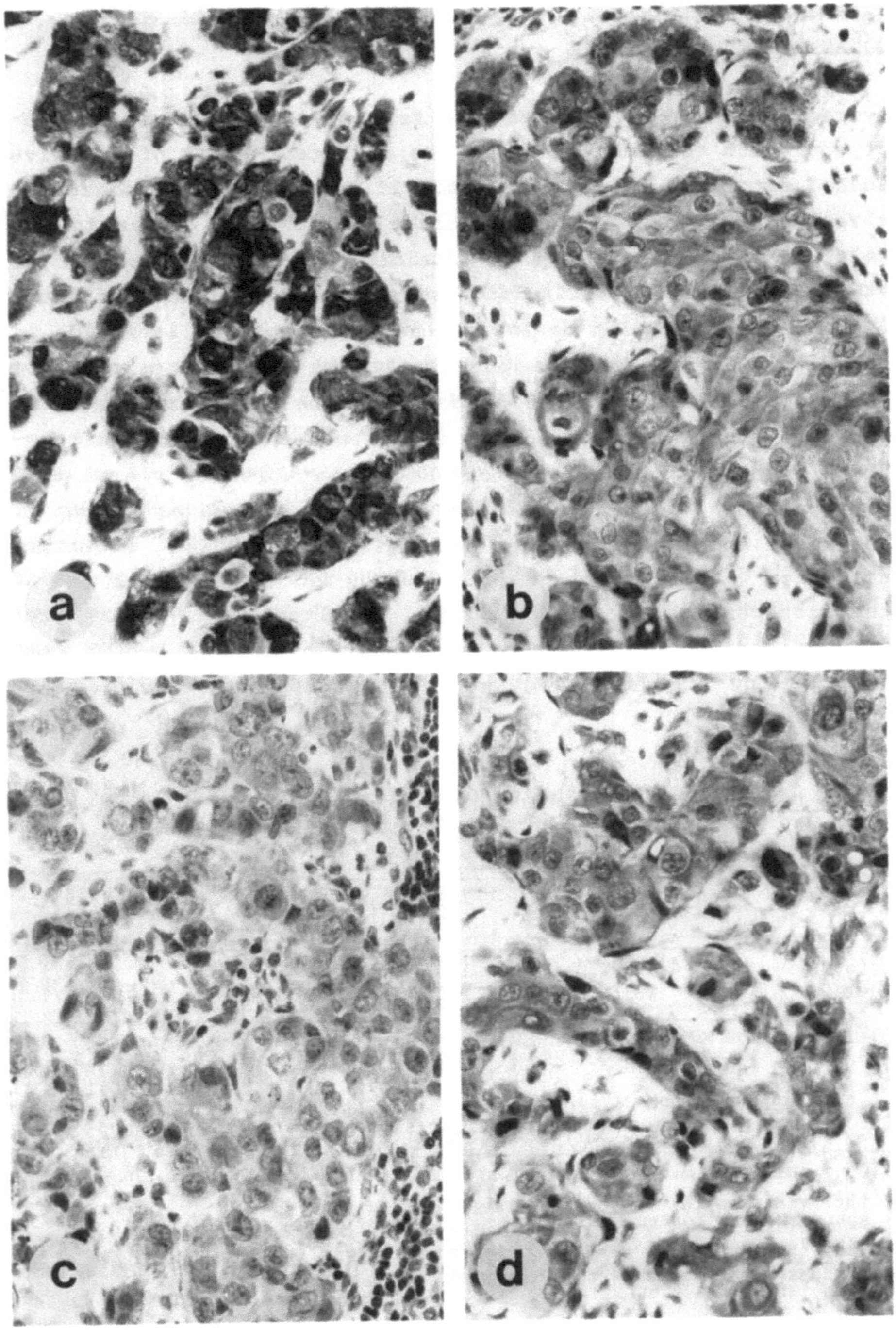
a
b
c
d

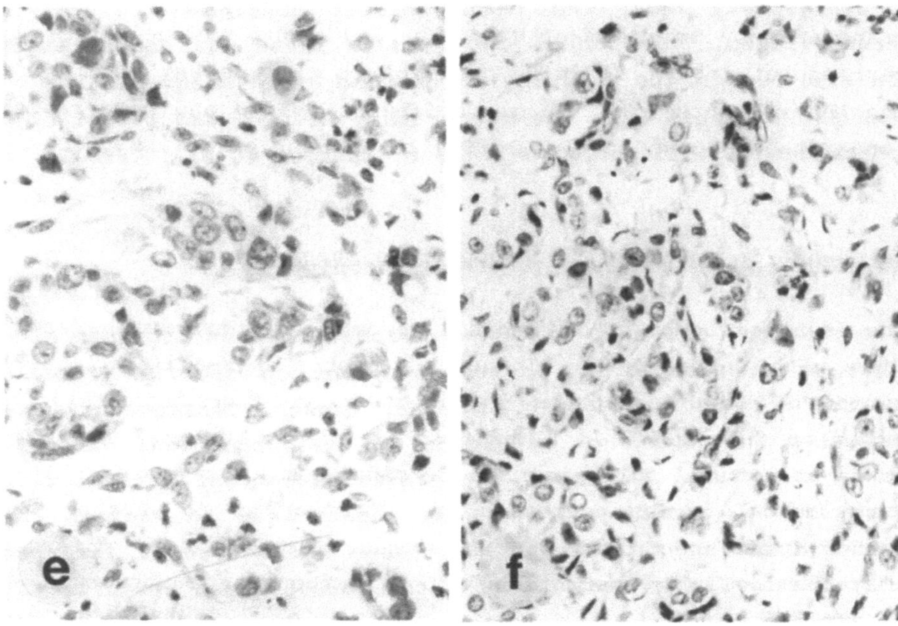

Fig. 1: Sections of a lymph node metastasis of an infiltrating ductal carcinoma after application of biotinylated lac-BSA (diazo) (a), biotinylated lac-BSA (thio) (b), biotinylated lac-BSA (red amin) (c), biotinylated gal-BSA (epi) (d) and after application of labelled as well as homologous unlabelled lactosylated lac-BSA (diazo) in the relation of 1:100 for demonstration of inhibition (e), ABC reagents and hematoxylin counterstaining and after performance of the standard protocol in the absence of an incubation with labelled neoglycoprotein (f); x 310.

Controls for specificity were carried out carefully by incubation with biotinylated, but non-glycosylated BSA (100 µg/ml) and by competitive inhibition experiments using the homologous unlabelled neoglycoprotein (Gabius et al., 1988a, b, 1989a, b). Further control reactions to assess the influence of glycosidases and glycosyltransferases on probe binding included prolonged incubation of the probes that might reduce binding due to the enzymatic reaction of a glycosidase and incubation in the presence of 20 mM uridine 5'diphosphate (UDP) and cytidine

5'diphosphate (CDP) that might inhibit binding to glycosyltransferases. Omission of the labelled probe was deliberately carried out in the staining protocol to exclude any binding of the ABC reagents like the glycoproteins avidin and horseradish peroxidase (Gabius et al., 1989a, b; Kuchler et al., 1990).

Relevance of subterminal carbohydrate moieties

The results, compiled in Table 1, prompt to focus further studies with carbohydrate ligands as probe on specific localization of sugar receptors in breast cancer. Consequently, we assessed the impact of alteration of the subterminal carbohydrate structure, following the terminal galactose moiety. This experimental series addresses the question as to whether the carbohydrate specificity of the tissue lectin can be inferred histochemically by variations of the carbohydrate structure of an applied marker.

Therefore, neoglycoproteins containing N-acetylglucosamine in β-1,3-linkage and β-1,4-linkage and N-acetylgalactosamine in β-1,3-linkage in the preterminal position were comparatively used as probes for localization of β-galactoside-specific lectins in sections of paraffin-embedded surgical biopsy specimens of 14 cases of malignant breast lesions and 7 lymph node metastases following a standard procedure (Gabius et al., 1991).

A N-acetylglucosamine residue in β-1,3-linkage proved to be a less favorable extension than this type of sugar in β-1,4-linkage or a N-acetylgalactosamine moiety in β-1,3-linkage (Table 2; Fig. 2). These histochemical results indicate the remarkable selectivity of protein-carbohydrate recognition in breast cancer tissue. Notably, the documented presence of endogenous lectins may be taken advantage of with respect to therapeutical implications.

Table 2: Binding of neoglycoproteins, of an endogenous lectin, of endogenous lectin-binding glycoproteins and of lectin-specific antibodies to tumor cells of malignant breast lesions.

case	gal-β1,4-glcNAc-Cete	gal-β1,3-glcNAc-Cete	gal-β1,3-galNAc-Cete	gal-β1,3-galNAc-(diazo)	lac (diazo)	anti-14 kDa	endogenous 14 kDa ligand	14 kDa
1	++/1	-/0	+/1	+++/1-2	+++/1-2	+++/3	+++/2-3	+++/2-3
1*	+++/2	-/0	+++/1	+++/2-3	+++/3-4	+++/2-3	+++/3-4	+++/3
2	++/1	-/0	+++/2	(+)/1	+++/2-3	+++/3	+++/3-4	+++/3
2*	(+)/1	-/0	+++/2	+++/1	+++/2	+++/2-3	+++/3-4	+++/3
3	+/2	+++/2	+++/3-4	+++/3-4	+++/2-3	+++/2-3	nd[a]	+++/4
3*	+++/3-4	+++/4	+++/3	+++/3	+++/3-4	+++/3-4	nd	+++/3-4
4	+++/2	(+)/1	(+)/1	++/2	+++/2	(+)/1	nd	+++/3
4*	+++/2	+++/3	++/2-3	++/2-3	+++/2	+++/3-4	nd	+++/3-4
5	+++/4	+++/1	+++/2	+++/2	+++/3-4	+++/3	+++/4	+++/3-4
6	-/0	-/0	+++/2	+++/1-2	+++/3	+++/3-4	+++/4	+++/3-4
7	+++/3	+++/2	+++/3	+++/3	+++/4	+++/2	nd	+++/3
8	+++/4	+++/2-3	+++/3-4	+++/3-4	+++/4	+++/4	nd	+++/4
9	++/1	(+)/1	++/1	++/1	+++/2-4	+++/2-3	nd	+++/4
10	(+)/1	(+)/1	++/2	++/1	(+)/1	+++/3	nd	+++/3
11	+++/2-3	+++/1-2	+++/2-3	+++/2	+++/3-4	+++/3-4	nd	+++/3-4
12	++/1	++/1	++/1	++/1	+++/2	+++/1-2	+++/3-4	+++/1-2
12*	+++/1-3	+/1	+++/2	+++/2	+++/2	+++/1-2	+++/3-4	+++/2-3
13	+++/2-3	+++/1-2	+++/2	+++/1-2	+++/2-3	+++/2-3	+++/3	+++/3-4
14	+++/2	+++/1	+++/2	+++/2	+++/3	+++/3	+++/4	+++/3
14*	+++/2-3	++/1	+++/1-2	+++/1-2	+++/3	++/3	+++/4	+++/3
15	+++/2	+++/1	+++/2	+++/2	+++/2	+++/1-2	nd	+++/2
16	(+)/1	(+)/1	+/1	(+)/1	++/1-2	+++/2-3	++/2	+++/1-3
average	3.1/2.0	2.4/1.2	3.5/2.0	3.5/1.9	3.8/2.6	3.8/2.7	3.9/3.4	4.0/3.0

[a] not determined

Cytoplasmic staining is evaluated by the percentage of positive cells, grouped into the categories: - (0 %), (+) (0-5 %), + (5-20 %), ++ (20-50 %) and +++ (50-100 %), and by the intensity of the individual staining reaction, grouped into the categories: 0 = no staining, 1 = weak, but significant staining, 2 = medium staining, 3 = strong staining and 4 = very strong staining. The tumor cases were classified histologically, as given in the section "Immuno- and glycohistochemical lectin detection".

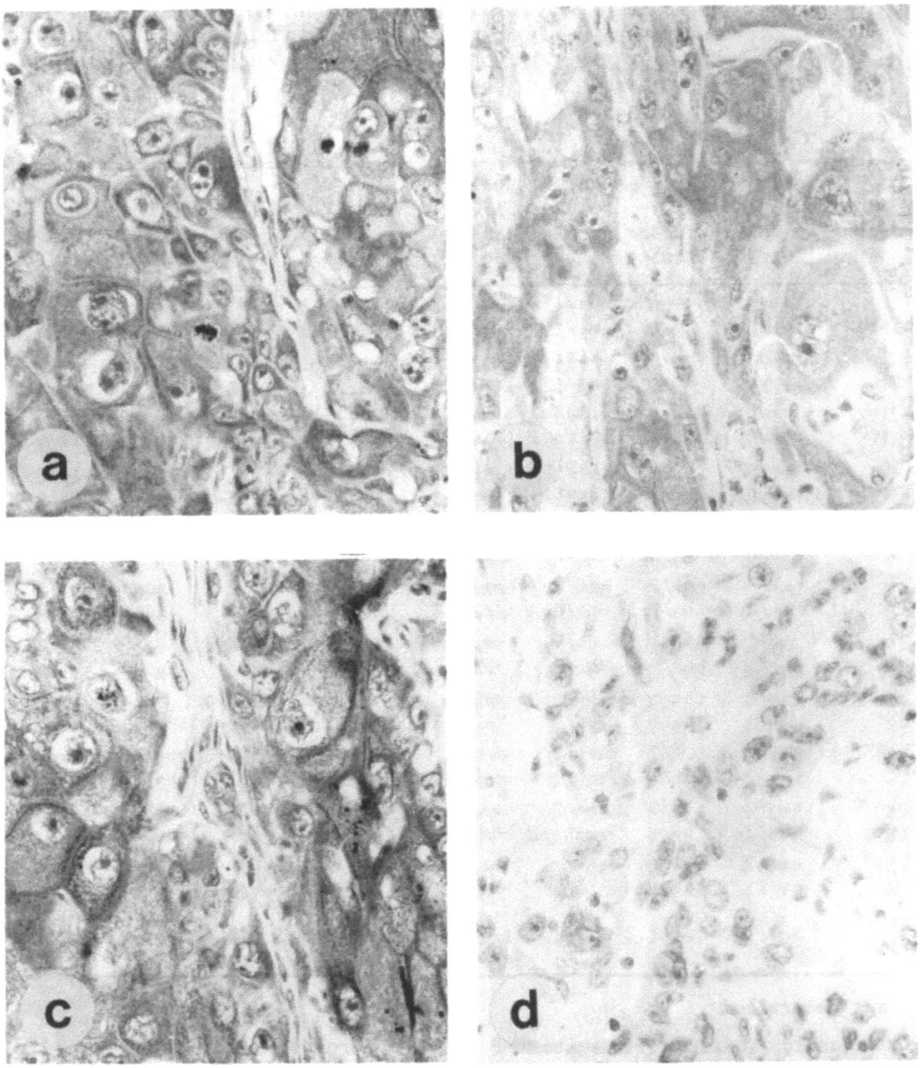

Fig. 2: Light micrographs of serial sections of a squamous cell carcinoma, referred to as case 14 in Table 2, after application of biotinylated neoglycoproteins, namely gal-β1,4-glcNAc-Cete-BSA (a), gal-β1,3-glcNAc-Cete-BSA (b), gal-β1,3-galNAc-Cete-BSA (c) and gal-β1,3-glcNAc-Cete-BSA in the presence of a 50fold excess of lactosylated BSA as well as 0.2 M lactose for demonstration of inhibition (d), ABC reagents and hematoxylin counterstaining; x 310.

Lectin-mediated drug targeting for breast cancer cells

In cancer therapy, cell surface determinants like lectins may be exploitable as binding sites for appropriately tailored conjugates within drug targeting (Gabius, 1988b; Monsigny et al., 1988). To gain evidence for the feasibility of this approach to breast cancer cells, we at first quantitatively analyzed the cell surface lectin expression by employing a chemically glycosylated enzyme.

This type of probe, termed neoglycoenzyme, was synthesized under mild, activity-preserving conditions by carbodiimide-mediated coupling of p-aminophenyl glycosides to β-galactosidase of E.coli (Gabius et al., 1989c). Since the same type of carbohydrate derivative can be coupled to the enzyme, all structural parameters except for the carbohydrate ligand are identical. Using an array of differently glycosylated neoglycoenzymes, incubated at an identical concentration with a constant number of cultured breast cancer cells (DU 4475), the surface expression of lectins with various specificities could be determined (Table 3).

Table 3: Determination of binding sites for carbohydrates on cells of the DU 4475 breast cancer cell line by glycosylated β-galactosidases[a].

sugar type	binding sites per cell
lactose	8.3×10^3
melibiose	8.7×10^3
β-D-galNAc	7.6×10^3
β-D-glcNAc	5.4×10^3
α-D-mannose	2.2×10^3
α-L-rhamnose	1.6×10^4
α-L-fucose	1.4×10^4

[a] Binding of neoglycoenzymes was performed at a concentration of 20 nM with 4×10^3 cells per assay.

To substantiate the proposal that these activities may serve as target for drug delivery, neoglycoproteins were further chemically modified by conjugation of a therapeutically active substance, the photoactive drug hematoporphyrin. The cultured carcinoma cells were then exposed to these hematoporphyrin-neoglycoprotein conjugates and the cytotoxic effect, initiated by irradiation, was quantitated.

Cell viability was reduced in correlation to the receptor density for neoglycoenzymes after irradiation, with non-glycosylated BSA-hematoporphyrin being uneffective. These results are in agreement with prior studies on retinoblastoma cells that similarly used toxin-carrying neoglycoproteins (Griegel et al., 1989).

Immuno- and glycohistochemical lectin detection

These results show that neoglycoproteins can detect tissue lectins that might prove useful for therapeutic implications. The localization of lectins in sections thus deserves further attention, aimed at refining this technique. Similarly, the glycohistochemical detection should be validated by the concomitant localization with lectin-specific antibody. Comparison of these two techniques, too, serves as an inherent control of the glycohistochemical reaction.

Consequently, serial sections of malignant breast lesions were analyzed by two different glycohistochemical probes and an immunohistochemical probe. Due to the presence of the β-galactoside-specific lectin of an apparent molecular weight of 14 kDa in breast cancer (Gabius et al., 1986) our experiments focussed on this lectin, which can be localized by lactosylated BSA. Besides lactosylated neoglycoprotein (lac-BSA (diazo)) human tissue lectin-binding glycoproteins, purified by affinity chromatography on resins with the immobilized β-galactoside-specific lectin (14 kDa), and a lectin-specific antibody to the 14 kDa-lectin were employed to answer the question as to whether differences occur in the capacity of these tools for lectin localization.

14 cases of primary tumor and 7 lymph node metastases including 11 cases of infiltrating ductal carcinoma, referred to as cases 1 - 11 in the table (Table 3), and 4 ipsilateral lymph node metastases from cases of this group, 1 case of tubular carcinoma with focal solid differentiation and its ipsilateral axillary lymph node metastasis, referred to as case 12 and 12* in the table, an axillary lymph node metastasis of an apocrine carcinoma, referred to as case 13 in the table, 1 case of squamous cell carcinoma with its ipsilateral axillary lymph node metastasis, referred to as case 14 in the table, and 1 case of intraductal non-infiltrating carcinoma, referred to as case 15 in the table, were studied. Additionally, serial sections of 1 case of benign breast lesion, an intracanalicular fibroadenoma, referred to as case 16 in Table 2, were analyzed. In glycohistochemistry the sections of paraffin-embedded biopsy specimens were dewaxed and rehydrated, processed with 1 % hydrogen peroxide to block endogenous peroxidase activity, subsequently incubated with 0.1 % periodate-treated BSA solution to saturate unspecific protein-binding sites, with 100 µg/ml biotinylated (neo)glycoprotein in phosphate-buffered saline, pH 7.4, containing 0.1 % BSA for 4 h at room temperature and then for 1 h with ABC reagents, after thorough rinses, as outlined elsewhere (Bardosi et al., 1989, 1990; Gabius et al., 1990). 3-Amino-9-ethylcarbazole served as chromogenic substrate. Counterstaining was carried out with hematoxylin. Specific polyclonal antibodies to the β-galactoside-specific lectin (14 kDa) from bovine heart that is closely related to its human counterpart were generated in rabbits, as described elsewhere (Bardosi et al., 1989). Within immunohistochemistry, normal goat serum was used for blocking unspecific binding sites, the antibody fraction was used at a 1:20 dilution to pick up low levels of expression and biotin-SP-goat anti-rabbit antibody (Dianova, Hamburg, FRG) was employed to detect specifically bound anti-lectin antibody.

Controls to ascertain the specificity of binding have already been outlined. Specificity of the immunohistochemical reactions was ascertained by omission of the antibody and by using an antibody preparation in the presence of a large excess of purified lectin to saturate the antigen-binding sites. Average staining intensity and quantity of stained tumor cells were determined semiquantitatively. They are compiled in Table 2.

The patterns of staining were qualitatively similar for neoglycoprotein, the biochemically purified lectin-binding glycoproteins and the lectin-specific antibody (Fig. 3). Control reactions underscored the specificity of the protein-carbohydrate interactions and the antigen-antibody reactions.

It is interesting to note that the tissue lectin itself, when appropriately labelled, exhibits a rather similar binding pattern (Gabius et al., 1991). Thus, further ultrastructural studies on the extent of colocalization of endogenous lectins and its accessible tissue ligands are encouraged. Equally notable is the possibility to extend the concept of applying a carrier-immobilized ligand for receptor detection, as performed herein for lectins.

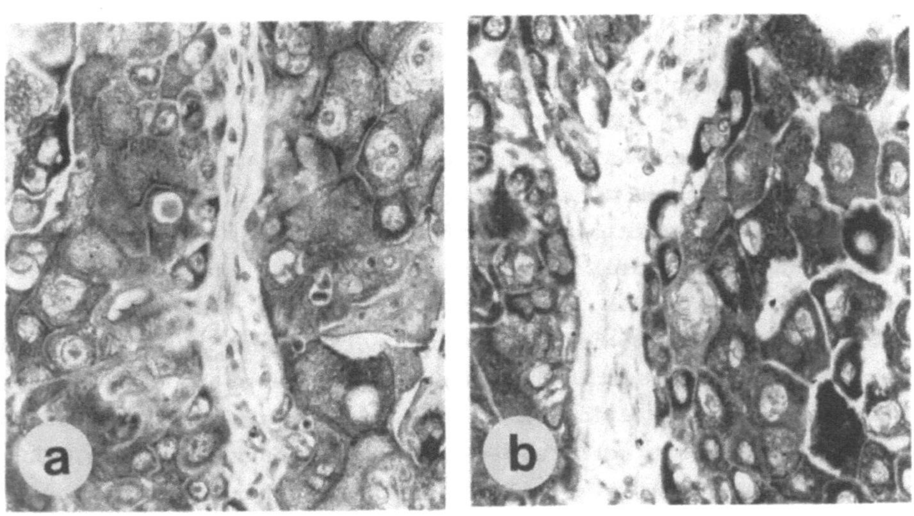

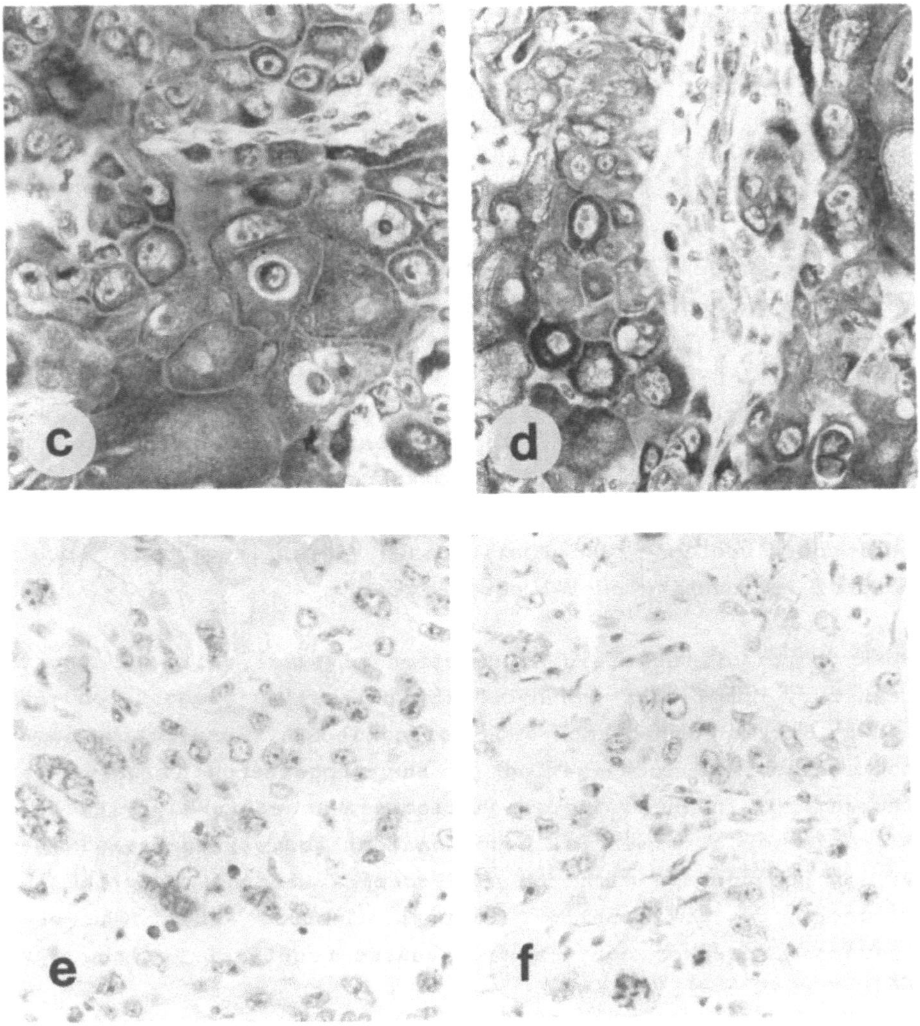

Fig. 3: Light micrographs of serial sections of a squamous cell carcinoma, referred to as case 14 in Table 2, after application of biotinylated lac-BSA (diazo) (a), of biotinylated lectin-binding glycoproteins (b), of lectin-specific antibodies and biotinylated goat-anti-rabbit antibodies (c), of biotinylated β-galactoside-specific lectin (14 kDa) (d) and of biotinylated lactosylated neoglycoprotein (e) or biotinylated β-galactoside-specific lectin (14 kDa) in the presence of a 50fold excess of unlabelled neoglycoprotein as well as 0.2 M lactose for demonstration of inhibition (f), ABC reagents and hematoxylin counterstaining; x 310.

Neohormoproteins as tools for receptor localization

An immobilized ligand as a probe for histochemical receptor detection may be applied to other receptor types than lectins. We herein illustrate that a steroid hormone-protein conjugate is effective as tool for receptor localization in breast cancer.

Progesterone 3-(O-carboxymethyl)oxime was synthesized via progesterone 3-(N-pyrolidyl)enamine and coupled to BSA at a molar ratio of 18:1 (Janoski et al., 1974; Mitsuma et al., 1987). Estradiol-17 β-hemisuccinate and the product of coupling to BSA with a molar ratio of 28:1 were synthesized, as described (Tsong and Koide, 1974). The progesterone-BSA-conjugate was biotinylated by biotinyl-N-hydroxysuccinimide ester, as performed for neoglycoproteins. Labelled progesterone-BSA was applied to sections of formalin-fixed, paraffin-embedded specimens of 29 invasive ductal carcinomas.

Specificity of the staining reaction in tumor cells was ascertained by homologous inhibition by unlabelled progesterone-BSA and by lack of inhibition with estradiol-BSA. Comparative immunohistochemical localization of the progesterone receptor in the same 29 invasive ductal carcinomas was achieved by the application of a commercial monoclonal antibody preparation (Let 64) in conjunction with the APAAP-complex according to the instruction of the supplier (Dianova, Hamburg, FRG). The progesterone receptor content was measured routinely by the dextran-coated charcoal assay.

Immunohistochemical demonstration and histochemical demonstration of progesterone receptors resulted in a similar staining pattern, as exemplarily shown in Fig. 4.

Discrepancies were noted, when the results of both types of analysis of 29 cases of invasive ductal mammary carcinoma were compared. 10 cases were found to be positive with the neohormoprotein. These cases, but also 3 further cases exhibited progesterone receptor, measured by the monoclonal antibody.

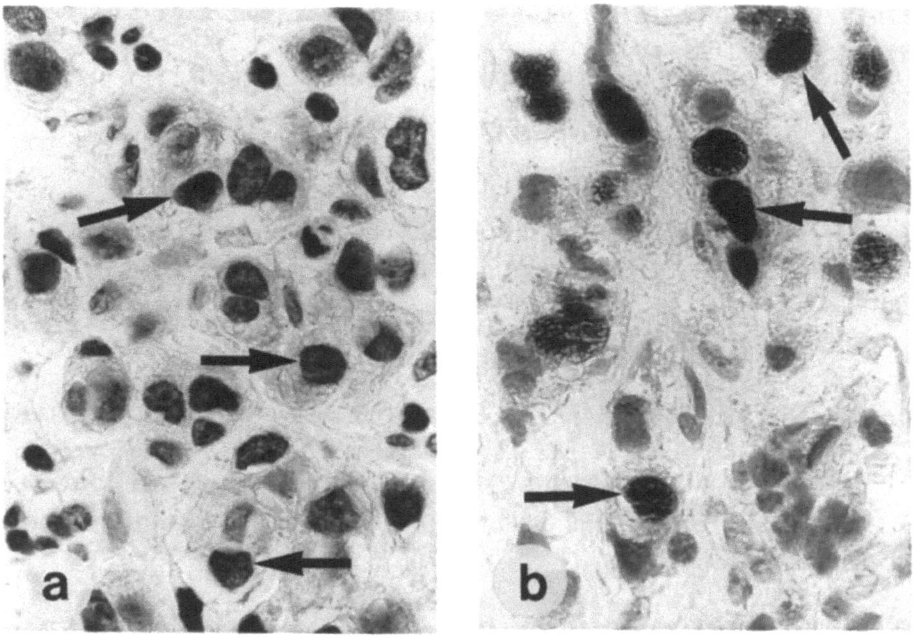

Fig. 4: Light micrographs of sections of an infiltrating ductal carcinoma, demonstrating histochemical detection of progesterone-binding sites with biotinylated progesterone BSA (a) and immunohistochemical detection of the progesterone receptor with a specific monoclonal antibody (b) after application of ABC or APAAP reagents and hematoxylin counterstaining. Arrows mark tumor cell nuclei with positive histochemical or immunohistochemical reaction; x 730.

It is at present unclear whether lack of detection may be accounted for by harm of tissue fixation or different levels of sensitivity of the methods. Among these 3 cases that exclusively originated from postmenopausal women, 1 case had no biochemically detectable progesterone receptor expression. From the viewpoint of biochemical results, only 1 case with a receptor content above 20 fmol/mg protein revealed no significant staining with both histochemical methods. This discrepancy may be due to differences of tissue sampling within biochemical and histochemical determination of the receptor content in a distinct tissue specimen, as has already been put forward (Bergqvist et al., 1984).

Conclusions

1. The type of synthesis of neoglycoproteins has primarily a quantitative influence on their capacity to react with tissue lectins in terms of staining intensity and number of stained cells.

2. Structural alterations of the preterminal sugar is a rational step to enhance the level of specificity of neoglycoproteins. This improvement may also be exploited in therapy when neoglycoproteins are used as carriers in lectin-mediated targeting.

3. Glycosylated enzymes can be employed to quantitatively assess the lectin density on cultured cancer cells, leading to a rational selection of certain types of sugar receptors for lectin-mediated targeting of neoglycoprotein-drug conjugates.

4. Labelled human lectin-binding glycoproteins, purified by affinity chromatography on resins with a distinct immobilized lectin, are valuable tools for histochemical localization of the respective lectin. The staining procedure with this probe mimics the putative interaction of endogenous lectins and respective glycoconjugates in situ. Comparison of histochemical staining with this type of probe, with the respective neoglycoprotein and with the antibodies to the lectin used for glycoprotein isolation yielded similar staining patterns in serial sections of malignant breast lesions, the

application of biochemically purified lectin-binding glycoproteins resulting in the most intense reactions.

5. The labelled β-galactoside-specific lectin (14 kDa) has been demonstrated to be effective for ligand localization. Thus, lectin-specific ligands and the endogenous lectin (14 kDa) are present in the same tumor cells.

6. The concept of application of an immobilized ligand has value beyond lectin localization, as examplarily substantiated for visualization of progesterone receptors by the use of a progesterone-protein-conjugate. This is another example of the applicability of "neoligandoproteins" for specific receptor detection.

References

Bardosi A, Dimitri T, Wosgien B, Gabius HJ (1989) Expression of endogenous receptors for neoglycoproteins, especially lectins, that allow fiber typing on formaldehyde-fixed, paraffin-embedded muscle biopsy specimens. J Histochem Cytochem 37: 989-998

Bardosi A, Bardosi L, Hendrys M, Wosgien B, Gabius HJ (1990) Spatial differences of endogenous lectin expression within the cellular organization of the human heart: A glycohistochemical, immunohistochemical and biochemical study. Am J Anat 188: 409-418

Barondes SH (1981) Lectins: Their multiple endogenous cellular functions. Ann Rev Biochem 50: 207-231

Bergqvist A, Carlström K, Ljungberg O (1984) Histochemical localization of estrogen and progesterone receptors: Evaluation of a method. J Histochem Cytochem 32: 493-500

McBroom CR, Samanen CH, Goldstein IJ (1972) Carbohydrate antigens: Coupling of carbohydrates to proteins by diazonium and phenyliosthiocyanate reactions. Methods Enzymol 28: 212-219

Gabius HJ (1988a) Mammalian lectins: Their structure and their glycobiological and glycoclinical roles. ISI Atlas of Science: Biochem 1: 210-214

Gabius HJ (1988b) Tumorlectinology: At the intersection of carbohydrate chemistry, biochemistry, cell biology and oncology. Angew Chem Int Ed Engl 27: 1267-1276

Gabius HJ (1991) Detection and function of mammalian lectins - with emphasis on membrane lectins. Biochim Biophys Acta 1071: 1-18

Gabius HJ, Brehler R, Schauer A, Cramer F (1986) Localization of endogenous lectins in normal human breast, benign breast lesions and mammary carcinoma. Virch Arch [Cell Pathol] 52: 107-115

Gabius HJ, Bodanowitz S, Schauer A (1988a) Endogenous sugar-binding proteins in human breast tissue and benign and malignant breast lesions. Cancer 61: 1125-1131

Gabius HJ, Kohnke B, Hellmann T, Dimitri T, Bardosi A (1988b) Comparative histochemical and biochemical analysis of endogenous receptors for glycoproteins in human and pig peripheral nerve. J Neurochem 51: 756-763

Gabius HJ, Debbage PL, Lang N, Lange W (1989a) Malignant and normally developing trophoblastic cells of human placenta display different characteristics defined by histochemical and biochemical mapping of endogenous lectins. Histochemistry 92: 283-289

Gabius HJ, Hellmann KP, Dimitri T, Bardosi A (1989b) Endogenous sugar receptor pattern in human glioblastomas and gangliocytomas studied by histochemical application of biotinylated (neo)glycoproteins and affinity chromatography. Histochemistry 91: 5-11

Gabius S, Hellmann KP, Hellmann T, Brinck U, Gabius HJ (1989c) Neoglycoenzymes: A versatile tool for lectin detection in solid-phase assays and glycohistochemistry. Anal Biochem 182: 447-451

Gabius HJ, Gabius S, Brinck U, Schauer S (1990) Endogenous lectins with specificity to β-galactosides and α- or β-N-acetylgalactosaminides in human breast cancer: Their glycohistochemical detection in tissue sections by synthetically different types of neoglycoproteins, their quantitation on cultured cells by neoglycoenzymes and their usefulness as targets in lectin-mediated phototherapy in vitro. Path Res Pract 186: 597-607

Gabius HJ, Wosgien B, Brinck U, Schauer A (1991) Localization of endogenous β-galactoside-specific lectins by neoglycoproteins, lectin-binding tissue glycoproteins and antibodies and of accessible lectin-specific ligands by a mammalian lectin in human breast carcinomas. Path Res Pract, in press

Gordon MR, Chida K, Takata I, Myrvik QN (1987) Macrophage migration inhibition induced by MDP, LPS, PMA and MIF/MAF: Reversal by macrophage migration enhancement factor, L-fucose, L-fucosyl BSA, D-mannose, and D-mannosyl BSA. J Leukocyte Biol 42: 197-203

Griegel S, Rajewsky MF, Ciesiolka T, Gabius HJ (1989) Endogenous sugar receptor (lectin) profiles of human retinoblastoma and retinoblast cell lines analyzed by cytological markers, affinity chromatography and neoglycoprotein-targeted photolysis. Anticancer Res 9: 723-730

Janoski AH, Shulman FC, Wright GE (1974) Selective 3-(O-carboxylmethyl) oxime formation in steroidal 3,20-diones for hapten immunospecificity. Steroids 23: 49-64

Kuchler S, Zanetta JP, Vincendon G, Gabius HJ (1990) Detection of binding sites for biotinylated neoglycoproteins and heparin (endogenous lectins) during cerebellar ontogenesis in the rat. Eur J Cell Biol 52: 87-97
Lotan R, Raz A (1988) Endogenous lectins as mediators of tumor cell adhesion. J Cell Biochem 37: 107-117
Mitsuma M, Kambegawa A, Okinaga S, Arai K (1987) A sensitive bridge heterologous enzyme immunoassay of progesterone using geometrical isomers. J Steroid Biochem 28: 83-88
Monsigny M, Roche AC, Modoux P, Kieda C, Mayer R (1988) Endogenous lectins of myeloid and tumor cells: Characterization and biological implications. In: Gabius HJ, Nagel GA (eds) Lectins and glycoconjugates in oncology. Springer, Berlin Heidelberg New York, pp 25-47
Schauer A, Rothe H, Balzer I, Bergholz M, Brehler R, Fiebig I, Rauschecker H, Nagel GA, Blossey HC, Weber K, Osborn M, Gabius HJ, Cramer F, Aus HM, Harms H, Sauer R, Scheurlen H, Schumacher M (1988) Monoclonal antibodies for improved evaluation of breast cancer prognosis. In: Rich MA, Hager JC, Lopez DM (eds) Breast cancer: Scientific and clinical progress. Kluwer Academic Puvlishers, Boston, pp 299-320
Sharon N, Lis H (1989) Lectins as cell recognition molecules. Science 246: 227-233
Tsong YY, Koide SS (1974) A study on the stability of an estrogen-protein conjugate in vivo and under in vitro conditions. Steroids 24: 825-837
Uhlenbruck G, Beuth HJ, Oette K, Schotten T, Ko HL, Roszkowski K, Roszkowski W, Lütticken R, Pulverer G (1986) Lektine und die Organotropie der Metastasierung. Dtsch med Wschr 111: 991-995

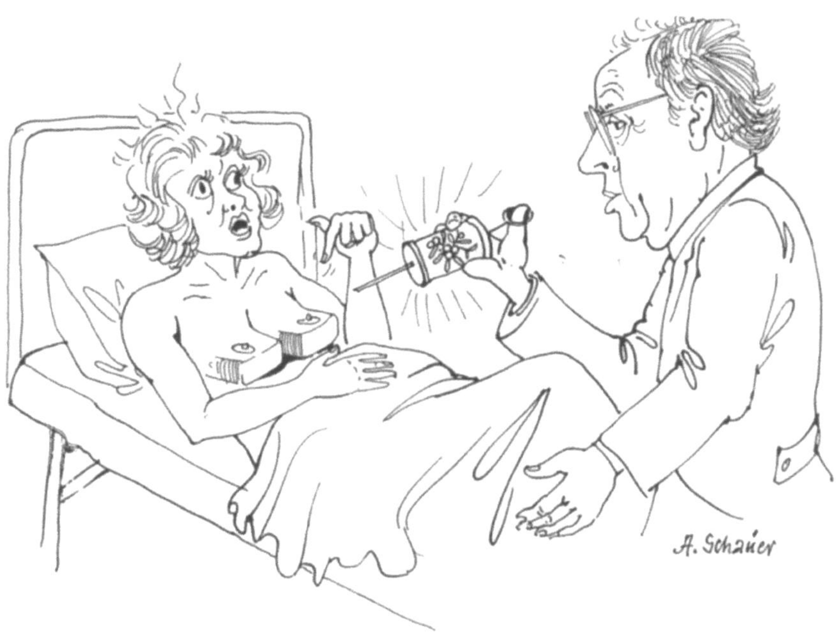

Endogenous receptors fore carbohydrate ligands in human renal cell carcinoma (RCC)

R. Heicappell[1], H. Buszello[1], S. Gabius[2], R. Ackermann[1], and H.-J. Gabius[3]

[1]Department of Urology
Heinrich-Heine University
Moorenstr. 5
4000 Düsseldorf
Germany

SUMMARY

The purpose of this study was to analyse expression of endogenous lectins in human renal cell carcinoma (RCC).
Frozen sections of 44 tumor specimens were analysed for endogenous lectins. In the majority of specimens, lectins with specificity for maltose and N-acetyl-galactosamine were detected. Specimens of both the primary tumor and metastasis were available in 10 cases.
When lectin expression of primary tumor and metastasis were compared, the patterns of lectin expression were similar with no clear gain or loss of certain lectins in the metastases.
Lectins were further analysed in a set of human RCC cell lines that were derived from the same surgical specimen by growth in nude mice. The cell lines were grown from a primary tumor in nude mice, from a liver metastasis thereof and malignant ascites; they thus represented different subpopulations present in the primary tumor. Patterns of lectin expression of the cell lines was heterogeneous.
Ligand binding analysis with neoglycoenzymes revealed that a lectin with specificity for maltose appeared to be modified in the cell derived from malignant ascites as compared to the parental cell line.
From our data, we conclude that lectins with specificity for maltose and N-acetyl-galactosamine are present on human RCC and their corresponding metastases. In the process of tumor progression, the maltose lectin may undergo modifications.

[2] Department of Internal Medicine, Division of Hematology/Oncology, Robert-Koch-Str. 40, 3400 Göttingen, Germany

[3] Max-Planck-Institute for Experimental Medicine, Department of Chemistry, Hermann-Rein-Str. 3, 3400 Göttingen, Germany

INTRODUCTION

Growth of tumor cells in distant organs represents one of the major problems in clinical oncology. In renal cell carcinoma, 5 years survival rate is approximately 95% for locally growing low grade tumors (Tosaka et al., 1990). At time of diagnosis, 25-40% of all patients have distant metastases (deKernion et al., 1978). Of all RCC metastases, no more than 50% can be diagnosed at time of initial presentation (deKernion et al., 1978), when the patient would have a good chance for longtime survival. Clearly then, there is need for early detection of distant metastases in renal cell carcinoma, because this would give the patients a better chance to survive. Renal carcinomas primarily metastasize to the lungs (75%), bones(41%), and lymph nodes (22%) (deKernion et al., 1978). Knowledge on the factors that influence metastasis or contribute to organ selectivity is limited (for review see Nicolson, 1988).

Tumor metastasis is a multistep process that is not random but directed and under positive and/or negative genetic control (for review see: Hart and Easty, 1991). In the process of metastasis, tumor cells interact with a variety of other cells in the primary tumor, in the circulation and in the target organ. Successfull completion of all steps of the metastatic cascade is a prerequisit for development of metastatic disease.

It is the physiologic role of white blood cells to travel in the circulation and evade the blood vessels for migration into lymph nodes, bone marrow or sites of inflammation. Several recognition systems for leukocyte homing have been described (for review see: Patarroyo and Makgoba, 1989). Some of them, e.g. binding of lymphocytes to hepatocytes (Kolb et al., 1978) or the homing receptor on mouse lymphocytes (Stoolman et al., 1984) involve interactions between carbohydrates and their receptors, the lectins.

For tumor cells, recognition systems similar to those described for white blood cells may exist, too (Raz and Ben-Ze'ev, 1987).

The purpose of this study was to analyse expression of lectins on human renal cell carcinoma. Both, primary tumors and - in 10 patients - a corresponding metastasis were analysed. In order to evaluate the effect of organ environment on lectin expression, human RCC cell lines were analysed that were derived from a xenotransplant of RCC in nude mice and its metastases.

PATIENTS, MATERIALS AND METHODS

Patients

Frozen sections of 44 patients were analysed for lectin expression. All patients had unilateral nephrectomy. The presence of renal cell carcinoma was confirmed by routine histology in all cases. In 10 of the 44 patients examined, specimens of both, the primary tumor and a metastasis were available.

In all patients, an additional specimen was obtained from normal renal tissue of the contralateral edge of the kidney.

Of the patients 32 were males and 12 females; mean age was 58.9 years.

Cell lines

The origin of the human renal cell carcinoma (h-RCC) cell lines SN12A1, SN12K1, and SN12L1 has been described in detail by Naito et al.(1986). Briefly, these cell lines have been derived from a single granular cell type h-RCC specimen, that was xenotransplanted under the renal subcapsule of a nude mouse. From the site of implantation, this tumor metastasized to several organs. Cell lines were derived from a primary tumor in the kidney (SN12K1), a liver metastasis thereof (SN12L1) and of metastatic tumor cells that grew in ascites (SN12A1). All cell lines were kindly made available by Dr.I.J.Fidler, Department of Cell Biology, M.D.Anderson Cancer Center, Houston, TX.

Cell lines were maintained in monolayer cultures on plastic tissue culture flasks in Eagles minimum essential medium (MEM; Gibco, Eggenstein, Germany) supplemented with 10% fetal bovine serum (FBS), sodium pyruvate, non-essential amino acids, L-glutamine, and a twofold vitamin solution (Gibco, Eggenstein, Germany).

Preparation of neoglycoproteins

Neoglycoproteins were prepared as described elsewhere (Gabius et al., 1991). Briefly, bovine serum albumin (BSA) was chemically glycosylated by diazotization (in the case of fucose, mannose, sialic acid, N-acetyl-glucosamine, and N-acetyl-galactosamine) or by reductive amination (in the case of lactose, melibiose, cellobiose and maltose) resulting in incorporation of 8-10 and 16-18 sugar moieties per molecule of BSA, respectively. Neoglycoproteins were then labelled with fluorescein-isothiocyanate (FITC).

Preparation of neoglycoenzymes

Glycosylation of ß-galactosidase from E.coli (Serva, Heidelberg, Germany) was performed under mild, activity-preserving conditions by carbodiimide-mediated coupling of p-aminophenyl glycosides, as previously described (Gabius S, et al. 1989). Briefly, 8 mg enzyme, dissolved in 1ml of 20mM phosphate-buffered saline (PBS, pH 7,4), was reacted with 16umol p-aminophenyl glycoside in the presence of 32umol 1-ethyl-3-(3-dimethylaminopropyl)carbodiimide for 16 hours at 4°C prior to dialysis against PBS and isolation of the modified enzyme by gel filtration. Fractions containing enzyme were pooled, dialyzed against PBS containing 60% glycerol and stored at -20°C. The sugar content of the enzyme fractions, quantitated for protein by application of dye-binding, was determined by a resorcinol-sulfuric acid micromethod to average 25 (±5) sugar moieties per monomeric subunit (Monsigny et al., 1988)

Expression of lectins on frozen sections of RCC

Specimens of the tumor and the corresponding normal kidney were snap-frozen in pentane, mounted on cork and stored at -20°C prior to use. For lectin analysis, 5um sections were cut on a cryostat. FITC-labelled neoglycoproteins were added at a concentration of 50ug/ml in PBS containing 1% BSA and 0.1% Na-azide for 30 minutes at 4° C. Specimens were washed three times with PBS and fluorescence was analysed using a Leitz fluorescence microscope equipped for FITC fluorescence. In the specimens of 19 patients, specificity of staining was analysed by addition of a 10 fold excess of the the respective non-labelled neoglycoprotein plus carbohydrate at 1.25mM.

Determination of extracellular and intracellular neoglycoprotein fluorescence in adherent cells

Tumor cells were grown on sterilized glass coverslips in 24 well tissue culture dishes at an initial density of 200,000 cells per well for 24 hours. Coverslips were then washed three times with phosphate buffered saline (PBS) supplemented with 1% BSA and 0.1% Na azide. Neoglycoproteins were added at 50ug/ml for 30 minutes at room temperature. For analysis of intracellular lectins, cells were permeabilized with methanol and nonidet P-40 (0.05% in PBS) prior to addition of neoglycoproteins and subsequently incubated at 4° C with neoglycoproteins for 30 minutes. Cultures were then washed three times and the coverslips were mounted using Kaiser's glycerol gelatine. Fluorescence was determined using a Leitz diaplan fluorescence microscope equipped for FITC-fluorescence.

Quantitative determination of carbohydrate binding sites

For ligand binding studies, 200,000 cells were seeded into 24-well culture plates and incubated for 24 hours at 37°C. Cells were then carefully washed off any serum components with

Hank's balanced salt solution (Gibco, Eggenstein, Germany) containing 20mM HEPES buffer (pH 7.5) (Biochrom, Berlin, Germany) and 0.05% sugar-free bovine serum albumin. Cells were incubated with increasing concentrations of neoglycoenzymes in a volume of 400ul for 120 minutes at 4°C. Cells were then washed three time in HBSS. Subsequently, 200ul of 100mM HEPES buffer (pH 7.0) containing 0.5% Triton X-100, 150mM NaCl, 2mM $MgCl_2$, 0.1% NaN_3, 0.1% BSA and 1.5mM chlorophenolred-ß-D-galactopyranoside (CPRG; Boehringer, Mannheim, Germany) was added for 60 minutes at 37°C. Substrate turnover by bound neoglycoenzymes was stopped by addition of 200ul 0.2M glycine (pH 10.5). Optical density of the colored reaction product was determined at 590nm in a microtiter plate reader (Dynatech Minireader II). For determination of non-specific binding, control cultures contained enzyme only. Transformation of the binding data for the neoglycoenzymes to obtain the dissociation constant (K_D) and to estimate the number of binding sites available at saturation was performed as described by Scatchard (Scatchard, 1949).

Biochemical analyis of lectins

Tumor specimens were trimmed off necrotic and connective tissue, snap-frozen in pentane and stored at -20° prior to analysis. After thawing, tumors were homogenized in 5-6 volumes of 25mM Tris-HCl buffer (pH 7.8) containing 1% Triton X-100, 0.1% sodium deoxycholate, 0.1mM phenylmethanesulfonyl fluoride, 1mM dithiotreitol, 5ug/ml leupeptin and antipain, and 0.2M KCl. Affinity chromatography was carried out with carbohydrates or glycoproteins immobilized on resin columns via cyanogen bromide or divinyl sulfone activation as described elsewhere (Gabius and Engelhardt, 1988). Samples were concentrated by ultrafiltration (Diaflo Ultrafiltration Model 50, Amicon, Witten, Germany). Detergent was removed by chloroform extraction and gel electrophoresis was performed. Gels were stained with a modified silver staining method, as decribed by Blum et al. (1987).

RESULTS

Tumor histology

In all specimens, renal cell carcinoma was confirmed by histopathology. Pathologic staging for the tumors were one T1 tumor, 23 T2 tumors and 20 T3 tumors. Pathologic grade was grade I in two cases, grade II in 19 cases and G3 in 12 cases.

Expression of lectins on frozen sections of RCC

Frozen sections of fourty four individual RCC specimens were incubated with neoglycoproteins (50ug/ml) for 30 minutes at 4°C. Fluorescence was analysed using a fluorescence microscope. Positive fluorescence was defined as intermediate to strong fluorescence of the whole specimen or strong fluorescence of spots in the specimen covering more than a third of the whole area.

In 19 cases, specifity of staining was checked. After incubation with neoglycoproteins these samples were incubated with a 10 fold excess of non-labelled neoglycoprotein and the respective sugar at 1.25mM.

The majority of specimens stained positive for N-actetyl-galactosamine and maltose (table 1). Binding of both carbohydrates was specific because addition of unlabelled neoglycoprotein at 10-fold excess plus free carbohydrate totally abolished fluorescence. Neoglycoproteins that were glycosylated with carbohydrates of similar structure exhibited different binding patterns. Thus, 32/44 (73 %) of the specimens stained positive for maltose; for cellobiose, which is structurally related, only 13 out of 44 specimens were positive. Similar findings were made for N-acetyl-galactosamine vs. N-acetyl-glucosamine.

Less than 30% of the specimens tested stained positive for monosaccharides (mannose, mannose-6-phosphate, fucose, xylose and sialic acid). Mannose and mannose-6-phosphate had a large proportion of non-specific staining.

For each tumor, a specimen of normal kidney (obtained from the contralateral edge of the tumor bearing kidney) was analysed for lectin expression. In some specimens certain lectins were visible after addition of FITC-labelled neoglycoproteins. In 19 specimens specificity of neoglycoprotein binding was analysed by addition of a 10 fold excess of nonlabelled neoglyprotein plus the respective carbohydrate at 1.25mM. No specific binding of neoglycoproteins could be detected in any of the 19 specimens examined.

Carbohydrate specificity	positive staining	specific positive staining
mannose	21 / 44	5 / 19
mannose 6-P	11 / 44	5 / 19
fucose	15 / 44	4 / 19
xylose	9 / 44	5 / 19
sialic acid	6 / 44	0 / 19
N-acetyl-glucosamine	15 / 44	3 / 19
N-acetyl-galactosamine	30 / 44	14 / 19
maltose	32 / 44	16 / 19
cellobiose	13 / 44	6 / 19
melibiose	3 / 44	0 / 19
lactose (red)	7 / 44	3 / 19
lactose (diaz)	15 / 44	3 / 19
control (BSA/FITC)	0 / 44	0/ 19

Table 1: Staining of frozen sections of human renal cell carcinoma (hRCC) with FITC-labelled neo-glycoproteins

Expression of lectins on frozen sections of RCC metastases

Twelve RCC metastases were available for detection of lectins (table 2). In 10 out of 12 specimens, metastasis and primary tumor were analysed. Sources of metastasis were lymph node (5/12), vena cava (4/12), adrenal gland (1/12), bone (1/12) and skin (1/12).

carbohydrate specificity	primary tumor	metastases
monosaccharides		
mannose	4 / 10	4 / 12
mannose 6-P	5 / 10	3 / 12
fucose	3 / 10	2 / 12
xylose	2 / 10	0 / 12
sialic acid	3 / 10	3 / 12
aminosugars		
N-acetyl-glucosamine	5 / 10	8 / 12
N-acetyl-galactosamine	6 / 10	9 / 12
disaccharides		
maltose	7 / 10	9 / 12
cellobiose	4 / 10	2 / 12
melibiose	2 / 10	3 / 12
lactose (red)	3 / 10	2 / 12
lactose (diaz)	6 / 10	4 / 12
control(BSA/FITC)	0 / 10	0 / 10

Table 2: Lectin patterns in pairs of primary tumor and metastasis

Lectin expression on the primary tumor of patients with metastatic disease was similar to those, in whom no metastases were present at time of surgery. Patterns of lectin expression on tumor metastases were similar to those found on primary tumors. No overall modifications e.g. gain or loss of specific lectins in metastases could be detected. Out of 10 pairs of primary tumors and metastases, in 7 specimens of metastases, more lectins were expressed than in the primary tumor. A lectin with specifity for N-acetyl galactosamine was detected on all (6/6) lymph node metastases.

Biochemical analysis of lectins on RCC

Tumor specimens of two patients were biochemically analysed for carbohydrate binding proteins. Following affinity chromatography on immobilized carbohydrate, SDS-PAGE revealed the following main bands:

Lactose:	14, 30, 56kD
Mannose:	56, 97 kD
Fucose:	56 kD
N-acetyl-glucosamine:	56 kD
N-acetyl-galactosamine	34 kD in one specimen two bands of 56 kD, 68 kD in the second specimen
Maltose:	56 kD, 64 kD
Sialic acid:	10 kD

Lectin expression on RCC cell lines

In order to obtain information on the influence of organ environment on the expression of lectins on human RCC, three RCC cell lines were analysed that were derived from outgrowths of the same primary tumor in different organs of nude mice. One cell line (SN12K1) was derived from a xenotransplanted primary tumor in the kidney of nude mice. A second cell line was obtained from liver metastases in nude mice of the same primary tumor. The third cell line (SN12A1) was established from malignant ascites after xenotransplantation of, again, the same primary tumor into the kidney of nude mice. All cell lines mentioned are established cell lines with stable growth *in vitro* and *in vivo* (nude mice).

These cell line were selected in order to find out, wether lectin expression would change in different organ environments *in vivo*. Cell lines were grown on glass coverslips and determination of lectin expression was carried out after addition of FITC-labelled neoglycoprotein using a fluorescence microscope. The lectins found in the majority of frozen sections were lectins with specificity for maltose and N-acetyl-galactosamine. These lectins were found on fixed cells of the cell line that was derived from a primary tumor in the kidney of nude mice as well (table 3).

	cell line					
carbohydrate specificity	kidney variant (SN12K1)		liver metastasis variant (SN12L1)		ascites variant (SN12A1)	
	M	I	M	I	M	I
monosaccharides						
mannose	+	-	-	+	-	+
mannose 6-P	(+)	-	-	(+)	+	-
fucose	+	-	-	-	(+)	-
xylose	+	-	-	-	+	+
sialic acid	-	-	-	-	-	-
aminosugars						
N-acetyl-glucosamine	-	-	-	+	-	-
N-acetyl-galactosamine	+	-	-	-	-	+
disaccharides						
maltose	+	-	+	-	-	-
cellobiose	-	-	-	-	(+)	-
melibiose	-	-	-	-	-	-
lactose (red)	-	-	-	-	-	-
lactose (diaz)	+	-	(+)	(+)	-	-
control						
(BSA/FITC)	-	-	-	-	-	-

Table 3: Expression of lectins on fixed and permeabilized RCC cell lines. Lectins were detected by FITC-labelled glycoproteins. Intracellular ("I") and membrane ("M") fluorescence was determined with a flurescence microscope. Fluorescence was rated as bright ("+"), weak ("(+)") or absent ("-").

Results indicate, that expression of lectins is not uniform on the cell lines examined. Rather, qualitative changes can be observed in those cell lines that were derived from tumor metastasis in the liver and from malignant ascites.

Quantitative analysis of lectins on RCC cell lines

Cell lines derived from the same primary tumor by selection in nude mice were examined for qualitative changes in lectin expression. Again, cell lines examined were from a primary tumor in the kidney (SN12K1), a liver metastasis (SN12L1) and malignant ascites (SN12A1), all of identical origin (table 4).

cell line		carbohydrate specificity			
		lac	mal	gluc NAc	fuc
Kidney variant (SN12K1)	B_{max}	1.40	0.73	0.97	0.62
	K_D	206	49	50	92
Liver metastasis variant (SN12L1)	B_{max}	1.80	0.55	1.02	0.94
	K_D	187	35	40	73
Ascites variant (SN12A1)	B_{max}	1.52	1.58	0.90	0.75
	K_D	220	162	36	65

Table 4: Determination of apparent affinity constant (K_D) and the apparent number of binding sites for neoglycoenzymes at saturation (B_{max}).

B_{max} is expressed in enzymes (x 10,000) per cell and K_D in nM. Abreviations used for carbohydrate specificities are:
lac: lactose, mal:maltose, glucNAc: N-acetyl-glucosamin and fuc: fucose.

Lectin expression on the primary tumor of patients with metastatic disease was similar to those, in whom no metastases were present at time of surgery. Patterns of lectin expression on tumor metastases was similar to the primary tumors. No overall modifications e.g. gain or loss of specific lectins in metastases could be detected. Out of 10 pairs of primary tumor and metastases, in 7 specimens of metastases, more lectins were expressed than in the primary tumor. A lectin with specifity for N-acetyl-galactosamine was detected on all (6/6) lymph node metastases.

DISCUSSION

Migration of tumor cells to distant organs is of predominant importance in clinical oncology. In renal cell carcinoma, factors that contribute to tumor metastasis have not been defined yet. Carbohydrate receptors are known to serve as recognition structures on the cell surface in a variety of physiological and pathological events (for review see Gabius, 1988).

Endogenous lectins have been described to be present on tumors of murine (e.g. S.Gabius, 1990) and human (e.g. Gabius and Kayser, 1989) origin. Tumor lectins may be involved in certain steps of tumor metastasis, specifically in cell adhesion (S.Gabius et al., 1990).

In renal cell carcinoma, few information on the biology of metastasis is currently available. Due to the relative inefficiency of current therapeutic strategies for disseminated disease (for review see Heicappell and Ackermann, 1990), it is of key interest to acquire information on the molecules involved in RCC metastasis.

In a first series of experiments we analysed frozen sections of human RCC for the presence of lectins. The majority of the specimens examined expressed lectins for maltose and N-acetyl-galactosamine. These lectins were not detectable on corresponding normal human kidney.

Galactoside binding lectins have been described in a variety of other tumor systems (Gabius and Kayser, 1989; Raz and

Lotan, 1987). Their function has been associated with adhesion in tumor metastasis (for review see Raz and Lotan, 1987). It may thus well be, that the galactoside binding lectin that is expressed on the majority of the primary tumors from human kidney is involved in tumor metastasis, too.

The role of the maltose binding lectin is much less clear. Maltose binding lectins have been found on other tumors as well. In the murine melanoma B16F1 the maltose lectin was the lectin with the highest number of receptors of all lectins examined (S.Gabius, 1990). Moreover, in the high metastatic variant B16F10, 40% less reseptors for maltose were found than in the low metastatic variant B16F1.

In 10 patients, pairs of primary tumors and metastases were examined for lectin expression. Patterns of lectins expression were inconsistent, however. There was no clear change in lectin expression, e.g. gain or loss of certain lectins in metastases as compared to the primary tumors. In primary tumors, clones with the propensity to metastasize may rapidly gain growth advantage over non-metastatic cells (Kerbel et al., 1988). The tumors examined were removed shortly after clinical detection. We have no information as to whether in the tumors under study metastatic clones have already gained growth advantage or not. The clinical observation, that the majority of RCC have visible or occult metastases at time of initial presentation suggest that the metastatic clones may have had growth advantage in the tumors we studied. If this was the case, it would not be surprising that lectin patterns did not differ very much between primary tumor and metastasis.

Thus, for analysis of the role of lectins in human tumor metastasis, a better defined system with defined transition from primary tumor to metastases would be preferable.

We therefore chose to select the SN12 system of human RCC tumor cell lines for further analysis. In this system, a set of cell lines has been established from a single RCC specimen (Naito et al., 1986). Naito injected tumor cells of a RCC specimen into the kidney of nude mice. There, the tumor metastasized into the liver and gave rise to malignant ascites. Cells of the primary tumor in the kidney of nude mice, the liver metastasis and the ascites was put into cell culture; subsequently three cell lines were established, designated as SN12K1 (primary tumor in the kidney), SN12L1 (liver metastasis), SN12A1 (ascites). Each of these cell lines presumably represents a subpopulation present in the primary tumor. For analysis of lectins, these cell lines were grown on glass coverslips, fixed and permeabilized and subsequently stained with FITC-labelled neoglycoproteins.

In the 3LL murine tumor system, Weiss and coworkers (Glaves et al., 1989; Vidal-Vanaclocha, 1990) have shown that there were qualitative and quantitative differences in lectin expression of the same tumor when grown in different organs of C57BL/6 mice.

We have presented indirect evidence that this might also be the case in human renal cell carcinoma:

i) We observed differences among the cell lines (SN12K1, SN12A1, SN12L1) in the pattern of lectins expressed. Lectins with specificity for maltose and N-acetyl-galactosamine were present in the cell line derived from the primary tumor in the kidney. These particular lectins were also present in the majority of frozen sections of human primary tumors, as described in more detail before.In the cell line derived from a liver metastasis, the maltose lectin was expressed, too, whereas the N-acetyl-galactosamine lectin was not visible with the method employed. In the cell line that was derived from malignant ascites, maltose lectin was not visible in fixed and permeabilized cells.

ii) When we analysed living cells with neoglycoenzyms, we found that apparently, the receptor for maltose was qualitatively different from the maltose receptors on the two other cell lines examined (primary tumor and liver metastasis). We base this assumption on the observation that the dissociation constant of SN12A1 was four times lower than on SN12L1 or SN12K1; moreover, B_{max} was two to three times higher than in the other cells lines.

It may thus be concluded, that a typical lectin pattern can de found on human renal cell carcinoma, but not on normal kidney cells. In metastases from renal cell carcinoma, there is no overall change in the pattern as compared to the primary tumor. Results from ligand binding experiments to human RCC cell lines representing different stages of tumor progression suggest that the receptor for maltose may be modified in the course of tumor progression.

REFERENCES

Blum H, Beier H, Gross HJ (1987) Improved silver staining of plant proteins, RNA and DNA in polyacrylamide gels. Electrophoresis 8: 93-99

deKernion, JB, Ramming KP, and Smith RB (1978) The natural history of metastatic renal cell carcinoma: a computer analysis. J.Urol., VI 120: 148-152

Fidler IJ, Kripke ML (1977) Metastasis results from pre-existing variant cells within a malignant tumour. Science 197: 893-897

Gabius H-J (1988) Mammalian lectins: their structure and their glycobiological and glycoclinical roles. ISI Atlas of Science: Biochemistry, 210-214

Gabius H-J, and Engelhardt R (1988) Sugar receptors of different types in human metastases to lung and liver. Tumor Biol 9:21-36

Gabius H-J and Kayser K (1989) Elucidation of similarities of sugar receptor (lectin) expression of human lung metastases from histogenetically different types of primary tumors. Anticancer Res. 9(6): 1599-1604

Gabius, H-J, and Bardosi, A (1991) Neoglycoproteins as tools in glycohistochemistry. Progr Histochem Cytochem 22(3),1-75, 1991

Gabius S, Hellmann KP, Hellmann T, Brinck U, and Gabius H-J (1989) Neoglycoenzymes: a versatile tool for lectin detection in solid phase assays and glycohistochemistry. Anal Biochem 182: 447-451

Gabius S, Schirrmacher V, Franz H, Joshi SS, and Gabius HJ (1990) Analysis of cell surface receptor expression by neoglycoenzyme binding and adhesion to plastic-immobilized neoglycoproteins for related weakly and strongly metastatic cell lines of murine tumor model systems. Int J Cancer 46: 500-507

Glaves D, Gabius HJ, and Weiss L (1989) Site-associated expression of endogenous tumor lectins. Int J Cancer 44: 506-511

Hart I and Easty D (1991) Identification of genes controlling metastatic behaviour. Br.J.Cancer 63: 9-12

Heicappell R and Ackermann R (1990) Rationale for immunotherapy of renal cell carcinoma.(Invited Review) Urol Res 18: 357-372

Kerbel RS, Waghorne C, Korczak B, Lagarde A and Breitman ML (1988) Clonal dominance of primary tumours by metastatic cells: genetic analysis and biological implications. Cancer Surveys 7: 597-629

Kolb H, Kolb-Bachhofen V, Kolb HA (1978) Possible mechanism of entrapment of neuraminidase-treated lymphocytes in the liver. Cell Immunol 40: 457-462

Monsigny M, Petit C, Roche AC (1988) Colorimetric determination of neutral sugars by a resorcinol sulfuric acid micromethod. Anal Biochem 175: 525-530

Naito S, von Eschenbach AC, Giavazzi R, and Fidler IJ (1986) Growth and metastasis of tumor cells isolated from a human renal cell carcinoma implanted into different organs of nude mice. Cancer Res 46: 4109-4115

Nicolson GL (1988) Organ specificity of tumor metastasis: role of preferential adhesion, invasion and growth of malignant cell at specific secondary sites. Cancer and Metastasis Rev 7: 143-188

Pataroyo M, and Makgoba MW (1989) Leukocyte adhesion to cells in immune and inflammatory responses. Lancet Nov 11: 1139-1142

Raz A and Ben-Ze'ev A (1987) Cell-contact and - architecture of malignant cells and their relationship to metastasis. Cancer and Metastasis Reviews 6: 3-21

Raz A, and Lotan R (1987) Endogenous galactoside-binding lectins: a class of functional cell surface molecules related to metastasis. Cancer and Metastasis Reviews 6: 433-452

Scatchard G (1949) The attractions of proteins for small molecules and ions. Ann NY Acad Sci 51: 660-672

Stoolman LM, Tenforde TS, and Rosen SD (1984) Phosphomannosyl receptors may participate in the adhesive interaction between lymphocytes and high endothelial venules. J Cell Biol: 1535-1540

Tosaka A, Ohya K, Yamada K, Ohashi H, Kitahara S, Sekine H, Takehara Y and Oka K (1990) Incidence and properties if renal masses and asymptomatic renal cell carcinoma detecte by abdominal ultrasonography. J Urol 144: 1097-1099

Vidal-Vanaclocha F, Barbera-Guillem E, Weiss L, Glaves D, and Gabius, H-J (1990) Quantitation of endogenous lectin expression in 3LL tumors, growing subcutaneously and in th kidney of mice. Int J Cancer 46: 908-912

ACKNOWLEDGEMENTS
This work was supported by a grants from DFG (He1362/5-1) and the Dr.Mildred Scheel Stiftung. The skillfull techical assistance of M.Lietzau, T.Hellmann, K.P.Hellmann and H.Franz is gratefully acknowledged.

Brain, nerves and muscles: Neoglycoproteins in the practice of neuromorphology and neuropathology

A. Bardosi[1], D. Brkovic[1], and H.-J. Gabius[2]

[1] Institut für Pathologie
Akademisches Krankenhaus der
Universität Münster
D-4930 Detmold, FRG

Introduction

The application of plant lectins and carbohydrate-specific monoclonal antibodies has provided a wealth of information on the spatial and temporal regulation of the expression of defined carbohydrate sequences in cellular glycoconjugates. However, the functional implication of the expression in such a controlled fashion for this particular kind of biochemical structure remains to be elucidated. It is a reasonable proposal to ascribe functional relevance to carbohydrate structures by a glycobiological interplay with respective receptors. Their experimental detection in histochemistry with appropriate laboratory tools may thus mimic an actual functional aspect. The implicitly suggested presence of endogenous receptors for the carbohydrate part of cellular glycoconjugates consequently needs to be substantiated. To obtain the desired information, we have taken advantage of procedures, using chemically derived appropriate tools. By covalently conjugating suitable carbohydrate ligands to an inert carrier, the ligands themselves are employed as detection devices for the receptors. Since the conjugate, termed neoglycoprotein (see also contribution of Lee and Lee, this volume), can additionally be

[2] Max-Planck Institut für experimentelle Medizin, Abt. Chemie, D-3400 Göttingen, FRG

conveniently labelled, e.g. biotinylated, routinely performec protocols for staining can thus be carried out. Consequently, an array of neoglycoproteins can be employed for the glycohistochemical search of specific receptors in fixed, paraffin-embedded sections of diverse tissue specimen. Appropriate controls, e.g. ascertaining the absence of non-specific binding of the labelled, but non-glycosylated carrier protein and the specificity of the involved protein-carbohydrate interaction, confirm the validity of the approach. This technique allows to reliably profile the expression of receptors for sugar sequences, fulfilling the prerequisite to critically evaluate whether its application can lead to progress in the typing of tumor classes and in understanding of their biological behavior. Moreover, the use of neoglycoproteins has proven valuable in other areas of neuropathology. It enables typing of muscle fibers in formaldehyde-fixed, paraffin-embedded tissue sections. The concomitant glycohistochemical and immunohistochemical analysis also substantiates that endogenous lectins can be localized by these types of probe, reinforcing the stringency of the control reactions. We herein demostrate its potential usefulness for normal and tumorous tissue within the peripheral and central nervous system, extending our previous compilation (Bardosi et al., 1988a).

Observations

Human brain

A panel of neoglycoproteins was applied to sections of different region of human brain to map the presence of sugar receptors. This comparative analysis will also help to delineate any pathological alterations.

A strong and specific staining with biotinylated neoglycoproteins, exposing different sugar moieties as ligands, indicated the presence of sugar receptors in the nuclei, neuronal pathways and accessory structures such as ependyma cells, plexus chorioideus, intra- and extracerebral vessels and leptomeninges, localized in the mesencephalon, in the pons, in the medulla oblongata and in the cerebellum. Examples of the staining reactions are given in Figs. 1-6. Significant differences in the extent of staining were seen for various neuroanatomical regions, like nerve cells in the basal and central regions of the nuclei pontis in the glycohistochemically detected level of expression of endogenous sugar receptors, in the glial cell of white matter, in neurons of hippocampal gyrus, basal ganglia and thalamus nuclei as well. Remarkably, among the neurons of five cortical laminae the binding of N-acetyl-D-galactosamine varied from strong to undetectable.

Exemplary glycobiochemical investigations on the medulla oblongata and cerebellum were performed to investigate the molecular nature of sugar receptors detected glycohistochemically. Despite notable overall similarities, carbohydrate-binding proteins of differing molecular weight can be isolated from these two regions of the central nervous system, namely in the case of receptors with specificity to ß-galactoside termini, to N-acetyl-D-galactosamine and N-acetyl-D-glucosamine and to D-xylose. These combined glycohistochemical and glycobiochemical results serve as a guideline for exploring the physiological relevance of the detected regional differences (Gabius and Bardosi 1990; 1991; Bardosi et al., 1990b).

Tumors of central nervous system

A specific staining with various neoglycoproteins indicated the presence of sugar receptors in different types of meningiomas, glioblastomas, gangliocytomas, well-differentiated and anaplastic oligodendrogliomas and ependymomas. All tumors of the CNS that had been examined showed significant binding of xylose-(BSA-biotin), lactose(diaz.)-(BSA-biotin), asialofetuin, fucose-(BSA-biotin), mannose-(BSA-biotin) and two further probes, carrying phosphate and acetyl groups. However, individual staining intensities were different. It is interesting to note that a significant reaction could be observed with cellobiose-(BSA-biotin) only in the cases of differentiated and anaplastic oligodendrogliomas with a medium staining intensity and in the case of malignant ependymoma with a weak staining intensity. Similar selectivity was observed for glioblastomas, when sialic acid-(BSA-biotin) was used. Characteristic examples of staining reactions from this tumor class are given in Figs. 7-12.

In comparison to the well-differentiated ependymomas, the anaplastic form of this tumor exibited a higher capacity to specifically bind the neoglycoproteins, containing α- or β-glucosides. Inverse intensity of the glycohistochemical reaction was observed with galactose-6-phosphate-, galactose-β(1,3)-N-acetyl-D-glucosamine, N-acetyl-D-glucosamine- and mannose-(BSA-biotin), respectively, when anaplastic and differentiated oligodendrogliomas were compared with each other. Dedifferentiated neurons, i.e. in gangliocytomas, showed a changed spectrum of endogenous sugar receptors in comparison to neurons of normal cerebral cortex.

Tumor type-dependent qualitative and quantitative alterations in the spectrum of endogenous sugar receptors could be observed among the subtypes of meningiomas. Receptors for N-acetyl-D-galactosamine were present only in the anaplastic form, while glucuronic acid-sepcific receptors were only found in the meningotheliomatous meningiomas. Overall, differences in the binding spectrum of neoglycoproteins suggest presence of a possible additional subtype of meningiomas, called submalignant meningioma.

Differences in the intracellular distribution of glycohistochemically detectable sugar-binding receptors were apparent even at the light microscopical level in some tumor types, such as in the whorle formations of submalignant meningiomas or cell islands, diagnostically useful for meningotheliomatous meningiomas.

These observations suggest that there are changes in the spectrum of tumor cell-associated, lectin-like proteins during malignant transformation. Moreover, labelled neoglycoproteins could be used routinely as diagnostic tools in neuro-oncology, as demonstrated in the case of submalignant meningioma. Recognitive interactions between carbohydrates and their endogenous receptors may be reflected in the arrangement of tumor cells, which can be characteristic for a given tumor type (Bardosi et al., 1988a, b, c; Bardosi at al., 1991; Dimitri et al., 1989; Gabius et al., 1989).

Human and pig sciatic nerve

As already seen in the CNS, nearly all glycohistochemical tools specifically bound to at least certain parts in the PNS, i.e. endoneurium, epineurium, myelin or axoplasm of the two types of peripheral nerve. Examples are shown in Figs. 13 and 14. Staining with the different probes for galactoside-specific receptors was consistently stronger in pig nerves than in human nerves, especially for endoneurium and axoplasm, emphasizing a dependence on the species for the extent of staining. Opposite quantitative differences, however, were detectable in epineurium and myelin for α-glucoside-specific receptors. Whereas application of BSA, exposing N-acetyl-D-galactosamine residues, detected stronger expression of appropriate endogenous receptors in pig nerve endoneurium, axoplasm, epineurium, and myelin, binding of N-acetyl-D-glucosamine, fucose, mannose, mannose-6-phosphate resulted in similar staining intensities for both types of mammalian peripheral nerve. Further differences could be noted with galactose-6-phosphate and the two carboxylated sugars, sialic acid and glucuronic acid, attached to BSA.

The assessment of sugar receptor distribution by application of neoglycoprotein binding among morphologically defined regions in nerves may hold promise in detecting developmental regulation and changes during nerve regeneration and degeneration after trauma or pathological changes (Gabius et al., 1988, Bardosi et al., 1989a, Gabius et al., 1991).

Tumors of peripheral nervous system

Both cell types of the neurinomas, namely Antoni A and B cells, exhibited an identical pattern of glycohistochemical expression, supporting the notion that both cell types originate from the same cell lineage. In contrast, the matrix cells (fibroblastoid cells) and Schwann cells that comprise the neurofibromas showed significant differences of glycohistochemical staining, especially in the cases of lactose(red.am.)-(BSA-biotin), N-acetyl-D-glucosamine-(BSA-biotin) and both phosphorylated carbohydrate-carrying probes (Bardosi et al.,1991).

These results suggest that analysis of the spectrum of endogenous sugar receptors can serve to distinguish different cell populations within a given tumor.

Human heart

Further fields of application within pathology can be opened, when these probes are exploited for analysis of muscle specimen like endomyocardial tissues, illustrated in Figs. 15 and 16. Differences in lectin expression were observed between the layers of endocardial tissue, myocardial cell constituents, connective tissue elements and vascular structures (Bardosi et al., 1990a). The endocardium proved to be positive with ß-galactoside-bearing probes, with neoglycoproteins carrying ß-xylosides, α-fucosides and galactose-6-phosphate moieties, and with probes containing a carboxyl group within the carbohydrate structure, namely sialic acid and glucuronic acid. In contrast, only fucose- and maltose-specific receptors were apparent in the elastic layers of the endocardium. Aside from ascertaining the specificity of the protein-carbohydrate interaction by controls, i.e. lack of binding of the probe in the presence of the unlabelled neoglycoprotein and lack of binding of the labelled sugar-free carrier protein, presence of sugar receptors was independently shown biochemically by using the histochemically effective carbohydrates as immobilized affinity ligands. Moreover, affinity chromatography using immobilized lactose as affinity ligand as well as the use of polyclonal antibodies against the abundant β-galactoside-specific lectin of heart demonstrated that the binding of the lactose-carrying neoglycoprotein was due to this lectin. Remarkably, the labelled endogenous lectin localized accessible ligand in tissue parts, where the lectin itself was detected glycohistochemically as well as immunohistologically. This demonstration of receptor-ligand presence in the same anatomical parts is a further step toward functional assignment of the recorded protein-carbohydrate interaction (Bardosi et al., 1990a). The analysis of the normal tissue will help to define pathological conditions of human heart such as idiopathic cardiomyopathies.

Human and rat skeletal muscles

The same analytical processing was applied to formaldehyde-fixed, paraffin-embedded muscle biopsy specimens, obtained from the human deltoid, quadriceps and biceps muscles, the tibial and quadriceps muscles of the rat and the masseteric muscle of the ox. α-L-Fucoside-, β-D-galactoside-, β-D-xyloside- and α-D-mannoside-specific binding, respectively, to muscle sections was detected, showing no species-specific differences (Bardosi et al., 1989b). Examples are given in Figs 17 and 18. Presence of receptors for the natural N-acetylated sugars and for sugars with a phosphate group, namely mannose-6-phosphate and galactose-6-phosphate, was demonstrated glycohistochemically. However, these binding specificities revealed species-specific differences, e.g. the absence of N-acetyl-D-galactosamine-specific receptors or galactose-6-phosphate-specific receptors in rat muscle. Other charged sugars included glucuronic acid and sialic acid, which either bound only to ox and rat muscle or failed to reveal the presence of respective receptors in all types of muscle. Positive glycohistochemical reaction developed within a sarcomere only at the level of A-bands. Granular staining could be observed in the sarcoplasm between the myofibrils and also in the subsarcolemmal regions. Notably, differences in expression of receptors were apparent between type 1, type 2A and type 2B fibers. The molecular properties of one type of glycohistochemically detectable sugar receptor was inferred immunohistochemically as well as biochemically, as already shown in heart tissue (Bardosi et al., 1990a). The glycohistochemical fiber typing corresponds to enzyme histochemical typing, what underscores the potential practical value of this method. Consequently, use of a panel of neoglycoproteins, could be a basis for accurate fiber typing on formaldehyde-fixed, paraffin-embedded biopsy specimens and thus aid in diagnosing neuromuscular disorders (Bardosi et al., 1989b).

Conclusions

These results support the notion that the expression of sugar receptors is not a uniform characteristic of the various cell types. In addition to its advantageous exploitation in neuropathology and neuromorphology, its continued assessment may uncover information on physiologically relevant processes. Neoglycoproteins thus provide information on the protein part of glycobiological interaction that is complementary to the results obtainded with plant lectins, carbohydrate-specific antibodies and, more recently, mammalian tissue lectins as tools to localize defined carbohydrate structures.

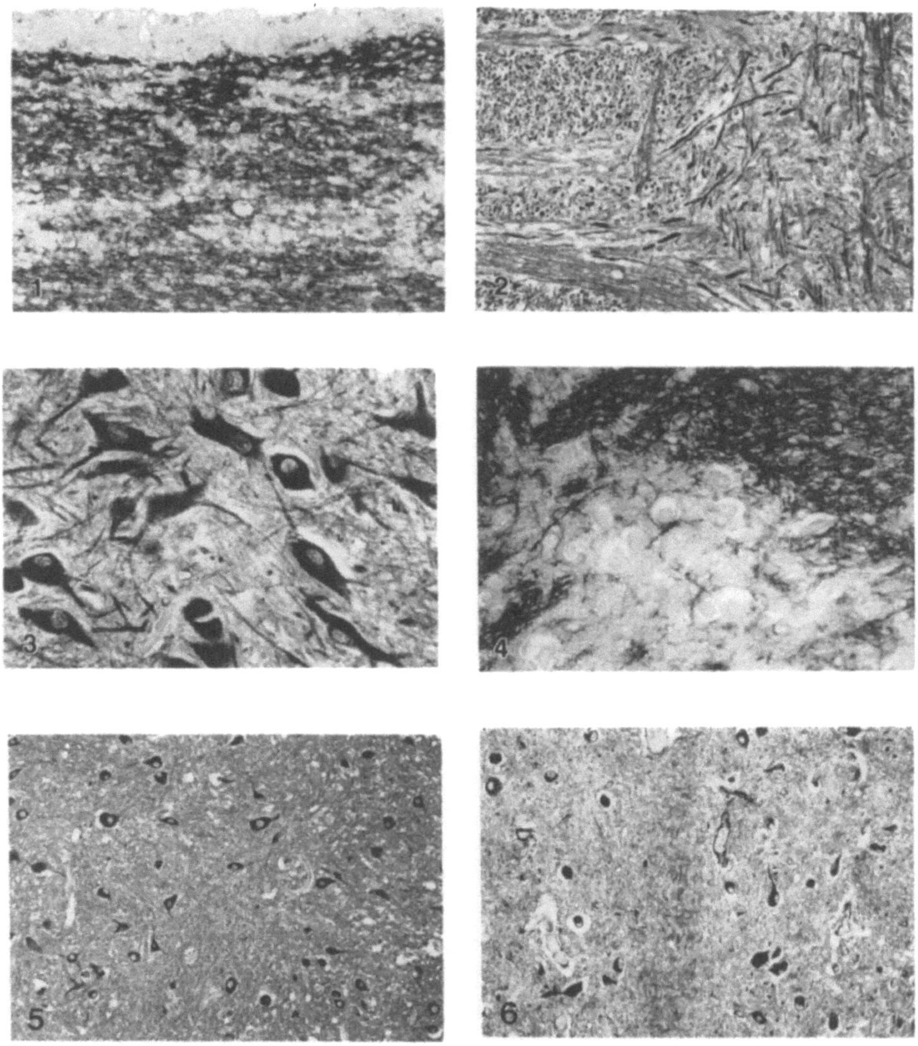

Figs. 1-6 Glycohistochemical staining of different regions of human adult brain. (1) No staining of the membrana limitans externa, while the fiber structures show an intense reaction with maltose-(BSA-biotin), X 250. (2) Fiber structures of the lemniscus medialis stained with lactose-(BSA-biotin), X 400. (3) Strong reaction of the neurons of substantia nigra with lactose-(BSA-biotin), X 680. (4) No staining of the neurons of the nci. pontis, but intense reaction of the fibrae pontocerebellares, using glucuronic acid-(BSA-biotin), X 680. (5) Intense staining of neurons within the nucleus of nc. mesencephalicus n. trigemini with asialofetuin-biotin, X 100. (6) Nucleus centralis superior stained with N-acetyl-D-glucosamine-(BSA-biotin), X 100.

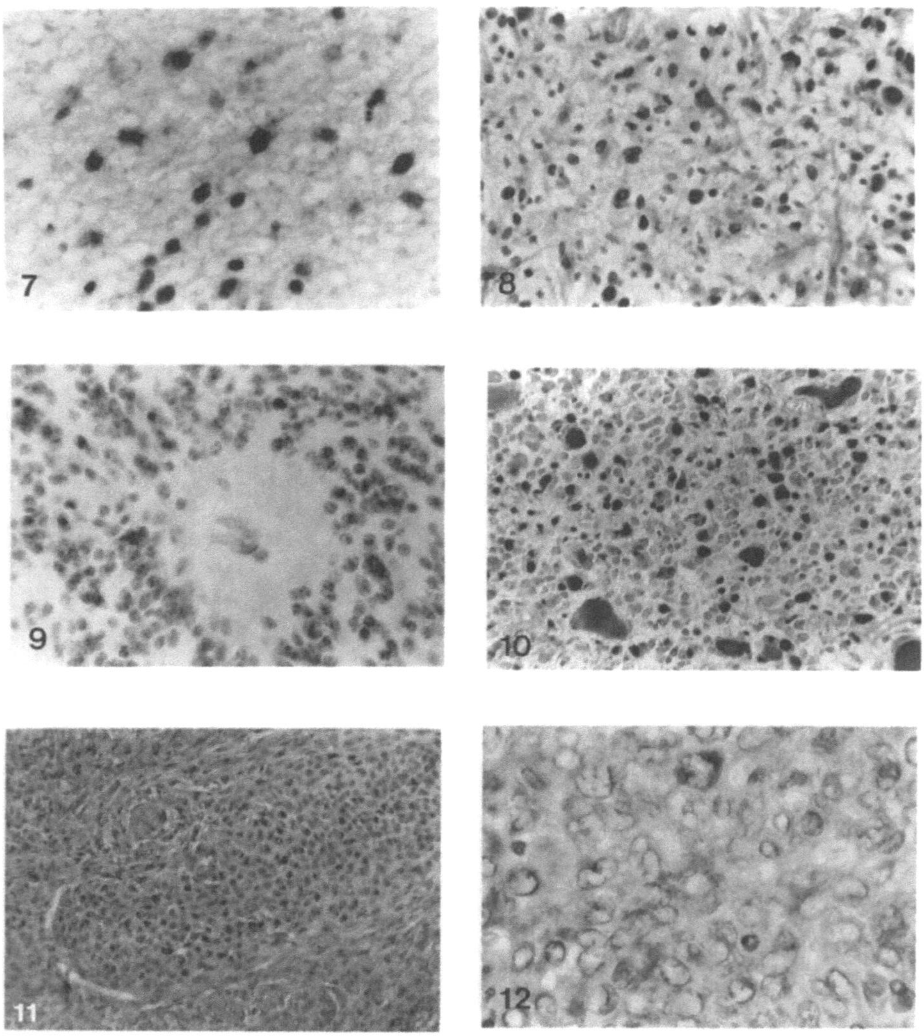

Figs. 7-12 Glycohistochemical staining of different tumors of the central nervous system. (7) Fibrillary astrocytoma stained with lactose-(BSA-biotin), X 220. (8) Glioblastoma stained with xylose-(BSA-biotin), X 220. (9) Well-differentiated ependymoma show patchy nuclear and strong nucleolar staining of tumor cells, using fucose-(BSA-biotin), X 333. (10) Tumor cells of the anaplastic ependymoma with intense cytoplasmic staining, using mannose-6-phosphate-(BSA-biotin), X 155. (11) Positive staining of tumor cells arranged in islands of meningotheliomatous meningioma, using mannose-(BSA-biotin), X 70. (12) Focal cytoplasmic reaction of tumor cells of malignant meningioma with N-acetyl-D-galactosamine-(BSA-biotin), X 500.

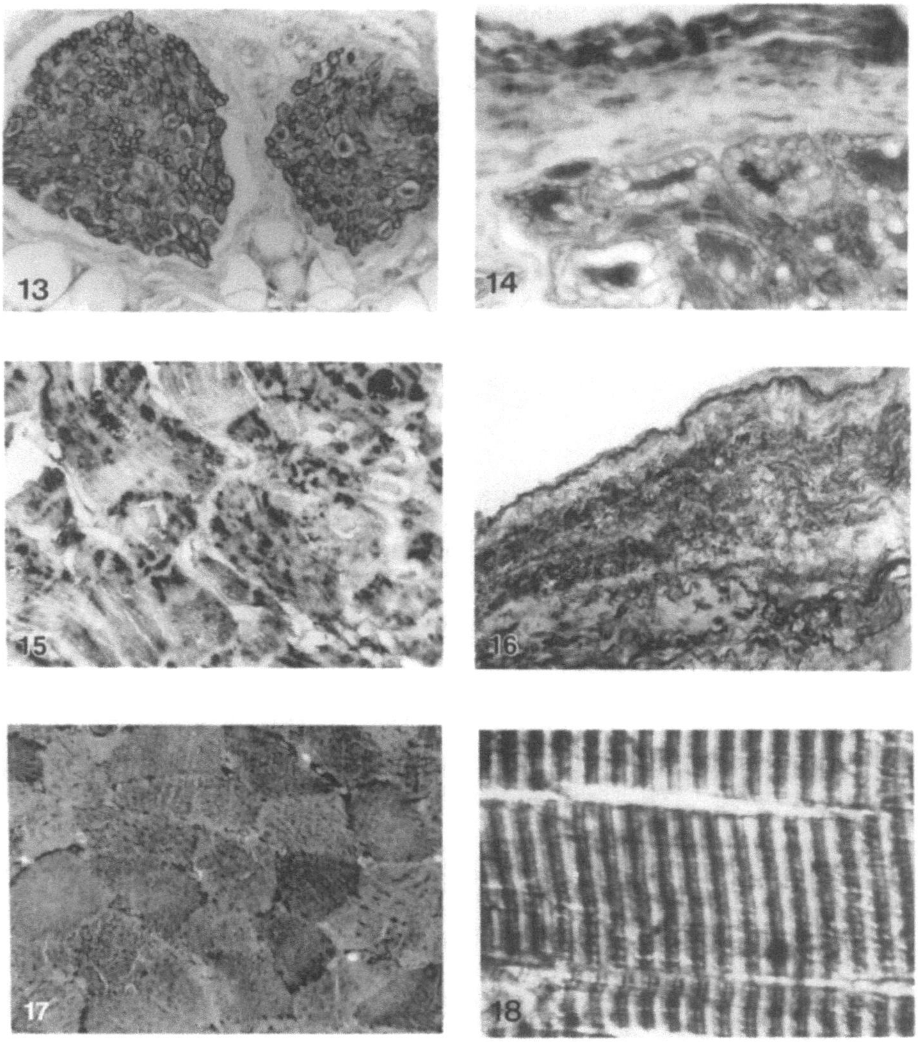

Figs. 13–18 Glycohistochemical staining of human peripheral nerves, heart and skeletal muscle. (13) Strong reaction of fascicles of human sciatic nerve stained with asialofetuin-biotin, X 260. (14) Patchy staining of the epi-perineurial border and strong reaction of axoplasm of pig peripheral nerve with N-acetyl-D-glucosamine-(BSA-biotin), X 665. (15) Strong granular activity of myocardial cells stained with cellobiose-(BSA-biotin), X 720. (16) Strongly stained collagen fibers of endocardium of human heart with xylose-(BSA-biotin), X 800. (17) Fiber typing on formaldehyde-fixed human skeletal muscle, using lactose-(BSA-biotin), 300. (18) Cross striation of human muscle fibers visualized with mannose-(BSA-biotin), X 1000.

Acknowledgements: We sincerely thank the Dr.-Mildred-Scheel-Stiftung für Krebsforschung and the Deutsche Forschungsgemeinschaft (Ba 916/3-1) for financial support and Mrs. Brodmerkel and Mrs. and Mr. Hellmann for their excellent technical assistance.

References

Bardosi A, Dimitri T, Gabius HJ (1988a) Endogenous carbohydrate-binding proteins in neuro-oncology. In: Gabius HJ, Nagel GA (eds.) Lectins and glycoconjugates in oncology. Springer Verlg. 1988, pp 143-152

Bardosi A, Dimitri T, Gabius HJ (1988b) Endogenous carbohydrate-binding proteins in oligodendrogliomas. Acta Neuropathol (Berl) 76:55-61

Bardosi A, Dimitri T, Gabius HJ (1988c) (Neo)glycoproteins as tools in neuropathology: histochemical patterns of the extent of expression of endogenous carbohydrate-binding receptors like lectins in meningiomas. Virch Arch B (Cell Pathol) 56:35-43

Bardosi A, Dimitri T, Behrends T, Autschbach D, Gabius HJ (1989a) Is part of the molecular basis of the perineurial barrier function the lack of endogenous carbohydrate-binding proteins? J Neurosci Res 22:65-73

Bardosi A, Dimitri T, Wosgien B, Gabius HJ (1989b) Expression of endogenous receptors for neoglycoproteins, especially lectins, that allow fiber typing on formaldehyde-fixed, paraffin-embedded muscle biopsy specimens. A glycohistochemical, immunohistochemical and glycobiochemical study. J Histochem Cytochem 37:989-998

Bardosi A, Bardosi L, Hendrys M, Wosgien B, Gabius HJ (1990a) Spatial differences of endogenous lectin expression within the cellular organization of the human heart: a glycohistochemical, immunohistochemical, and glycobiochemical study. Am J Anat 188:409-418

Bardosi A, Bardosi L, Lindenblatt R, Gabius HJ (1990b) Detection and mapping of endogenous receptors for carrier-immobilized constituents of glycoconjugates (lectins) by labelled (neo)glycoproteins and by affinity chromatography in human adult mesencephalon, pons, medulla oblongata and cerebellum. Histochemistry 94:285-291

Bardosi A, Brkovic D, Gabius HJ (1991) Localization of endogenous sugar-binding proteins (lectins) in tumors of the central and peripheral nervous system by biotinylated neoglycoproteins. Anticancer Res, in press

Dimitri T, Bardosi A, Gabius HJ (1989) Mapping of receptors for carbohydrate constituents of glycoconjugates in well-differentiated and malignant ependymomas: a glycohistochemical study. Neuropath Appl Neurobiol 15:121-134

Gabius HJ, Bardosi A (1990) Regional differences in the distribution of endogenous receptors for carbohydrate constituents of cellular glycoconjugates, especially lectins, in cortex, hippocampus, basal ganglia and thalamus of adult human brain. Histochemistry 93:581-592

Gabius HJ, Bardosi A (1991) Neoglycoproteins as tools in glycohistochemistry. Progr Histochem Cytochem 22(3), 1-66

Gabius HJ, Kohnke B, Hellmann T, Dimitri T, Bardosi A (1988) Comparative histochemical and biochemical analysis of endogenous receptors for glycoproteins in human and pig peripheral nerve. J Neurochem 51:756-763

Gabius HJ, Hellmann KP, Dimitri T, Bardosi A (1989) Endogenous sugar receptor pattern in human glioblastomas and gangliocytomas studied by histochemical application of biotinylated (neo)glycoproteins and affinity chromatography. Histochemistry 91:5-11

Gabius HJ, Wosgien B, Hendrys M, Bardosi A (1991) Lectin localization in human nerve by biochemically defined lectin-binding glycoproteins, neoglycoprotein and lectin-specific antibody. Histochemistry 95:269-277

Site-associated differences in endogenous lectin expression by mouse colon carcinoma cells

D. Glaves, L. Weiss, and F. Vidal-Vanaclocha[1]

Department of Experimental Pathology
Roswell Park Cancer Institute
Elm and Carlton Streets
Buffalo, New York 14263
U.S.A.

Organ patterns of arterial metastatic involvement are different in patients with histories of upper and lower rectal carcinoma. Analyses of autopsy data (Weiss et al, 1981;1984) suggest that these patterns are partially imposed by the step-wise nature of metastasis. As shown in Figure 1, in upper rectal carcinomas, seeding mainly via the portal venous system results in secondary metastases in the liver; cells from these lesions seed tertiary lung lesions via the inferior vena cava, which in turn generalize via the arterial system to form widespread quaternary arterial metastases. In contrast, primary carcinomas of the lower rectum predominantly first generate secondary lung metastases via the systemic venous system, and these in turn generate tertiary arterial metastases in other organs. Both groups of primary tumors arose in a common site, the rectum; both groups were of the same histologic type (adenocarcinomas), both were of the same stage, and both used similar seeding routes from the lungs to generate the arterial metastases (Weiss, et al 1981, 1984). The only obvious difference between the two groups to account for the difference in arterial metastatic patterns, is that those from the primary upper rectal lesions were mainly generated directly from tertiary lung metastases via secondary lesions in the liver, whereas the patterns

[1]Department of Cell Biology and Morphological Sciences
University of the Basque Country
48940 Leioa
Viscaya
Spain

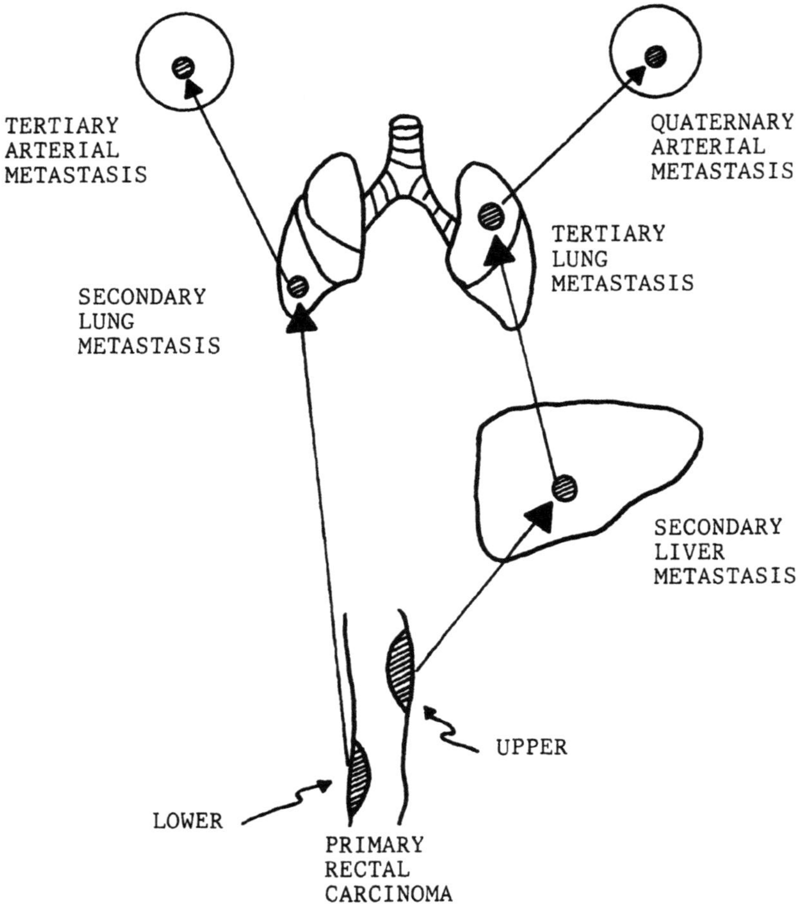

FIGURE 1. Patterns of metastasis from adenocarcinomas of the upper and lower rectum.

from the primary lower rectal cancers were mainly generated from secondary lung lesions, which were seeded by cells with no prior history of growth in the liver. Therefore, the question arises of whether cancer cells growing in different organs, namely the liver and lungs, develop site-associated differences after seeding.

The results of experiments using a murine model of colorectal carcinoma are consistent with the previous interpretation of human autopsy data, since the patterns of spontaneous and artificial metastases were different following injection of colon 26 cells with a history of growth in the liver, lungs, or liver-then-lungs (Weiss & Ward, 1988). These studies imply that growth-site associated changes do occur in cancer cell populations which may influence metastatic pattern formation.

Since the expression of endogenous lectins by cancer cells appears to be associated with a variety of metastasis-related interactions (e.g. Kieda & Monsigny, 1986; Raz & Lotan, 1987; Vavasseur et al, 1990), we have evaluated endogenous lectins as potential markers for changes occurring in cancer cells populations as a consequence of growth in different organs using colon 26 carcinoma cells growing in anatomically relevant sites as a model for certain aspects of hematogenous metastasis from human colorectal carcinoma.

Colon 26 tumors were grown as primary or metastatic tumors in the paracecal region, spleen or subcutis, and as spontaneous metastases in the liver or lungs. Liver or lung colonies were generated by direct injection of carcinoma cells into the portal vein or tail vein, respectively. A panel of biotinylated glycoproteins (neoglycoproteins), with carbohydrate residues representing those commonly found in naturally-occurring glycoconjugates, was used in ABC-peroxidase histochemical assays to locate carbohydrate binding sites in paraffin-embedded tissues. Computerized interactive image analysis and quantitative microdensitometry were used to quantitate peroxidase reactions. On each tissue section, pixel-intensities were measured by microdensitometry in a prescribed area (Fig. 2A) and following image reversal (Fig. 2B), pixel-intensity corresponded to peroxidase staining intensity. Non-specific stromal staining and

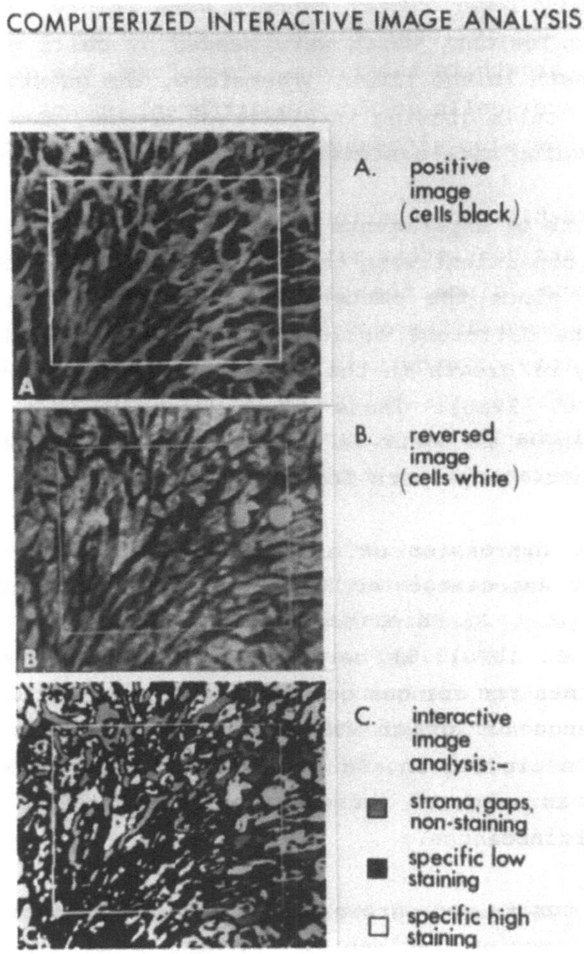

FIGURE 2. Computerized image analysis of ABC peroxidase reaction of a tissue section from a liver metastasis treated with biotinylated rhamnose. The designated field outlined in white in C measures approximately 2800 μm^2.

tissue gaps were eliminated by interactive image analysis (Fig. 2C). Numerical limits were used to classify non-staining, low-staining and high-staining intensities, and the relative areas occupied by cells falling into these three categories were determined in each field. Staining intensities for 6 of the 17 neoglycoprotein probes exceeded the maximum extinction levels obtained with the other 11 probes. Therefore, the incident illumination was increased and intensity thresholds (low- medium-high-staining) re-defined for this group (Group II neoglycoproteins). The results were analyzed separately from those obtained with the other neoglycoproteins (Group I neoglycoproteins). Integrated staining-intensity areas for Group I neoglycoproteins reacted with colon 26 tumors are given in Table 1 and for group II neoglycoproteins in Table 2.

Complex patterns of lectin expression were observed with both groups of neoglycoproteins. Attempts to clarify reactivity patterns were made in which the staining-intensity areas were ranked and compared using Scheffe's multiple comparison test (Pollard, 1977). The results of this ranking (Tables 3 and 4) show significantly different expression among tumors in up to four different sites for the all the neoglycoproteins examined. In general, the lectins expressed at "high" levels for group I neoglycoproteins were those with specificities for N-acetylated carbohydrates, α-fucosides and α-mannosides and most of the highest ranked staining-intensity areas were detected in the liver, followed by cecal then subcutaneous sites. Spleen and lung tumors generally expressed the lowest levels of lectins. However, for the strongly reactive group II neoglycoproteins, liver and cecal tumors were outranked in several staining classes by lung tumors and even spleen or subcutaneous tumors.

Spontaneous metastases in the liver and lungs exhibit clear-cut differences in endogenous lectins. For example, reactivity of liver metastases was greater than that of lung metastases with lactose, rhamnose and N-acetyl-D-galactosamine but lower in liver lesions than lung lesions with asialocasein and the sulphated polysaccharides, heparin and fucoidan. Overall, statistically significant differences between liver and lung metastases were detected in 44 of 62 individual intensity-area classes. In

Table 1. Integrated staining-intensity areas with group I neoglycoproteins reacted with sections of colon 26 tumors.

Tissue Source (Staining Intensity)[a]	Neoglycoproteins:										
	α D-Gluc	Fucose	GalgalNAc	Melibiose	GalNAc	Mannan	Maltose	Galactose	GlcNAc	Sialic Acid	Mannose
Cecum:											
High	0	0	0	0	0	0	0	0	0	0	0
Low	0	85.0±2.5[b]	6.8±1.3	4.4±3.2	4.3±0.4	87.0±0.1	6.5±0.8	9.9±2.4	59.3±3.0	55.3±1.9	87.0±2.3
Non-staining	87	2	80.3	82.6	82.8	0	80.5	77.1	27.4	31.8	0
Liver Metastases:											
High	0	0.9±0.5	0	0	0	0	0	0	9.7±3.3	0	4.4±3.1
Low	28.8±1.9	83.4±0.5	41.7±3.6	16.6±2.2	69.0±4.4	1.4±0.2	0.3±0.1	11.3±1.7	74.6±1.1	35.8±5.9	79.9±3.5
Non-Staining	55.5	0	42.6	67.7	15.3	82.9	84.0	73.1	0	48.6	0
Liver Colonies:											
High	0	25.3±6.9	0	0	14.2±6.4	0	0	0	19.8±5.0	0	5.1±3.8
Low	0	61.4±4.4	20.3±5.1	22.0±5.1	68.5±1.7	46.4±3.0	29.7±4.8	52.3±8.4	62.9±6.3	28.6±3.4	77.5±1.6
Non-staining	81.7	0	62.4	60.6	0	36.3	53.0	53.0	0	54.0	0
Lung Metastases:											
High	0	0	0	0	0	0	0	0	0	0	0
Low	0	32.5±5.4	0	0	0	0	0	0	22.8±0.1	27.8±0.1	11.5±4.6
Non-staining	90.7	58.3	90.7	90.7	90.7	90.7	90.7	90.7	67.9	63.0	79.3
Lung Colonies:											
High	0	0	0	0	0	0	0	0	0	0	0
Low	0	38.0±5.7	0	6.3±2.4	10.0±1.8	0	0	0	62.1±7.0	0	13.0±1.5
Non-staining	85.1	47.1	85.1	78.8	75.1	85.1	85.1	85.1	23.0	85.1	72.1
Subcutaneous:											
High	0	0.3±0.1	0	0	0	0	0	0	0	0	0
Low	19.4±2.7	81.5±2.6	19.1±1.8	0	37.1±3.6	0	3.0±0.1	0.8±0.1	76.4±2.9	13.0±0.8	80.7±1.4
Non-staining	64.8	2.4	64.1	84.2	47.1	84.2	81.2	83.5	7.8	71.2	3.5
Spleen:											
High	0	0	0	0	0	0	0	0	0	0	0
Low	0.2±0.0	26.3±2.3	4.8±0.2	5.2±0.8	16.4±3.5	0.4±0.3	4.1±0.0	0	54.9±7.5	0.5±0.1	53.6±4.0
Non-staining	88.5	62.7	84.1	83.9	72.6	88.6	84.9	89.0	34.1	88.5	33.4

a - Areas of tissue gaps and stroma given in Table 2.
b - Percent area (Mean ± SE)

Table 2. Integrated staining-intensity areas with group II neoglycoproteins reacted with sections of colon 26 tumors.

Tissue Source (Staining Intensity)	Neoglycoproteins:						Tissue gaps and Stroma
	Heparin	Fucoidan	ATF	Lactose	Rhamose	ASC	
Cecum:							13.1±0.1
High	2.0±0.0[a]	84.4±1.4	0	0	0	5.7±0.0	
Medium	25.9±4.8	2.6±0.0	35.0±6.7	35.0±6.7	31.5±4.1	40.5±6.9	
Low	59.0±5.9	0	51.9±8.2	51.9±8.2	55.5±4.2	40.7±2.5	
Liver Metastases:							15.7±1.0
High	0	3.8±1.1	0	6.8±1.9	15.7±5.0	0	
Medium	28.1±2.8	80.5±1.1	5.6±0.0	66.5±5.3	66.3±3.4	0	
Low	56.2±0.9	0	78.1±1.8	11..0±5.3	2.3±3.0	84.3±1.0	
Liver Colonies:							17.3±2.0
High	23.8±4.5	45.4±1.2	2.9±1.3	0	38.9±5.4	0	
Medium	47.1±4.8	37.2±2.5	59.8±8.2	82.7±7.8	43.8±2.9	82.7±0.9	
Low	11.7±4.4	0	19.9±5.2	0	0	0	
Lung Metastases:							9.3±1.9
High	16.4±1.3	34.0±1.8	0	0	0	22.3±2.5	
Medium	71.0±3.7	56.7±1.4	19.9±3.3	18.6±2.0	36.9±5.1	68.4±1.1	
Low	3.3±2.2	0	70.8±4.6	72.1±2.1	53.8±6.7	0	
Lung Colonies:							14.9±3.3
High	27.2±3.6	55.7±4.1	0	0	29.7±3.8	31.6±1.5	
Medium	57.2±2.2	29.4±3.7	28.9±6.2	29.5±6.4	56.1±3.1	53.5±2.6	
Low	0	0	56.2±3.8	55.6±4.5	0	0	
Subcutaneous:							15.8±2.0
High	0	11.5±2.6	0	0	0	29.9±3.3	
Medium	22.1±4.3	72.7±3.6	10.3±0.0	9.1±1.4	59.6±4.9	54.3±2.9	
Low	62.1±1.6	0	73.9±2.0	75.1±0.0	24.6±7.2	0	
Spleen:							11.3±1.0
High	1.8±0.0	31.9±3.7	0	0	0	7.4±2.3	
Medium	25.9±4.8	56.8±3.9	2.5±0.0	17.7±3.4	41.9±11.2	81.3±1.1	
Low	59.0±5.9	0	86.1±0.1	71.0±2.3	46.9±11.9	4.6±0.0	

[a] - Percent area (Mean ± SE)

Table 3. Ranking of integrated staining-intensity areas with group I neoglycoproteins reacted with sections of colon 26 tumors.[a]

Tissue Source (Staining Intensity)	Neoglycoproteins:										
	α D-Gluc	Fucose	GalgalNAc	Melibiose	GalNAc	Mannan	Maltose	Galactose	GlcNAc	Sialic Acid	Mannose
Cecum:											
High	-[c]	5.5[b]	-	-	-	-	-	-	5	-	5
Low	4.5	1.5	4	4	6	1	2	2.5	5	1	1
Liver Metastases:											
High	-	3	-	-	-	-	-	-	1.5	-	2
Low	1	1.5	1	2	1.5	3.5	5	2.5	2	2	2.5
Liver Colonies:											
High	-	2	-	-	-	-	-	-	1.5	-	1
Low	4.5	4	2.5	1	1.5	2	1	1	3.5	3.5	4
Lung Metastases:											
High	-	5.5	-	-	-	-	-	-	5	-	5
Low	4.5	6	6.5	6.5	7	6	6.5	6.5	7	3.5	6.5
Lung Colonies:											
High	-	5.5	-	-	-	-	-	-	5	-	5
Low	4.5	5	6.5	4	5	6	6.5	6.5	3.5	7	6.5
Subcutaneous:											
High	-	1	-	-	-	-	-	-	5	-	5
Low	2	3	2.5	6.5	3	6	3.5	4.5	1	5	2.5
Spleen:											
High	-	5.5	-	-	-	-	-	-	5	-	5
Low	3	7	5	4	4	3.5	3.5	4.5	6	6	5

a. - Calculated from data in Table 1.
b. - Where rank numbers are different, differences are significant at the 5% level.
c. - No staining at this level in any tissue; ranking impossible.

Table 4. Ranking of integrated staining-intensity areas with group II neoglycoproteins reacted with sections of colon 26 tumors.[a]

Tissue Source (Staining Intensity)	Neoglycoproteins:					
	Heparin	Fucoidan	ATF	Lactose	Rhamose	ASC
Cecum:						
High	4.5[b]	1	4	4	5.5	5
Medium	6	7	2	3	7	6
Low	2	4	6	5	1	2
Liver:						
High	6	7	4	1	3	6.5
Medium	5	1	6	2	1	7
Low	3	4	2	6	5	1
Liver Colonies:						
High	2	3	1	4	1	6.5
Medium	3	5	1	1	4	1.5
Low	5	4	7	7	6.5	5.5
Lung Metastases:						
High	3	4	4	4	5.5	3
Medium	1	3.5	4	5.5	6	3
Low	6	4	4	2.5	2	5.5
Lung Colonies:						
High	1	2	4	4	2	1
Medium	2	6	3	4	3	4.5
Low	7	4	5	4	6.5	5.5
Subcutaneous:						
High	6	6	4	4	5.5	2
Medium	7	2	5	7	2	4.5
Low	1	4	3	1	4	5.5
Spleen:						
High	4.5	5	4	4	5.5	4
Medium	4	3.5	7	5.5	5	1.5
Low	4	4	1	2.5	3	3

a - Calculated from data in Table 2.
b - Where rank numbers are different, differences are significant at the 5% level.

addition to ***metastases***, we examined lectin expression in liver and lung ***colonies*** which were seeded by direct injection of cells into portal or tail veins. Examples of similarities and differences in lectin expression in liver and lung colonies are illustrated in Figure 3. Reactivities with heparin, fucoidan and asialocasein were higher in lung than liver colonies, as they were with lung versus liver metastases. Indeed, differences between the two types of colonies were similar in direction to those between the two types of metastases, even though the magnitudes of the differences were not necessarily the same. For both colonies and metastases, expression of individual lectins was independently variable, changing in different directions in the two sites, which emphasizes the potential biologic significance of these alterations. This is underscored by the results of biochemical analyses which showed differences in lectins expressed by xenogeneic transplants of liver and lung metastases from human colon carcinoma (Gabius et al, 1989).

It was of considerable interest to find that endogenous lectin expression differed not only between metastases and colonies in different sites but also between metastases and colonies in the same organ, liver or lung. Thus, liver colonies expressed significantly greater reactivity than liver metastases for 10 of the 17 neoglycoprotein probes but less reactivity for α-D-glucose and galgalNAc. Similarly, lung colonies showed more reactivity than lung metastases with 5 neoglycoproteins and less reactivity with asialotransferrin and sialic acid. The differences in lectin expression between lung colonies and liver colonies indicate that site-associated differences develop in cancer cell populations after delivery to the target organ. However, the differences in expression between colonies and metastases in both organs, indicates that differences already exist before delivery, between circulating cancer cells spontaneously disseminated from primary tumors growing in different sites. This might have been expected from our data showing shifts in lectin expression among colon 26 "primary" tumors in the various sites. Since the post-dissemination phases of spontaneous metastasis and colony formation are the same, differences may reflect population selection associated with the invasion/intravasation stages of metastasis (Hart, 1979).

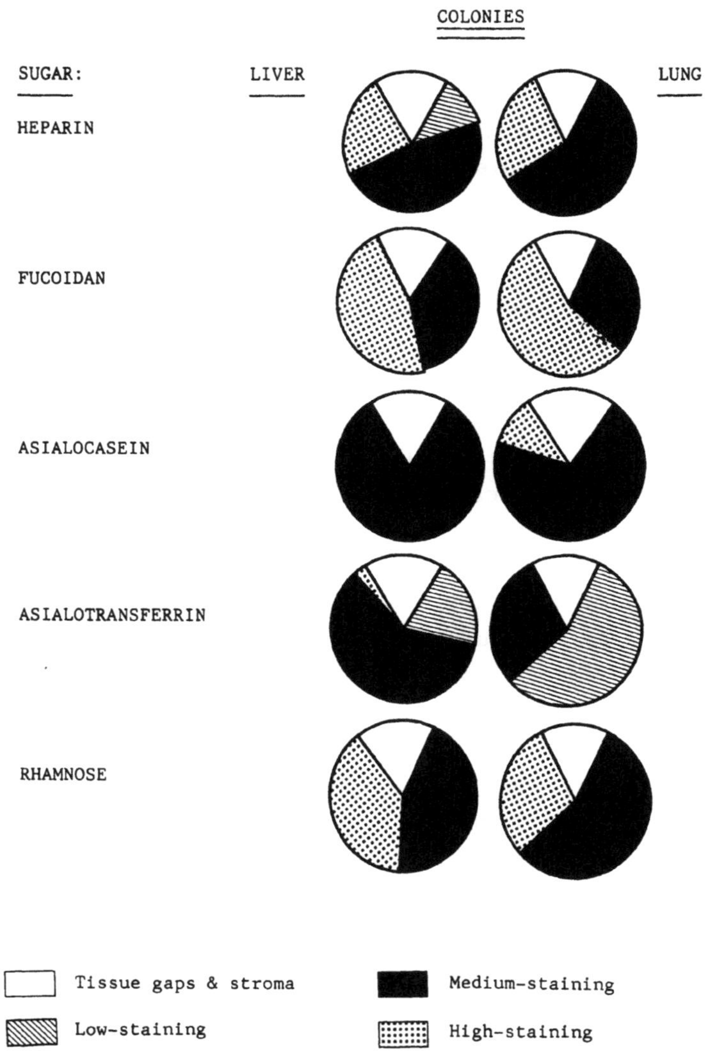

FIGURE 3. Comparison between relative staining areas in tissues from liver and lung colonies reacted with selected neoglycoproteins. Areas calculated from Table 2.

Site-associated selective changes occur in other cancer cell parameters. For example, electrokinetic studies identified subpopulations of mouse Lewis lung carcinoma cells with different surface properties, one or other of which was dominant in different growth sites (Harlos & Weiss, 1983). Interestingly, we detected differences in endogenous lectin expression in Lewis lung carcinomas growing in the same sites (Glaves et al, 1989). As far as metastasis-related properties are concerned, site-associated generation of cancer cell homing to specific organs has been described (Nicolson et al, 1985) and other studies support the concept that local factors at the growth site of a tumor can influence the metastatic behavior of that tumor(Meyvisch & Mareel, 1985; Price et al, 1988). Similarly, metastasis and organ colonization by colon 26 carcinoma cells are modified by prior growth in the liver or lungs and, in addition,cancer cells with a prior history of growth in the liver then lungs show different colonization patterns from cells grown in the lungs alone. (Weiss & Ward, 1988). We are currently investigating whether or not parallel changes occur in endogenous lectin expression.

In the present context, endogenous lectins have been operationally defined as tissue constituents with different affinities for selected carbohydrates. As with the mirror image exogenous lectins more commonly used to investigate metastasis-associated properties of cancer cells, the specific functions of their receptors are for the most part uncertain. However, lectin-carbohydrate interactions have been used to monitor differences in of cancer cells which may be associated with their tumorigenic or metastatic properties (e.g. Kellokompu, 1986; Lang et al, 1988; Nicolson, 1988; Dennis et al, 1989; Hakomori, 1989; Berrada et al, 1990). By the same token, the present studies do not assume a causal relationship between endogenous lectins and cancer cell behavior at the population level. These studies do, however, show that endogenous lectins can be used as markers for changes in cancer cell populations.

Sometimes (Fidler, 1973), but not invariably (Weiss et al, 1982; Chambers et al, 1984; Vaage, 1988; Milas et al, 1983), differences occur in both the colonization and metastatic potentials, and organ patterns of colonies and metastases

(Nicolson, 1988) between cells derived from "primary" cancers and their "metastases" in laboratory animals. In general terms, these differences have been attributed to the generation of metastases in non-random manner, uniquely from a "metastatic phenotype" or, alternatively by a random process involving a "transient metastatic compartment." In general, both of these non-exclusive hypotheses attempt to explain events leading to the seeding process and in terms of the "seed and soil" hypothesis, they emphasize the role of the cancer cell "seed" (Weiss, 1985). In the present experiments, differences between metastases and colonies, formed by intravenous injections of cells, also point to metastasis-relevant differences in cell populations before seeding. However, our studies show that quantitative, but not qualitative differences, can occur between cancer cells growing in different sites AFTER seeding has occurred. This emphasizes the role of the "soil".

Although quantitative histochemistry may be valuable in analyzing site-associated changes in cancer cell populations, this approach should be relevant to assessment of clinical or experimental materials for other markers using molecular biologic, exogenous lectin and immunologic probes. Such an approach has the added advantage that the topographic relationships of cancer cell populations to each other, normal cells, stroma and blood vessels are preserved. Objective quantitation is desirable when using probes for any markers because, in general, differences in cell constituents which have been used as parameters for progression, or to distinguish non-metastatic and metastatic cells, are quantitative rather than qualitative. Our own studies of endogenous lectins are no exception and they highlight the feasibility of using quantitative histochemistry to develop multi-marker profiles of cancer cell populations which may differ in subtle and independently variable properties.

ACKNOWLEDGEMENTS

The biotinylated neoglycoprotein probes used in these studies were synthesized and made available by Dr. Hans-Joachim Gabius, Max-Planck-Institut fur Experimentelle Medizin, Gottingen, Federal Republic of Germany.

REFERENCES

Berrada A, Vavasseur F, Le Pendu J, Meflah K (1990) Correlation between cell surface oligosaccharides and tissue target-selective adhesion of two rat adenocarcinoma cell livers. Invas Met 10:352-371

Chambers A F, Harris J F, Ling V, Hill R P (1984) Rapid phenotype variation in cells derived from lung metastases from KHT fibrosarcoma. Invas Met 4:225-237

Dennis V W, Laferte S, Yager S, Breitman M L (1989) Asparagine-linked oligosaccharides associated with metastatic cancer. Cancer Cells 1:87-92

Fidler I J (1973) Selection of successive tumor lines for metastasis. Nature 242:148-149

Gabius H-J, Ciesiolka T, Kunze E, Vehmeyer K (1989) Detection of metastasis-associated differences for receptors of glycoconjugates (lectins) in histomorphologically unchanged xenotransplants from primary and metastatic lesions of human colon adenocarcinomas Clin Exptl Met 7:571-584

Gabius H-J, Engelhardt R (1988) Sugar receptors of different types in human metastases to lung and liver. Tumor Biol 9:21-36

Glaves D, Gabius H-J, Weiss L (1989) Site-associated expression of endogenous tumor lectins. Int J Cancer 44:506-511

Hakomori S (1989) Aberrant glycosylation in tumors and tumor-associated carbohydrate antigens. Adv Cancer Res 52:257-331

Harlos J P, Weiss L (1983) Differences in the peripheries of Lewis lung tumor cells growing in different sites in the mouse. Int J Cancer 32:745-750

Hart I R (1979) The selection and characterization of an invasive variant of the B16 melanoma. Am J Pathol 97:587-600

Kellokompu I H (1986) Differences in lectin reactivities of cellular glycoconjugates between primary human tumors and their metastases. Cancer Res 46:4620-4625

Kieda C, Monsigny M (1985) Involvement of membrane sugar receptors and membrane glycoconjugates in the adhesion of 3LL cell subpopulations to cultured pulmonary cells. Invas Met 6:347-366

Lang E, Schirrmacher V, Altevogt P (1988) Molecular identification of lectin binding sites differentiating related low and high metastatic murine lymphomas. Clin Exptl Met 6:61-72

Meyvisch C, Mareel M (1985) Site-induced differences in spontaneous metastasis of MO_4 mouse fibrosarcoma cells. Invas Met 5:185-192

Milas L, Peters L J, Ito H (1983) Spontaneous metastasis, random or selective? Clin Exptl Met 1:309-315

Nicolson G L (1988) Cancer metastasis: tumor cell and host-organ properties important in metastasis to specific secondary sites. Biochim Biophys Acta 948:175-224

Nicolson G L, Dulski K, Basson C, Welch D R (1985) Preferential organ attachment and invasion in vitro by B16 melanoma cells selected for differing metastatic colonization an invasive properties. Invas Met 5:144-158

Pollard J H (1977) Handbook of numerical and statistical techniques, Cambridge University Press, London, p.191

Price J E, Naito S, Fidler I J (1988) Growth in an organ microenvironment as a selective process in metastasis. Clin Exptl Met 6:91-102

Raz A, Lotan R (1987) Endogenous galactoside-binding lectins: a new class of functional tumor cell surface molecules related to metastasis. Cancer Met Rev 6:433-452

Sargent N S E, Oestreicher M, Haidvogl H, Madwick H M, Burger M M (1988) Growth regulation of cancer metastasis by their host organ. Proc Natl Acad Sci 85:7251-7255

Vaage J (1988) Metastasizing potentials of mouse mammary tumors and their metastases. Int J Cancer 41:855-858

Vavasseur F, Berrada A, Heuze F, Jotereau F, Meflah K (1990) Fucose and galactose receptor and liver recognition by lymphoma cells. Int J Cancer 45:744-751

Vidal-Vanaclocha F, Barbera-Guillem E, Weiss L, Glaves D, Gabius H-J (1990) Quantitation of endogenous lectin expression in 3LL tumors growing subcutaneously and in the kidneys of mice. Int J Cancer 46:908-912

Vidal-Vanaclocha F, Glaves D, Barbera-Guillem E, Weiss L (to be published) Quantitative microscopy of mouse colon 26 cells growing in different metastatic sites. Br J Cancer

Weiss L, Bronk J, Pickren J W, Lane W W (1981) Metastatic patterns and target organ arterial blood-flow. Invas Met 1:126-135

Weiss L, Mayhew E, Glaves-Rapp D, Holmes J C (1982) Metastatic inefficiency in mice bearing B16 melanomas. Br J Cancer 45:44-53

Weiss L, Voit A, Lane W W (1984) Metastatic patterns in patients with carcinomas of the lower esophagus and upper rectum. Invas Met 4: 47-60

Weiss L, Ward P M (1988) Effects of metastatic cascades on metastatic patterns: Studies on colon-26 carcinomas in mice. Int J Cancer 41:450-455

Weiss L, Ward P M, Harlos J P (1986) The stability of kidney-induced selection of Lewis lung tumor cell populations and their metastasis-related behavior. Int J Cancer 38:207-213

D. Glaves

F. Vidal-Vanaclocha

Differentiation-associated modulation of lactoside binding lectins in cancer cells

R. Lotan

Department of Tumor Biology
University of Texas M. D. Anderson Cancer Center
Houston, TX 77030
USA

INTRODUCTION

Normal development, growth and differentiation of normal and malignantly-transformed cells are accompanied by interactions among cells and between cells and exogenous macromolecules (e.g. extracellular matrix components). These interactions are mediated via homotypic or heterotypic protein-protein, carbohydrate-carbohydrate, or protein-carbohydrate interactions (Brandley et al., 1990; Brandley and Schnaar, 1986; Edelman, 1985; Kleinman et al., 1985; Lotan and Raz, 1988a; Stoolman, 1989). The physiological consequences of such interactions include recognition, adhesion, directed cell migration, positional information, and transduction of signals, which result in changes in the expression of genes involved in cell growth and differentiation (Barcellos-Hoff et al., 1989; Brandley and Schnaar, 1986; Edelman, 1985; Kleinman et al., 1985; Morrison-Graham and Weston, 1989; Rutishauser and Jessell, 1988; Schuetz et al., 1988). The pathological consequences of such interactions include enhanced metastatic properties (Raz and Lotan, 1987). Qualitative and quantitative alterations in the molecules which mediate cellular interactions may occur during differentiation (Avner et al., 1985; Dennis et al., 1987; Feizi, 1985; Fukuda, 1985; Muramatsu, 1988) as well as after malignant transformation and during tumor progression (Alhadeff, 1989; Dennis et al., 1987; Hakomori, 1989; Muramatsu, 1988) and result in aberrant cellular interactions that can lead to anchorage-independent growth and enhanced tumor cell invasion, embolisation in the circulation, and organ colonization and metastasis

(Dennis et al., 1987; Gabius, 1987; Monsigny et al., 1983; Nicolson, 1988; Raz and Lotan, 1987).

Although protein-protein interactions are a well-established mechanism for interactions of cell surface receptors with cell-surface-associated and extracellular matrix components (Humphries, 1990) the importance of protein-carbohydrate interactions is gaining recognition (Brandley et al., 1990; Drickamer, 1989; Gabius, 1991; Lotan and Raz, 1988a; Paulson, 1989; Sharon and Lis, 1989; Stoolman, 1989). The different sugar residues that are present in vertebrate cell surface glycoconjugates are linked in specific sequences and types of anomeric linkages that form specific codes that could be deciphered by carbohydrate-binding proteins (lectins) and play a role in cell recognition and adhesion (Brandley et al., 1990; Brandley and Schnaar, 1986; Gabius, 1987; 1991; Lotan and Raz, 1988a; Paulson, 1989; Rutishauser and Jessell, 1988; Sharon and Lis, 1989; Stoolman, 1989).

Here I describe the presence of lactoside-specific lectins in cancer cells, their modulation during malignant transformation and differentiation of tumor cells, and some ideas on the possible utilization of such changes for the design of new approaches to cancer therapy.

PROPERTIES OF LECTINS FOUND IN NORMAL CELLS

The changes in the glycosylation of specific glycoconjugates, which occur during embryonal development are thought to be important in the regulation of cell growth, differentiation, and intercellular interactions (Avner et al., 1985; Feizi, 1985; Fukuda, 1985; Hakomori, 1989; Lotan and Raz, 1988a; Muramatsu, 1988; Paulson, 1989). Carbohydrate-binding proteins may be involved in some of the carbohydrate-mediated processes as complementary partners in specific interactions. Thus, such processes could be regulated either via changes in carbohydrate structure, for example by alterations in glycosyltransferases (Amos et al., 1990; Deutsch et al., 1983; Paulson, 1989; Schachter, 1986), or via modulation of the expression of carbohydrate-binding proteins, as will be described here, or both.

Carbohydrate-binding proteins, which are neither antibodies nor enzymes are called lectins (Barondes, 1988). Lectins exhibiting affinity for different sugars were found in various vertebrate tissues (Gabius,

1987; 1991) and a group of lectins that bind lactosides was studied most extensively in recent years (Allen et al., 1987; Barondes, 1984; 1986; 1988; Carding et al., 1985a; 1985b; Caron et al., 1987; Catt and Harrison, 1985; Cerra et al., 1984; 1985; Couraud et al., 1989; Crittenden et al., 1984; Harrison and Chesterton, 1980; Hirabayashi et al., 1987; Hirabayashi and Kasai, 1988; Hynes et al., 1990; Jia and Wang, 1988; Lotan et al., 1991; Lotan and Raz, 1988a; 1988b; Merkle et al., 1989; Monsigny et al., 1983; 1988b; Ohyama et al., 1986; Regan et al., 1986; Sparrow et al., 1987). These studies have demonstrated the presence of lactoside-binding lectins in various tissues including spleen, thymus, heart, lung, bone marrow, skin, intestine, pancreas, liver, placenta, and muscle. Some of these lectins have been purified by affinity chromatography and found to constitute two distinct classes of polypeptides; one exhibiting a molecular weight of about 14,500 dalton and the other includes lectins ranging in molecular weight from 29,000 to 35,000 dalton, depending on the species of origin. These two types of lectins are encoded by two distinct genes but share some sequence homology in a region that is presumed to be part of the lactoside-binding sites (Couraud et al., 1989; Hirabayashi and Kasai, 1988; Jia and Wang, 1988; Raz et al., 1987a; 1988).

The carbohydrate-binding specificities of the two classes of lectins are distinct in that the 14.5-kDa and 29-kDa (or 34-kDa) lectins bind well to oligosaccharides containing terminal nonreducing *N*-acetyllactosamine (Galβ1-4GlcNAc) groups. Substitution with *N*-acetylneuraminic acid (NeuAc) residue in an α2-3 linkage to Gal has only a small effect on binding compared to lactose, whereas substitution with NeuAc in an α2-6 linkage to Gal is detrimental for binding to the lectins. Differences were found in fine carbohydrate binding specificity among 14.5-kDa lectins from different species, as well as between the 14.5-kDa lectins and the 29-kDa lectins within the same species. In addition, it was found that the 29-kDa lectins bind β1-6-branched poly-*N*-acetyllactosaminoglycans consisting of repeating (3Galβ1-4GlcNAcβ1) units, with a much greater affinity than they bind lactosamine. The 14.5-kDa lectins also bind the lactosaminoglycans better than they bind lactosamine, but less than the 29-kDa lectins (Abbott et al., 1988; Barondes, 1988; Childs and Feizi, 1979; Leffler and Barondes, 1986; Sparrow et al., 1987). The finding that the major ligand for the chick embryonic skin 14 kDa lectin in embryonic chick skin is a polylactosamino-proteoglycan indicates that this is an important complementary structure for the function of the lectin in vivo (Oda and Kasai, 1984). Polylactosaminoglycans have been found in laminin (Arumugham et al., 1986) and the binding of endogenous lectins to laminin has been demonstrated (Cooper and Barondes, 1990; Woo et al., 1990; Zhou and Cummings, 1990). Interestingly, the laminin receptor has been found to share antigenic epitopes with the 14.5-kDa lectin and to bind lactose (Mecham et al., 1989). These results implicate laminin as an endogenous complementary molecule for lactoside-specific lectins.

One of the preliminary indirect approaches to elucidate the function of lectins is to determine their cellular and subcellular distribution. The lactoside-specific lectins were found primarily in the cytoplasmic compartment of various normal cells, suggesting some intracellular function(s). However, they were detected also in the cell nucleus, on the cell surface as well as in the extracellular compartment, where they appear to be associated with the extracellular matrix. Based on these locations, various functions have been proposed for the lectins including mediation of intercellular recognition and adhesion, control of cellular proliferation, and organization of glycoprotein constituents of the extracellular matrix (Agrwal et al., 1989; Allen et al., 1990; Barondes, 1984; 1986; 1988; Catt and Harrison, 1985; Cerra et al., 1984; Cooper and Barondes, 1990; Cowles et al., 1989; Gabius, 1991; Harrison and Chesterton, 1980; Hinek et al., 1988; Joubert et al., 1987; 1988; Monsigny et al., 1983; Moutsatsos et al., 1986; 1987; Seve et al., 1986).

Because their levels are developmentally regulated in several tissues it is thought that the lectins play some role during embryonal development (Barondes, 1984; 1986; 1988; Cook et al., 1979; Cooper and Barondes, 1990; Harrison and Chesterton, 1980; Hynes et al., 1990; Joubert et al., 1988; Milos and Zalik, 1983; 1986; Regan et al., 1986). The finding that the sequences of lectins cloned from different tissues and from different species (e.g. electric eel, chick embryo skin, mouse 3T3 fibroblasts, muscle cell line, rat lung and brain, bovine fibroblasts, human placenta) have been conserved during evolution and exhibit a considerable homology also implies that the lectins play some important roles in fundamental cellular processes.

PROPERTIES OF LECTINS FOUND IN MALIGNANT CELLS

In vitro-transformed cells and cells derived from tumors which were induced by viruses, chemical agents, or developed spontaneously in vivo, were found to contain lactoside-binding lectins (Allen et al., 1990; Carding et al., 1985b; Couraud et al., 1989; Gabius, 1987; 1991; Godsave et al., 1981; Hass et al., 1990; Irimura et al., 1991; Lotan and Raz, 1988b; Monsigny et al., 1983; 1988a; Raz and Lotan, 1981; 1987) as well as lectins expressing different carbohydrate-binding specificities (Gabius, 1987; 1991; Gabius and Engelhardt, 1988; Gabius et al., 1986; 1987a; 1987b; 1987c; 1988; 1989; 1990; Haas et al., 1990; Kieda and Monsigny, 1986; Magnusson and Standahl, 1988; Monsigny et al., 1983). Comparisons have been made between lactoside-binding lectins found in normal cells and in malignant cells and it appears that they are similar in sugar binding specificity, molecular size, cDNA and amino acid sequences, and antigenicity. There are no indications for gene rearrangement or mutations in lectins from tumor cells. For example, the nucleotide sequence of the 14.5-kDa lectin cDNA from human placenta is identical to that from HL-60 cells (Couraud et al, 1989). Likewise, the sequence of the cDNA of the 35 kDa lectin from mouse muscle cell line C2, 3T3 fibroblasts and mouse fibrosarcoma UV-2237 are similar if not identical (Cooper and Barondes, 1990; Raz et al., 1988; Wilson et al, 1989). There are , however, some differences between lectins in normal and tumor cells; although usually undetected on the surface of most normal cells, lactoside-binding lectins were found on the surface of many tumor cells (Gabius et al., 1987a; 1987b; 1990; Lotan and Raz, 1988b; Lotan et al, 1989a; Raz and Lotan, 1987; Raz et al., 1984) suggesting that they can participate in cellular interactions. The process of malignant transformation has been shown to be accompanied by changes in lectin level and cellular distribution. For example, the level of cell surface lectins increased after certain normal cells were transformed by transfection with certain oncogenes or by retroviruses, or when cells transformed with a temperature-sensitive viral mutant were switched from growth at the non-permissive to the permissive temperature, where they expressed the transformed phenotype (Raz et al., 1987b; Raz and Lotan, 1987). In particular, the level of the 34 kDa lectin increased both at the protein and mRNA levels (Agrwal et al., 1989; Raz et al., 1987a; Raz et al., 1987b). Another interesting finding was that among tumor cells differing in metastatic propensity, those exhibiting a higher metastatic potential express higher levels of surface lectins (Raz et al., 1986). Indirect indications that cell surface lectins may be involved in the maintenance of the transformed state and in metastasis are based on the finding that anti-lectin antibodies were able to inhibit the anchorage-independent growth of tumor cells in semi-solid medium (Lotan et al., 1985) and to suppress the ability of certain tumor cells to form lung colonies after intravenous injection into syngeneic mice (Meromsky et al., 1986). Furthermore, overexpression of the 35 kDa lectin gene transfected into poorly metastatic fibrosarcoma cells enhanced their metastatic potential and increased growth in semi-solid medium of untransformed fibroblasts (Raz et al., 1990).

The clinical relevance of the above observations begins to be unravelled with the recent analysis in our laboratory of lectin expression in specimens of colorectal carcinoma from 46 patients with known Dukes' stages. This study established a relation between lectin expression and spontaneous progression to metastatic stages in cancer patients (Irimura et al, 1991; Lotan et al, 1991). Specifically, the levels of a 31 kDa lectin in colorectal cancer specimens from primary tumors of patients having distant metastases (Dukes' stage D) were significantly higher than in those from patients without detectable metastases (Dukes' stages B1 and B2). Likewise, the level of the 31 kDa lectin correlated significantly with the serum level of carcinoembryonic antigen (CEA) in the same patients. In contrast, the variation in the level of the 14.5-kDa lectin among the different specimens was smaller and there was no correlation between the amount of the 14.5-kDa lectin and cancer stage or CEA levels. The results indicate that the relative amount of the 31-kDa lectin increases as the colorectal cancer progresses to the more malignant stage.

MODULATION OF LECTIN EXPRESSION IN TUMOR CELLS BY DIFFERENTIATION INDUCING AGENTS

Cancer cells often exhibit aberrant differentiation and express features of normal fetal cells due to a block in differentiation or maturation. The reports that lactoside-binding lectins are modulated during embryonal development (Barondes, 1984; 1986) raised the question as to whether induction of differentiation of cancer cells by various agents in vitro might also lead to changes in the patterns of lectin expression. Indeed, several agents known to induce differentiation and suppress the transformed phenotype of tumor cells were found to alter the level of lactoside-binding lectins in various cultured murine and human tumor cell lines (Lotan et al., 1989a; Lotan et al., 1989b; Ohanessian et al., 1989). The analysis revealed the presence of two lectins of 14.5-kDa and 34-kDa in the murine melanomas K-1735P, B16-F1, and embryonal carcinoma F-9, whereas only the 14.5-kDa lectin was detected in neuroblastoma S20, and normal mouse embryo lung fibroblasts. The human breast carcinoma MDA-MB 175 contained two lectins of 14.5-kDa and 31-kDa, whereas the breast carcinoma MDA-MB 453 and the leukemias HL-60 and THP-1 contained only the 14.5-kDa lectin. In contrast, only the 31-kDa lectin was detected in the human colon carcinoma KM12P. Differentiation and growth inhibition induced in the tumor cells by exposure to one or more of the differentiation inducing agents dibutyryl cAMP, β-all-trans retinoic acid, dimethylsulfoxide, or butyrate resulted in altered expression of the lectins, which was specific for each of the cell types. Treatment of the K-1735P and B16-F1 cells with either retinoic acid or dibutyryl cAMP enhanced melanocytic differentiation and resulted in a selective decrease of 70-90% in the level of the 34-kDa lectin without a concurrent effect on the level of the 14.5-kDa lectin. In contrast, the growth was inhibited by retinoic acid and the level of the 14.5-kDa lectin decreased by 75%, whereas the level of the 31-kDa lectin was not altered in MDA-MB 175 breast carcinoma cells. Similarly, the induction of differentiation of other cells resulted in suppression of the expression of the 14.5-kDa lectin. For example, the expression of the 14.5-kDa lectin decreased by 60% following induction of neuronal differentiation of the S20 neuroblastoma cells by dibutyryl cAMP and by about 50% after induction of granulocytic differentiation in HL-60 cells by retinoic acid or dimethylsulfoxide, and after monocytic differentiation was induced in THP-1 cells by retinoic acid. In contrast, retinoic acid-induced differentiation of F-9 embryonal carcinoma cells into endoderm like cells was accompanied by about a 300-400% increase in the expression of the 14.5-kDa lectin and a marked suppression of the level of the 34-kDa lectin. Similar results were observed when the F-9 cells were treated with other differentiation inducers including difluoromethylornithine and hexamethylenebis-acetamide (Lotan et al., unpublished results). The differentiation of F-9 cells into endodermal cells is accompanied by the production of laminin, which contains polylactosaminoglycan chains to which the lactoside-specific lectins bind avidly. Double immunofluorescence labeling demonstrated that the 14.5-kDa lectin co-localizes with laminin in retinoic acid-treated F-9 cells, suggesting that the lectin mediates cell surface binding to extracellular laminin fibrils (Lotan et al., unpublished results). Similar results were reported for the concurrent expression of lactoside-specific lectins and their complementary glycoconjugates in rat sensory neurons (Regan et al., 1986) and the co-localization of the 14.5-kDa lectin and laminin on the surface of myotubes (Cooper and Barondes, 1990).

It is noteworthy that the differentiation-inducing agents inhibited the growth of those tumor cells that were induced to differentiate. However, growth inhibition per se is not sufficient to alter lectin levels; for example, suppression of the growth of K-1735 melanoma cells by depletion of serum factors did not alter lectin levels (Lotan et al., 1989a) and treatment of human colon carcinoma KM12P cells with retinoic acid suppressed anchorage-independent growth but had no effect on lectin expression (Lotan et al., unpublished).

Butyrate is an established inducer of differentiation of various normal and tumor cell types in vitro and in vivo and is one of a few that is known to affect colon cancer cells. Butyrate treatment of several cultured colon carcinoma cell lines resulted in both growth inhibition and induction of the expression of a 14.5-kDa lectin. The treatment with butyrate also resulted in an increase in the level of carcinoembryonic antigen, which is a marker of differentiation for colorectal carcinoma cells (Ohanessian et al., 1989). The induction of the 14.5-kDa lectin by butyrate was dependent on the butyrate dose and was detected after a 24-h treatment with 2 mM butyrate, before any effect on cell growth could be detected.

Changes in the expression of lectins with other sugar specificities after induction of differentiation of tumor cells in vitro have been reported by other investigators (Gabius et al., 1989; 1990; Hass et al., 1990; Magnusson and Standahl, 1988). Treatment of cultured human colon carcinoma COLO 205 cells with butyrate, retinoic acid, or tetradecanoyl phorbol acetate (TPA) resulted in alterations in the binding of neoglycoproteins carrying different saccharides to the cell surface (Gabius et al., 1990). For example, retinoic acid and TPA increased binding of *N*-acetylgalactosamine-containing neoglycoprotein and both butyrate and retinoic acid increased the binding of fucosylated neoglycoprotein. Likewise, treatment of HL-60 and U937 leukemia cells with either dimethylsulfoxide or TPA resulted in increased binding to their surface of neoglycoproteins with galactose, lactose, fucose and ganglioside (Magnusson and Stendahl, 1988). Using affinity chromatography it was demonstrated that the differentiation of U937 induced by TPA was accompanied by induction of a 24-kDa lactose-binding lectin and a 18-kDa mannose-binding lectin, whereas a 20-kDa fucose-binding lectin expression is suppressed (Hass et al., 1990). A similar method revealed that induction of differentiation of HL-60 cells by treatment with dimethylsulfoxide resulted in the increase in a 29 kDa lactoside-binding lectin, the appearance of several *N*-acetylgalactosamine-binding lectins and several heparin-binding lectins (Gabius et al., 1989).

The altered lectin levels may result in changes in the interactions between adjacent cells as well as in interactions between cells and the extracellular matrix. Some of these interactions may suppress the malignant properties of the tumor cells.

CONSIDERATIONS ON THE USE OF LECTIN-TARGETED THERAPEUTIC APPROACHES

It has been proposed that the lectins present on the surface of tumor cells can be a target for therapeutic approaches. For example, treatment with anti-lectin antibodies can suppress growth of tumor cells in agarose (Lotan et al., 1985) and inhibit lung colonization in vivo (Meromsky et al., 1986; Raz and Lotan, 1987) suggesting a potential clinical application. A more effective approach may be to target anti-lectin antibodies conjugated with drugs or toxins to the tumor cells. The finding that lectins present on the surface of tumor cells are capable of binding exogenous carbohydrate-containing molecules and internalize them by endocytosis can be used for several treatment strategies. For example, a combination of two neoglycoproteins, one bearing fucose residues and the other glucose residues, inhibited lung colonization by Lewis lung carcinoma cells (Monsigny et al., 1988b) and lactose derivatives were shown to suppress in vivo lung colonization by metastatic melanoma cells (Oguchi et al., 1990), presumably by blocking the function of surface lectins (Kieda and Monsigny, 1986). Another approach has been to use neoglycoproteins as carriers to target drugs (e.g. methotrexate, 5-

fluorodeoxyuridine, daunomycin, etoposide, or *cis*-dichlorodiammine-platinum) or toxins (e.g. gelonin) to tumor cells. The neoglycoproteins used included lactosylated serum albumin and polylysine to which the drugs were linked via a proteolytically-cleavable bridge (Gabius, 1991; Gabius et al., 1987b; 1987c; Monsigny et al., 1983; 1984; 1988b; Roche et al., 1983). The antitumor activity of such conjugates is based on their internalization by the cells by endocytosis, degradation of the carrier in the lysosomes, and release of the free drug or toxin into the cytoplasm, where it then exerts its toxic effects. The distribution of radiolabeled neoglycoproteins injected intravenously into mice bearing solid Ehrlich tumor indicated that for certain carbohydrates there is a significant localization in the tumor compared to blood and muscle levels (Kojima and Gabius, 1988).

Liposomes containing glycolipids (e.g. *N*-palmitoyl dihydrolacto-cerebroside) can bind to HeLa cells presumably via cell surface lectins and it has been proposed that entrapment of chemotherapeutic agents such as L-asparaginase in such glycosylated liposomes may increase their efficiency of delivery to tumor cells (Bussian and Wriston, 1977). HL-60 and U937 human leukemia cells were also reported to bind liposomes displaying specific carbohydrates (Magnusson and Standahl, 1988). Lectin-specific targeting of an enzyme entrapped in glycosylated liposomes to different liver cells was demonstrated in vivo (Das et al., 1985). At present there are no convincing results of the success of any of the methods directed at tumor cell surface lectins for treatment of tumors in vivo. Improvement in the methods may come after elucidation of the precise oligosaccharide sequence that best fits the lectin's sugar binding site.

The recent availability of recombinant lectins (Couraud et al., 1989; Hirabayashi et al., 1989; Frigeri et al., 1990) indicates that they can be produced in sufficient quantities and used like plant lectins for tumor imaging (Kojima and Jay, 1986; Shysh et al., 1985) or as carriers of drugs and toxins for targeting to complementary molecules on the surface of tumor cells (Monsigny et al., 1980). Furthermore, lactoside-specific lectins can stimulate immune responses either by acting as mitogens for lymphocytes (Lipsick et al., 1980; Pitts and Young, 1981) or by stimulating release of cytotoxins by macrophages (Kajigawa et al., 1986).

One of the possible limitations on the success of lectin-targeted therapy is the number of cell surface lectin molecules. A treshold may exist below which the drug conjugates may be ineffective. The ability of certain differentiation inducing agents to increases lectin expression in tumor cells can be used in combination with antilectin antibodies conjugated with toxins or drugs; neoglycoprotein-drug or neoglycoprotein-toxin conjugates; or glycosylated, drug-containing liposomes to augment responsiveness.

REFERENCES

Abbott WM, Hounsell EF, Feizi, T (1988) Soluble bovine galactose-binding lectin: cDNA cloning reveals the complete amino acid sequence and an antigenic relationship with the major encephalitogenic domain of myelin basic protein. Biochem J 252:283-287

Agrwal N, Wang JL., Voss PG (1989) Carbohydrate binding protein 35: levels of transcription and mRNA accumulation in quiescent and proliferating cells. J Biol Chem 264:17236-17242

Alhadeff JA (1989) Malignant cell glycoproteins and glycolipids CRC Crit Rev Oncol Hematol 9:37-107

Allen HJ, Karakousis C, Piver MS, Gamarra M, Nava H, Forsyth B, Matecki B, Jazayeri A, Sucato D, Kisailus E, DiCioccio R (1987) Galactoside-binding lectin in human tissues. Tumor Biol 8:218-229

Allen HJ, Sucato D, Woynarowska B, Gottstine S, Sharma A, Bernacki RJ (1990) Role of galaptin in ovarian carcinoma adhesion to extracellular matrix in vitro. J Cell Biochem 43:43-57

Amos BD, Lotan D, Lotan R (1990) Increased fucosylation of high molecular weight glycoproteins accompanies retinoic acid induced differentiation of F-9 embryonal carcinoma cells. Int J Cancer 46:86-94

Arumugham RG, Hsieh T. C-Y, Tanzer ML, Laine RA (1986) Structures of the asparagine-linked sugar chains of laminin. Biochim Biophys Acta 883:112-126

Avner P, Arnaud D, Blaineau C, Le Pendu J (1985) Developmentally regulated cell surface structures on mouse and human embryonal carcinoma cell lines. Cell Diff 17:13-20

Barcellos-Hoff MH, Aggeler J, Ram TG. Bissell MJ (1989) Functional differentiation and alveolar morphogenesis of primary mammary epithelial cells cultured on reconstituted basement membrane. Development 105:223-235

Barondes SH (1984) Soluble lectins: A new class of extracellular proteins. Science 223:1259-1264

Barondes SH (1986) Vertebrate lectins: Properties and functions. In: Liener IE, Sharon N, Goldstein IJ (eds) The Lectins: Properties, Functions, and Applications in Biology and Medicine. Academic Press, Orlando, pp 437-466

Barondes SH (1988) Bifunctional properties of lectins: lectins redefined. Trends Biochem Sci 13:480-482

Brandley BK, Swiedler SJ, Robbins PW (1990) Carbohydrate ligands of the LEC cell adhesion molecules. Cell 63:861-863

Brandley BK, Schnaar RL (1986) Cell-surface carbohydrates in cell recognition and response. J Leukocyte Biol 40:97-111

Bussian RW, Wriston JC, Jr (1977) Influence of incorporated cerebrosides on the interaction of liposomes with HeLa cells. Biochim Biophys Acta 471:336-340

Carding SR, Childs RA, Thorpe R, Spitz M, Feizi T (1985a) Multiple proteins related to the soluble galactose-binding animal lectin revealed by a monoclonal antilectin antibody. Biochem J 228:147-153

Carding SR, Thorpe SJ, Thorpe R, Feizi T (1985) Transformation and growth related changes in levels of nuclear and cytoplasmic proteins antigenically related to mammalian β-galactoside-binding lectin. Biochem Biophys Res Commun 127:680-686

Caron M, Joubert R, Bladier D (1987) Purification and characterization of a β-galactoside-binding soluble lectin from rat and bovine brain. Biochim Biophys Acta 925:290-296

Catt JW, Harrison FL (1985) Selective association of an endogenous lectin with connective tissues. J Cell Sci 73:347-359

Cerra RF, Gitt MA, Barondes SH (1985) Three soluble rat β-galactoside-binding lectins. J Biol Chem 260:10474-10477

Cerra RF, Haywood-Reid PL, Barondes SH (1984) Endogenous mammalian lectin localized extracellularly in lung elastic fibers. J Cell Biol 98:1580-1589

Childs RA, Feizi T (1979) Calf heart lectin reacts with blood group Ii antigens and other precursor chains of the major blood group antigens. FEBS Lett 99:175-179

Cook GMW, Zalik SE, Milos N, Scott,V (1979) A lectin which binds specifically to β-D-galactoside groups is present at the earliest stages of chick embryo development. J Cell Sci 38:293-304

Cooper DN, Barondes SH (1990) Evidence for export of a muscle lectin from cytosol to extracellular matrix and for a novel secretory mechanism J Cell Biol 110:1681-1691

Couraud P-O, Casentini-Borocz D, Bringman TS, Griffith J, McGrogan M, Nedwin GE (1989) Molecular cloning, characterization, and expression of a human 14-kDa lectin. J Biol Chem 264:1310-1316

Cowles EA, Moutsatsos IK, Wang JL, Anderson RL (1989) Expression of carbohydrate binding protein 35 in human fibroblasts: comparisons between cells with different proliferative capacities. Exp Gerontol 24:577-585

Crittenden SL, Roff CF, Wang JL (1984) Carbohydrate-binding protein 35: Identification of the galactose-specific lectin in various tissues of mice. Molec Cell Biol 4:1252-1259

Das PK, Murray GJ, Zirow GC, Brady RO, Barranger JJ (1985) Lectin-specific targeting of β-glucuronidase to different liver cells via glycosylated liposomes. Biochem Med 33:124-131

Dennis,JW, Laferte S, Waghorne C, Breitman ML, Kerbel RS (1987) β1-6 Branching of Asn-linked oligosaccharides is directly associated with metastasis. Science 236:582-585

Deutsch V, Lotan R (1983) Stimulation of sialyltransferase activity of melanoma cells by retinoic acid. Exp Cell Res 149:237-245

Drickamer K (1988) Two distinct classes of carbohydrate-recognition domains in animal lectins. J Biol Chem 263:9557-9560
Edelman GM (1985) Cell adhesion and the molecular processes of morphogenesis. Ann Rev Biochem 54:135-169
Feizi T (1985) Demonstration by monoclonal antibodies that carbohydrate structures of glycoproteins and glycolipids are onco-developmental antigens. Nature 314:53-57
Frigeri LG, Robertson MW, Liu F-T (1990) Expression of biologically active recombinant rat IgE-binding protein in Escherichia coli. J Biol Chem 265:20763-20769
Fukuda M (1985) Cell surface glycoconjugates as onco-differentiation markers in hematopoietic cells. Biochim Biophys Acta 780:119-150
Gabius H-J (1987a) Endogenous lectins in tumors and the immune system. Cancer Invest 5:39-46
Gabius H-J (1987b) Targeting of neoglycoprotein-drug conjugates to human tumor cells via endogenous lectins. Ann N Y Acad Sci 507:337-338
Gabius H-J (1991) Detection and functions of mammalian lectins-with emphasis on membrane lectins. Biochim Biophys Acta 1071:1-18
Gabius H-J, Engelhardt R (1988) Sugar receptors of different types in human metastases to lung and liver. Tumor Biol 9:21-36
Gabius H-J, Bandlow G, Schirrmacher V, Nagel GA, Vehmeyer K (1987a) Differential expression of endogenous sugar-binding proteins (lectins) in murine tumor model systems with metastatic capacity. Int J Cancer 39:643-648
Gabius H-J, Bokemeyer C, Hellmann T, Schmoll H-J (1987b) Targeting of neoglycoprotein-drug conjugates to cultured human embryonal carcinoma cells. J Cancer Res Clin Oncol 113:126-130
Gabius H-J, Engelhardt R, Hellmann T, Midoux P, Monsigny M, Nagel GA, Vehmeyer K (1987c) Characterization of membrane lectins in human colon carcinoma cells by flow cytofluorometry, drug targeting and affinity chromatography. Anticancer Res 7:109-112
Gabius H-J, Engelhardt R, Rehm S, Barondes SH, Cramer F (1986) Presence and relative distribution of the endogenous β-galactoside-specific lectins in different types of rat tumors. Cancer J 1:19-21
Gabius H-J, Vehmeyer K, Gabius S, Nagel GA (1988) Clinical application of various plant and endogenous lectins to leukemia. Blut 56:147-152
Gabius S, Hellmann KP, Ciesiolka T, Nagel GA, Gabius H-J (1989) Lineage- and differentiation-dependent alterations in the expression of receptors for glycoconjugates (lectins) in different human hematopoietic cell lines and low grade lymphomas. Blut 59:165-170
Gabius S, Yamazaki N, Hanewacker W, Gabius H-J (1990) Regulation of distribution, amount and ligand affinity of sugar receptors in human colon carcinoma cells by treatment with sodium butyrate, retinoic acid and phorbol ester. Anticancer Res 10:1005-1012
Godsave SF, Harrison FL, Chesterton CJ (1981) Molecular properties of a galaptin and its activity in normal and leukemic erythroid tissue. J Supramolec Struct Cell Biochem.17:223-230
Hakomori S (1989) Aberrant glycosylation in tumors and tumor-associated carbohydrate antigens. Adv Cancer Res 52:257-331
Harrison FL, Chesterton CJ (1980) Factors mediating cell-cell recognition and adhesion. FEBS Lett 122:157-165

Hass R, Köhler L, Rehfeldt W, Lebmann V, Müller W, Resc K, Goppelt-Struebe M (1990) Alterations in glycosylation and lectin pattern during phorbol ester-induced differentiation of U937 cells. Cancer Res 50:323-327

Hinek A, Wrenn DS, Mecham RP, Barondes SH (1988) The elastin receptor: A galactoside-binding protein. Science 235:1539-1541

Hirabayashi J, Ayaki H, Soma G-I, Kasai K-I (1989) Production and purification of recombinant human 14 kDa β-galactoside-binding lectin. FEBS Lett 250:161-165

Hirabayashi J, Kasai K.-i (1988) Complete amino acid sequence of a β-galactoside-binding lectin from human placenta. J Biochem 104:1-4

Hirabayashi J, Kawasaki H, Susuki K, Kasai K-i (1987) Complete amino acid sequence of 14 kDa β-galactoside-binding lectin of chick embryo. J Biochem 101:775-787

Humphries MJ (1990) The molecular basis and specificity of integrin-ligand interactions. J Cell Sci 97:585-592

Hynes MA, Gitt M, Barondes SH, Jessell T, Buck LB (1990) Selective expression of an endogenous lactose-binding lectin gene in subsets of central and peripheral neurons. J Neurosci 10:1004-1013

Irimura T, Matsushita Y, Sutton RC, Carralero D, Ohannesian DW, Cleary KR, Ota DM, Nicolson GL, Lotan R (1991) Increased content of an endogenous lactose-binding lectin in human colorectal carcinoma progressed to metastatic stages. Cancer Res 51:387-393

Jia S, Wang JL (1988) Carbohydrate binding protein 35. Complementary DNA sequence reveals homology with proteins of the heterogeneous nuclear RNP. J Biol Chem 263:6009-6011

Joubert R, Caron M, Bladier D (1987) Brain lectin-mediated agglutinability of dissociated cells from embryonic and postnatal mouse brain. Develop Brain Res 36:146-150

Joubert R, Caron M, Bladier D (1988) Distribution of β-galactoside specific lectin activities during pre- and post-natal mouse brain development. Cell Molec Biol 34:79-87

Kajikawa T, Nakajima Y, Hirabayashi J, Kassai K, Yamazaki M (1986) Release of cytotoxin by macrophages on treatment with human placenta lectin. Life Sci 39:1177-1181

Kieda C, Monsigny M (1986) Involvement of membrane sugar receptors and membrane glycoconjugates in the adhesion of 3LL cell subpopulations to cultured pulmonary cells. Invasion Metastasis 6:347-366

Kleinman HK, Cannon FB, Laurie GW, Hassell JR, Aumailley M, Terranova, VP, Martin GR, DuBois-Dalcq M (1985) Biological activities of laminin. J Cell Biochem 27:317-325

Kojima S, Gabius H-J (1988) Biodistribution of neoglycoproteins in mice bearing solid Ehrlich tumor. J Cancer Res Clin Oncol 114:468-472

Kojima S, Jay M (1986) Application of lectins to tumor imaging radiopharmaceuticals. Eur J Nucl Med 12:385-389

Leffler H, Barondes SH (1986). Specificity of binding of three soluble rat lung lectins to substituted and unsubstituted mammalian β-galactosides. J Biol Chem 261:10119-10126

Lipsick JS, Beyer EC, Barondes SH, Kaplan NO (1980) Lectins from chicken tissues are mitogenic for Thy-1 negative murine spleen cells. Biochem Biophys Res Commun 97:56-61

Lotan R (1991) β-Galactoside-binding vertebrate lectins-synthesis, molecular biology, function, In: Allen HJ, Kisailus E (eds), Glycoproteins: Composition, Structure, and Function, Marcel Dekker, New York, in press

Lotan R, Raz A (1988) Lectins in cancer cells. Ann N Y Acad Sci 551:385-398

Lotan R, Raz A (1988) Endogenous galactoside-specific lectins as mediators of tumor cell adhesion. J Cell Biochem 37:107-117

Lotan R, Carralero D, Lotan D, Raz A (1989) Biochemical and immunological characterization of K-1735 melanoma galactoside-binding lectins and their modulation by differentiation inducers. Cancer Res 49:1261-1268

Lotan R, Lotan D, Raz A (1985) Inhibition of tumor cell colony formation in culture by a monoclonal antibody to endogenous lectins. Cancer Res 45:4349-4353

Lotan R, Lotan D, Carralero D (1989) Modulation of galactoside-binding lectins in tumor cells by differentiation-inducing agents. Cancer Lett 48:115-122

Lotan R, Matsushita Y, Ohannesian D, Carralero D, Ota DM, Cleary KR, Nicolson GL, Irimura T (1991) Lactose-binding lectin expression in human colorectal carcinomas. Relation to tumor progression. Carb. Res 210:in press

Magnusson K, Stendahl O (1988) Development of distinct clonal patterns of carbohydrate-binding activity in human promyelocytic HL60 cells and histiocytic U937 cells during DMSO- or PMA-induced differentiation. Glycoconjugate J 5:271-281

Mecham RP, Hinek A, Griffin GL, Senior RP, Liotta LA (1989) The elastin receptor shows structural and functional similarities to the 67-kDa tumor cell laminin receptor. J Biol Chem 264:16652-16657

Merkle RK, Zhou Q, Schultz TK, Harper WB, Cummings RD (1989) Characterization of an S-type lectin purified from porcine heart. Arch Biochem Biophys 274:404-416

Meromsky L, Lotan R, Raz A (1986) Implications of endogenous tumor cell surface lectins as mediators of cellular interactions and lung colonization. Cancer Res 46:5270-5275

Milos N, Zalik SE (1983) Calcium-independent adhesion of extra-embryonic endoderm cells from the early chick blastoderm is inhibited by the blastoderm β-D-galactoside-binding lectin and by β-galactosidase. Cell Diff 12:341-347

Milos N, Zalik SE (1986) Release of β-D-galactoside-binding lectins into the cavities of aggregates of chick extraembryonic endoderm cells. Cell Diff 18:1-7

Monsigny M, Kieda C, Roche A-C (1983) Membrane glycoproteins glycolipids and membrane lectins as recognition signals in normal and malignant cells. Biol Cell 47:95-110

Monsigny M, Kieda C, Roche A-C, Delmotte F (1980) FEBS Lett 109:181-186

Monsigny M, Roche AC, Kieda C, Midoux P, Obrenovitch A (1988b). Characterization and biological implications of membrane lectins in tumoral, lymphoid and myeloid cells. Biochimie 70:1633-1649

Monsigny M, Roche A-C, Midoux P (1984) Uptake of neoglycoproteins via membrane lectin(s) of L1210 cells evidenced by quantitative flow cytofluorometry and drug targeting. Biol Cell 51:187-196

Monsigny M, Roche A-C, Midoux P (1988) Endogenous lectins and drug targeting. Ann N Y Acad Sci 551:399-413

Monsigny M, Roche A-C, Kieda C, Midoux P, Obrenovitch A (1988) Characterization and biological implications of membrane lectins in tumor, lymphoid and myeloid cells. Biochimie 70:1633-1649

Morrison-Graham GK, Weston JA (1989) Mouse mutants provide new insights into the role of extracellular matrix in cell migration and differentiation. Trends Genet 5:116-121

Moutsatsos IK, Davis JM, Wang JL (1986) Endogenous lectins from cultured cells: Subcellular localization of carbohydrate-binding protein 35 in 3T3 fibroblasts. J Cell Biol 102:477-483

Moutsatsos IK, Wade M, Schindler M, Wang JL (1987). Endogenous lectins from cultured cells. Nuclear localization of carbohydrate-binding protein 35 in proliferating 3T3 fibroblasts. Proc Natl Acad Sci USA 84:6452-6456

Muramatsu T (1988) Developmentally regulated expression of cell surface carbohydrates during mouse embryogenesis. J Cell Biochem 36:1-14

Nicolson GL (1988) Organ specificity of tumor metastasis: role of preferential adhesion, invasion and growth of malignant cells at specific secondary sites. Cancer Metastasis Rev 7:143-188

Oda Y, Kasai K-i. (1984) Photochemical cross-linking of β-galactoside-binding lectin to polylactosamino-proteoglycan of chick embryonic skin. Biochem Biophys Res Commun 123:1215-1220

Oguchi H, Toyokuni T, Dean B, Ito H, Otsuji E, Jones V, Sadozai K, Hakomori S (1990) Effect of lactose derivatives on metastatic potential of B16 melanoma cells. Cancer Comm 2:311-316

Ohannesian D, Nicolson GL, Fidler IJ, Carralero D, Lotan D, Lotan R (1989) The 14.5-kDa lectin is induced in human colon carcinoma cells by butyrate treatment J Cell Biol 109:179a

Ohyama Y, Hirabayashi J, Oda Y, Ohno S, Kawasaki H, Suzuki K, Kasai K-i (1986) Nucleotide sequence of chick 14K β-galactoside-binding lectin mRNA. Biochem Biophys Res Commun 134:51-56

Paulson JC (1989) Glycoproteins: what are the sugar chains for? Trends Biochem Sci 14:272-276

Pitts MJ, Yang DCH (1981) Mitogenicity and binding properties of b-galactoside-binding lectin from chick-embryo kidney. Biochem J 195:435-439

Raz A, Lotan R (1981) Lectin-like activities associated with human and murine neoplastic cells. Cancer Res 41:3642-3647

Raz A, Lotan R (1987) Endogenous galactoside-binding lectins: a new class of functional tumor cell surface molecules related to metastasis. Cancer Met Rev 6:433-452

Raz A, Avivi A, Pazerini G, Carmi P (1987a) Cloning and expression of cDNA for two endogenous UV-2237 fibrosarcoma lectin genes. Exp Cell Res 173:109-116

Raz A, Carmi P, Pazerini G (1988) Expression of two different endogenous galactoside-binding lectins sharing sequence homology. Cancer Res 48:645-649

Raz A, Meromsky L, Lotan R (1986) Differential expression of endogenous lectins on the surface of nontumorigenic, tumorigenic and metastatic cells. Cancer Res 46:3667-3672
Raz A, Meromsky L, Carmi P, Karkash R, Lotan D, Lotan R (1984) Monoclonal antibodies to endogenous galactoside-specific tumor cell lectins. EMBO J 3:2979-2984
Raz A, Meromsky L, Zvibel I, Lotan R (1987b) Transformation-related changes in the expression of endogenous cell lectins. Int J Cancer 39:353-360
Raz A, Zhu D, Hogan V, Shah N, Raz T, Karkash R, Pazerini G, Carmi P (1990) Evidence for the role of 34-kDa galactoside-binding lectin in transformation and metastasis. Int J Cancer 46:871-877
Regan LJ, Dodd J, Barondes SH, Jessell TM (1986) Selective expression of endogenous lactose-binding lectins and lactoseries glycoconjugates in subsets of rat sensory neurons. Proc Nat Acad Sci USA 83:2248-2252
Roche A-C, Barzilay M, Midoux P, Junqua S, Sharon N, Monsigny M (1983) Sugar-specific endocytosis of glycoproteins by Lewis lung carcinoma cells. J Cell Biochem 22:131-140
Rutishauser U, Jessell TM (1988) Cell adhesion molecules in vertebrate neural development. Physiol Rev 68:819-857
Schachter H (1986) Biosynthetic controls that determine the branching and microheterogeneity of protein-bound oligosaccharides. Biochem Cell Biol 64:163-181
Schuetz EG, Li D, Omiecinski CJ, Muller-Eberhard U, Kleinman HK, Elswick B, Guzelian PS (1988) Regulation of gene expression in adult rat hepatocytes cultured on a basement membrane matrix. J Cell Physiol 134:309-323
Seve A-P, Hubert J, Bouvier D, Bourgeois C, Midoux P, Roche A-C, Monsigny M (1986) Analysis of sugar-binding sites in mammalian cell nuclei by quantitative flow microfluorometry. Proc Natl Acad Sci USA. 83:5997-6001
Sharon N, Lis H (1989) Lectins as cell recognition molecules. Science 246:227-234
Shysh A, Eu SM, Noujaim AA, Suresh MR, Longenecker BM (1985) Radioimmunodetection of murine mammary adenocarcinoma (TA3/Ha) lung and liver metastases with radioiodinated PNA. Int J Cancer 35:113-119
Sparrow CP, Leffler H, Barondes SH (1987) Multiple soluble β-galactoside-binding lectins from human lung. J Biol Chem 262:7383-7390
Stoolman LM (1989) Adhesion molecules controlling lymphocyte migration. Cell 56:907-910
Wilson TJG, Firth MJ, Powell JT, Harrison FL (1989) The sequence of the mouse 14 kDa β-galactoside binding lectin and evidence for its synthesis on free cytoplasmic ribosomes. Biochem J 261:847-852
Woo H-J, Shaw LM, Messier JM, Mercurio AM (1990) The major non-integrin laminin binding protein of macrophages is identical to carbohydrate binding protein 35 (Mac2). J Biol Chem 265:7097-7099
Zhou Q, Cummings R D (1990) The S-type lectin from calf heart tissue binds selectively to the carbohydrate chains of laminin. Arch Biochem Biophys 281: 27-35

R. Lotan

Change in metastasis and organotropism of mouse lymphoma variants associated with changes in expression of adhesion molecules and chemotactic responsiveness

V. Schirrmacher, A. Kowitz, G. Kadmon, R. Benke, and P. Altevogt

Institut für Immunologie und Genetik,
Deutsches Krebsforschungszentrum,
D-6900 Heidelberg.

Introduction

We have previously reported on the isolation of a plastic adhesive variant, ESb-MP, derived from the suspension growing highly metastatic murine lymphoma ESb (1,2). Animals inoculated subcutanesously with ESb-MP cells developed large primary tumors and survived three to four times as long as animals inoculated with the non-adhesive ESb cells. The two lines were similar in expression of cell surface differentiation antigens and tumor antigens but differences could be seen in reactivity to certain lectins (2) and in expression of adhesion molecules. More information on cytogenetic aspects, changes in cell surface molecules, chemotactic responsiveness in vitro and of changes in the in vivo growth and organotropism will be reviewed.

Selection of adhesion variants from cloned and uncloned ESb lines and cytogenetic analysis.

A detailed cytogenetic analysis of the original ESb line (non-cloned standard batch 289) revealed two populations of cells, a minor one, characterized by a distinct caryotype composed of a number of specific marker chromosomes and a major population which differed in particular marker chromosomes (3). The comparison of the marker chromosomes revealed that the major subpopulation must have been derived from the minor one by chromosome translocations involving the chromosomes 12 and 17. The original plastic adhesive variant ESb-MP was in its caryotype very similar to the minor subpopulation (5). This indicated that the plastic adhesive lines which we could reproducibly isolate from the standard batch were derived from a pre-existing variant in the mixed population which possibly had a slightly enhanced tendency to stick to plastic. Upon further selective growth on plastic these cells could eventually be turned into a strongly plastic adhesive cell line.

Changes in adhesiveness affecting speed of dissemination, organotropism and mode of metastatic growth.

When the adhesive variants were tested in syngeneic animals they showed a greatly reduced overall malignancy as seen by the greatly increased survival of tumor bearing mice (4). The increased survival, however, did not mean that the animals were devoid of metastases. A detailed kinetic analysis of the dissemination and spread of the adhesive variants from the local (s.c.) inoculation site revealed various steps of differences in comparison to the non-adhesive cells. The adhesive variants showed a longer latency period before the tumors became palpable, the release of single tumor cells from the local tumor into the blood was much delayed and the extent to which disseminated tumor cells could be found in the blood was reduced. Colonization of internal organs such as liver was much delayed and the metastases which developed were of focal type whereas the non-adhesive cells metastasized more diffusely and caused death of the animals within a very short time interval. Thus there was an overall retardation of tumor cell dissemination by the adhesive variants and because of this the animals developed much larger primary tumors. There was however, also a change in the type of organs involved, i.e. in the organotropism. The adhesive variants showed a particular affinity for the kidney and for the meninga of the spinal cord. The great majority of animals inoculated subcutaneously or intravenously with the adhesive variants developed a hind-leg paralysis syndrome under very reproducible conditions (4). This tumor system might thus be an interesting model for this type of metastases and it will be interesting in the future so see whether these changes may be associated with distinct changes in specific membrane molecules.

Selective chemotactic responsiveness to organ derived chemoattractants correlating with Kidney metastases.

Two subclones of the adherent variant (Esb-M2 A8, ESb-M2 H6) showed a reduced malignant potential in vivo very similar to the ESb-MP cell line, but only the clone ESb-M2 A8 infiltrated the Kidney. Cell motility and chemotaxis of the various tumor lines was tested in a modified Boyden chamber assay. ESb-MP cells always migrated about 45 µm into a nitrocellulose filter of 45 µm pore size during 4h incubation whereas ESb cells and the two subclones migrated only to a distance of 12-

25µm. There was no apparent correlation between in vitro motility and malignancy in vivo.

The addition of several factors known to be chemo-attractants for granulocytes (C5a, FMLP) or for tumor cells (FMLP, laminin, fibronectin) did not effect the migration of ESb or ESb-MP cell lines. Enhanced migration of the ESb MP cell line was, however, observed when Kidney conditioned medium (KCM) was used as attractant in the lower chamber. This was not true for ESb-cells. Additionally, within the subclones of the ESb-MP cell line, there was a remarkable correlation between Kidney metastases in vivo and responsiveness to KCM in vitro. The motility stimulating effect of KCM was due to a chemotactic activity. ESb cells, which did not respond to KCM, responded to other factors derived from inflammatory exudate fluid.

We conclude that the stimulation of tumor cell motility by an organotropic factor was highly selective for certain tumor lines. The correlation between responsiveness in vitro and Kidney metastasis in vivo suggests that it is not so much the random migration of tumor cells which plays a role in metastasis but the motility following selective stimulation by tissue-derived factors and chemo-attractants (5).

A monoclonal antibody (MAb 12-15) detecting increased expression of mouse CD2 and immunolocalization of liver micrometastases.

In order to look for distinct changes in cell surface membrane molecules we were able to select monoclonal antibodies which reacted preferentially with the adhesive variants. Our studies concentrated on one of these antibodies, MAb 12-15 which was able to immunoprecipitate a membrane glycoprotein from the adhesive variants of a molecular weight range between 65 and 69 kd. Meanwhile, the molecules have been identified as CD2 (6) and Fcγ receptor proteins (alpha, beta 1 and beta 2) (7). There was only a low reactivity of this monoclonal with the suspension growing ESb cells and with normal lymphoid tissue. The antibody could be easily purified and labelled with ^{125}J so that quantitative binding studies could be done both in vitro and in vivo. The antibody showed particularly interesting immunolocalization properties in vivo (8). There was a clear cut difference in the stability of the corresponding glycoprotein on the tumor cells and on the normal lymphoid cells which seemed to be an important pre-requisit for the in vivo immunolocalization of tumor cells at the primary site or from metastases in the liver.

Changes in expression of adhesion molecules L1, LFA-1, Mac1 and ICAM-1.

Esb-MP cells were shown by immunofluorescence and cyto-fluorographic analyis to have increased expression of the two integrin molecules LFA-1 and Mac-1. We also found increased expression of ICAM-1 which functions as a ligand for LFA-1 and belongs to the immunoglobulin super family. When we screened for expression of neuronal adhesion molecules we made the unexpected observation that ESb-MP lymphoma cells expressed the neuronal adhesion molecule L1 but did not express N-CAM. In further experiments it could be shown that the L1-glycoprotein could be immunoprecipitated and was identical in the core-protein (150 Kd) to authentic L1 molecules. Due to differences in glycosylation the molecular wheights of cell surface expressed glycoproteins vary between 190 and 210 Kd.

Comparison of L1 negative and L1 positive clones of ESb-MP with regard to homotypic aggregation and growth and malignancy in vivo.

L1 can mediate calcium independant homotypic aggregation and could thereby lead to increased stickyness of the tumor cells to each other. To test the contribution of L1 to the in vivo growth-behaviour of ESb-MP cells we isolated L1 negative variants. This was achieved with the monoclonal antibody 324 in indirect immunofluorescence by sorting spontaneous L1 negative variants by means of the flourescence activated cell sorter (FACS). The last sorting was followed by cloning, expansion and retesting. While the expression of L1 on these selected clones was reduced about 7-fold, there was no difference in the expression of other adhesion molecules.

The L1 mediated aggregation was tested in 24 well plates following rotation at 85 rpm for 3 - 5 h at 37° (9). Four clones were selected which showed no aggregation (clone 5), weak aggregation (clone 12), medium aggregation (clone 5-9) and strong aggregation (clone 12-16). The homotypic aggregation correlated to the intensity of L1 cell surface fluorescence. When these four clones were tested for growth and metastasis in vivo it was found that the animals injected with the L1 negative clone showed the shortest survival while those with the strongest expression and the best aggregation showed the longest survival. The increased tendency of the clones to aggregate with each other was thus correlated with increased survival and reduced spread and metastasis in vivo.

Differences in glycosylation and sialylation of membrane components; Molecular identification of CD45 (T200) SBA lectin binding sites.

A number of fluoresceine-labelled plant lectins with defined carbohydrate specificity were used to compare their reactivity with non-adhesive or adhesive tumor variants. As reported before (2), Soybean agglutinin (SBA), reacted with the adhesive cells but not with the non-adhesive parental line. However, when the parental line was treated with sialidase, cryptic binding sites for SBA became exposed so that the cells now stained positively. Adhesive variants treated with sialidase also showed an increased reactivity with SBA.

Quantitative lectin binding studies performed with iodinated lectins revealed two types of binding sites which differed in their affinity. At low lectin concentrations (10-25 ug/ml) only the adhesive variants bound SBA thus indicating high affinity receptor sites. At high lectin concentrations (greater 100 ug/ml) both, the parental line and the adhesive variants, bound SBA thus indicating the presence of low affinity binding sites.

The distinction between high and low affinity binding sites could be corroborated by immunochemical studies (10).

We were particularly interested in the characterization of the high affinity binding sites because these were more distinct on the adhesive variants as compared to the non-adhesive cells. According to their molecular weight we speculated that they might represent the common leukocyte antigen T 200 or CD45. It is known that members of this family of molecules are expressed on B lymphocytes, T lymphocytes and macrophages. The molecular weights of the gene products in these cells are different. It is also known that these molecules are coded by one gene (12). B cells, T cells and macrophages could be shown to carry corresponding mRNAs of different size. Thus a differential splicing might be involved in the expression of related molecules in these cells (13).

We could show with biochemical techniques that the major SBA-binding cell surface component of ESb-MP cells is the T 200 glycoprotein. In ESb-cells, T 200 antigens bind SBA only after sialidase treatment. Enzymatic studies suggested that glycans detected by the lectin with or without sialidase treatment are different. Inhibition of N-glycosylation by tunicamycin and biosynthetic labelling revealed two T 200 chains for ESb-MP cells that were larger in size than the single chain detected in ESb-cells. Our data suggest that the molecules differ in expression of O-linked glycans that can be recognized by SBA. Additional O-linked sugars on ESb-MP T200 molecules seemed to be expressed in particular after trimming of the second T200 precursor chain. Thus, altered glycosylation as detected by lectin binding can depend on structural modifications in a single glycoprotein as shown here for the CD45 protein. It is possible that such subtle glycosylation changes may effect cellular interactions as well as migration and homing of normal or neoplastic lymphoid cells.

REFERENCES

1. Fogel, M., Altevogt, P., Schirrmacher, V., (1983) Metastatic potential severely altered by changes in tumor cell adhesiveness and cell surface sialylation. In: J. Exp. Med. 157, 371-376.

2. Altevogt, P., Fogel, M., Cheingsong-Popov, R., Dennis, J., Robinson, P., and Schirrmacher, V., (1983) Different patterns of lectin binding and cell surface sialylation detected on related high- and low-metastatic tumor lines. In: Cancer Res. 43, 5138-5144.

3. Pflüger, E., Lang, E., Benke, R., Heckl-Östreicher, Altevogt, P., and Schirrmacher, V. (1988) Generation of adhesive tumor variants: chromosomal changes, reduction in malignancy and increased expression of a distinct membrane glycoprotein. Clin. Exp. Metastasis 6, 485-499.

4. Benke, R., Lang, E., Komitowski, D., Muto, S., Schirrmacher, V. (1988) Changes in tumor cell adhesiveness affecting speed of dissemination and mode of metastatic growth. Invasion and Met., 8, 159-176.

5. Benke, R., and Schirrmacher, V., (1991) Change in organotropism of mouse lymphoma variants associated with selective chemotactic responsiveness to organ derived chemoattractants. Clin. Exptl. Metastasis, in press.

6. Altevogt, P., Kohl, U., von Hoegen, P., Lang, E., and Schirrmacher, V. (1989)
Antibody 12-15 cross-reacts with mouse Fcγ receptors and CD2: study of thymus expression, genetic polymorphism and biosynthesis of the CD2 protein.
Eur. J. Immun. 19, 341-346.

7. Altevogt, P., Heckl-Östreicher, B., Lang, E., Kohl, U., Kratzin, H. and Schirrmacher, V. (1988)
Murine Fcγ receptor proteins: identification of a previously unrecognized molecule with a monoclonal antibody (12-15).
Eur. J. Immunol., 18, 677-683.

8. Matzku, S., Kirchgeßner, H., Schirrmacher, V. (1988)
Antibody targeting to the murine lymphoma ESb-MP: Increased accumulation due to reduced internalization by lymphoma cells as compared to normal lymphoid cells.
Int. J. Cancer 41, 108-114.

9. Kadmon, G., Kowitz, A., Altevogt, P., Schachner, M. (1990)
The neural cell adhesion molecule N-CAM enhances L1-dependent cell-cell interactions.
J. Cell Biol. 110, 193.

10. Lang, E., Schirrmacher, V., and Altevogt, P. (1988)
Molecular identification of lectin binding sites differentiating related low and high metastatic murine lymphomas.
Clin. and Exptl. Met. 6, 61-72.

11. Lang, E., Kohl, U., Schirrmacher, V., Brossmer, R., and Altevogt, P. (1987)
Structural basis for altered SBA lectin binding between a murine metastatic lymphoma and an adhesive low malignant variant.
Exptl. Cell Res., 173, 232-243,

12. Saga, Y., Tung, J.-S., Shen, F.-W., and Boyse, E.A. (1986)
Sequences of Ly-5 cDNA: Isoform-related diversity of Ly-5 mRNA.
Proc. Natl. Acad. Sci. USA, 83, 6940-6944.

13. Lefrancois, L., Thomas, M.L., Bevan, M.J., Trowbridge, I.S. (1986)
Different classes of T lympohcytes have different mRNAs for the leukocyte-common antigen, T200
In: J. Exp. Med., 163, 1337-1342.

SCHIRRMACHER

Biomodulation: An integrated approach to access and manipulate biological information

P. L. Mann* **, D. Eshima*, D. M. Bitner*, R. H. Griffey**, R. Wenk**, J. L. Born*, and N. A. Matwiyoff** ***

* College of Pharmacy
** Center for Non-Invasive Diagnosis
*** Cell Biology
University of New Mexico
B-64e Radiopharmacy
Albuquerque New Mexico USA 87131

The concept of altering host/tumor responsiveness (surveillance) as a specific tool in the management of neoplastic disease has been the subject of both basic science and clinical studies for many years. One of the first practical applications of this approach was described by Von Leyden and Blumenthal in 1902. In this study autologous tumor cells were injected into a patient as part of an intervention protocol. In 1908 Paul Ehrlich predicted the existence of a host immuno-dependent protective mechanism. Burnet formalized this concept in 1970. The first element of Burnet's proposal is the existence of unstable cellular clones with neoplastic potential. These clones are a normal part of mammalian existence and are important in the maintenance of surveillance mechanisms. The second element is the proposal that the host immune system is designed to hold these clones in stasis. These two inter-related elements laid the conceptual groundwork for the theory that the "induction" of the immune system could be beneficial in cases where the normal stasis mechanisms have failed and a neoplasm has expressed. Thus, from its inception these two elements were given equal weight. The first is termed by us immuno-regulatory, with the implication of recognition and regulation [both proliferative and differentiative] components. The second is termed immuno-defensive, with its implication of effector functions involved in lytic actions. In recent years these separate concepts have been apparently collapsed into the search for enhanced specificity in immuno-defensive functions. Thus the modern day applications of Burnet's concept have evolved and the field of *BIOLOGICAL RESPONSE MODIFIERS* [BRMs] was created. This emphasis on the immuno-defensive aspect of the concept limits the original broadly based idea of regulation and defense. The simple defense-based BRM approach fails to recognize the dynamic regulatory nature of the immune system as well as the integration of the immune system within the overall framework of the functional entity as a whole.

These BRMs have many pseudonyms [most of them implying that the primary target is the immune system], but in essence the definition by Chirigos (1984) covers the sense of the field: "agents which can modify a host's cellular immune response to tumor cells with resultant therapeutic benefit". Inherent in this definition and much of the work on BRMs is that there is a direct linear relationship between our ability to modify the host immune response and the tumor's development. Again, this places the emphasis on the immuno-defensive aspects and de-emphasizes the immuno-regulatory functions as proposed by Burnet. A more subtle feature of this definition is the assumption that it is practical to "modify" these immuno-defensive responses directly. By definition many of these responses are at or near the terminal end of the differentiation spectrum. It would seem less likely that an effective direct access route to alter the phenotypically specific defensive or lytic component of the immune system would be successful than would be an approach based on modulation of the broadly based immuno-regulatory function(s). This is one significant area where the current work, which will be described below, differs from the standard BRM approach. The immuno-regulatory aspect of the original Burnet's hypothesis appears to be more amenable to modulation and is the focus of our approach. This immuno-regulatory system is a continuously connected [from simple/primitive to complex/specific] information management system. The control over the very primitive regulatory stage is hypothetically under the control of very simple molecules termed by us as *BIOMODULATORS*. The different term is employed to draw attention to the purported difference in access point along the continuum of the cellular differentiation pathway. We believe that the BRMs described in the literature are examples of *BIOMODULATORS* which have a highly defined, specific access point. Thus there is continuity from the most primitive to the most specific elements of control. The control system is capable of self-adjustment. The advantage of incorporating the immuno-regulatory aspects of the theory into the approach is that the purported primitive [in terms of differentiation] access point provides both a broader diversity of accessible responses and offers the biological system an opportunity to adjust [reduce toxicity] to the perturbation.

Although it is not the intent here to review the extensive BRM literature, there are several common threads running through this work which may help to place the *BIOMODULATION* studies into a historical perspective. The evolution of BRMs began in ernest after the pioneering work of Georges Mathe (1969) with BCG. BCG as a non-specific immunotherapy for tumors has been the subject of hundreds of studies. It activates macrophage-dependent processes especially the cytostatic effects. It also potentiates both humoral and cellular responses and induces suppressor T-cell populations in a time and dose-dependent manner.

C parvum is the other notable mycobacterium bovis BRM with demonstrated reticulostimulatory activity. Its anti-tumor activity and general BRM properties are discussed extensively by Tuttle and Cantrell (1981). A differential effect has been observed between the polysaccharide fraction with its reticulostimulatory activity and the peptidoglycan fragment with its local, anti-tumor activity. The implications of this work regarding the specificity of sub-components of the stimulant are very important. It also shows the changing trend in the BRM studies towards control over specificity of BRM producer and target.

Thymosin, as a complex mixture of natural product BRMs prepared from thymic tissue as well as a number of characterized sub-components [thymosin alpha1, beta4, and thymulin], has been shown to have a variety of biological consequences. These have been reviewed by Low and Goldstein (1984). There are specific effects of these materials on different stages of the T-cell differentiation.This is additional evidence for the concept that it is possible to interact with cellular differentiation at phenotypically specific levels.

The BRMs described above fall generally under the category of ill-defined natural products, although characterized sub-components have been identified in some cases. The next series of BRM's are also natural products but have been characterized, purified, and/or synthesized. Tuftsin [Thr-Lys-Pro-Arg] is a chemotactic/activator of both macrophage and the granulocytic differentiation series. Its anti-tumor activity has been shown in a variety of models [Catane et al, 1981].

Another class of immunoadjuvant molecules [synthetic BRMs] is represented by the polycarboxylates. One of the most studied examples is maleic anhydride divinyl ether co-polymer [MVE-2]. This material appears to possess anti-viral, immuno-adjuvant activities towards natural killers, T and B-lineage cells. Its toxicity is related to molecular weight and perhaps its high affinity for divalent cations [see Baird and Kaplan, 1980].

TABLE 1 BIOLOGICAL RESPONSE MODIFIERS

GENERAL CATEGORY	*EXAMPLES*	*TARGET TISSUE*	*COMMENTS*
NATURAL PRODUCTS	THYMOSIN	T-DEPENDENT IMMUNITY	BROAD SPECTRUM INITIATION
DEFINED NATURAL PRODUCTS	TUFTSIN	MACROPHAGE, GRANULOCYTE, HI, CMI	SELF-REGULATO RY
SYNTHETICS	MVE-2	T, B, NK, HI, CMI	TOXICITY
ACTIVATION PRODUCTS	IFNs	T, B, NK, BONE MARROW, MACROPHAGE	INTEGRATED CIRCUIT

The fourth and final class of BRMs consist of cellular activation products. Overall, these products are the result of cellular proliferation. The interferons are the most extensively characterized representatives of this group. Isaacs and Lindenmann first described these materials with respect to their anti-viral activity in 1957. More recent reviews are beginning to demonstrate a much broader diversity of IFN-induced activities [see Vilcek and De Maeyer, 1984 and Hooks and Detrick, 1985]. As a group these complex natural products have specificity both at the level of the cellular producer and target. Alpha IFN used in single dose regimens has been used successfully in hairy cell leukemias, and to a lesser extent in early CML. The beta IFN is similar to alpha and has shown some utility in brain, while the immune form [gamma IFN] has a number of unique features and has shown utility in renal cell carcinoma, and malignant melanoma [see Balmer, 1990]. Again, the trend towards specificity control *via* both effector molecule and target designation is clear.

Table 1 shows a simple tabular representation of the evolution of BRMs. It appears that the BRM field has evolved from its natural product [BCG] beginnings in two related directions. First, the BRMs themselves appear to be moving towards the development of a synthetic "drug". The underlying assumption is that control over specificity lies in the development of a unique structural entity. The desired effect of the BRMs seems to have evolved also. Originally, a generalized action on the host immune system was the goal of the BRM treatment. It appears now that absolute control over specificity of reaction is the desired outcome. The assumption is that the observed toxicity and lack of generic applicability of the BRM approach is the result of our inability to discover or design the appropriate BRM. The assumption is that these highly specific BRMs will interact with the biological system in such a specific manner so as to affect the tumour destruction without interfering with any normal cellular mechanisms. This is not, as yet, supported by the experimental data on BRMs.

The *BIOMODULATOR* [BM] approach proposes the hypothesis that control over cellular differentiation is a continuous process, connected in some very complex way from the simplest [primitive] to the most sophisticated [specific]. The general assumptions under which BRMs were developed still apply to this hypothesis, the exception is the point of access for the cellular differentiation mechanism itself. The traditional BRM approach tacitly assumes that the access point is close to or at the phenotypically specific lymphocyte. In contrast our hypothesis posits that the access point should be the most primitive level of differentiation possible. This hypothesis assumes that; BMs are non-cell-lineage specific; BMs are differentiative in nature; BMs interact at a primitive phenotypic level; and there is some level of connectivity between this "primitive" differentiative control and cell specific control.

TABLE 2 BIOMODULATOR CLASSES

GENERAL CATEGORY	*EXAMPLES*	*STRUCTURE*
NATURAL PRODUCTS	PWM	25 - 35 KDa GLYCOPROTEIN
DEFINED NATURAL PRODUCTS	Sw	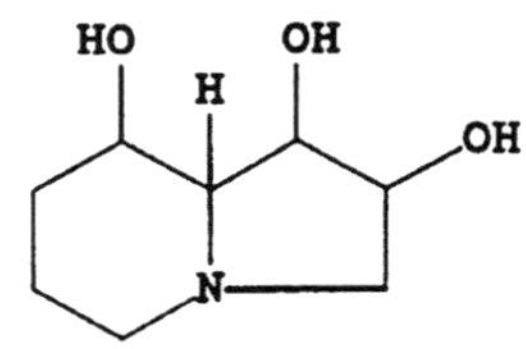
SYNTHETICS	HHD	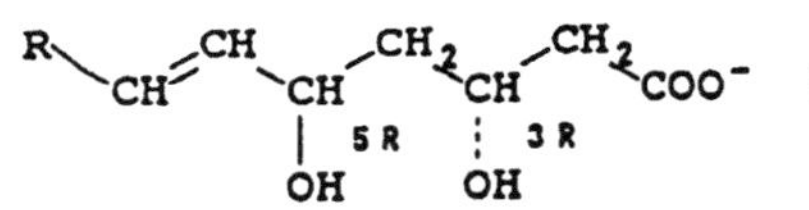

Table 2 shows some of the pertinent characteristics of the BMs which have been used in these studies. The large molecular weight glycopeptide BMs are represented by PokeWeed Mitogen [PWM] which is a mixture of five iso-mitogens, and have had a long history of scientific investigation [Qintans and Lefkovits 1975; Gmelig-Meyling, UytdeHaag and Ballieux 1977; Janossy et al 1977; and Mann 1983]. The emphasis in the literature has been on PWM's proliferative capacity, especially as it relates to leukocyte stimulation, with relatively few references to true cellular differentiation. There is an element of phenotypic expression which is associated with the proliferating leukocyte, but this aspect of differentiation appears to be resistant to the modulation. The non-proliferative-dependent modulation of cellular differentiation is central to the definition of BMs. Our studies have shown that cellular proliferation and differentiation can be separated [Mann 1983]. The BM property of PWM is a function of the differentiation component of the molecule. If both activities are present the proliferative capacity is dominant and inhibitory to the differentiative capacity. In most of the experimental data presented herein we have used what we refer to as a "crude" preparation of PWM. This is primarily differentiative in nature and therefore substantially different from most commercially available PWM preparations, which content large amounts of the proliferative agents. In separate experiments we have found that the Pa-1 and Pa-5 IEF fractions of PWM are differentiative and therefore can be used in these experiments. The expression of the differentiative capacity is one of the central definitions of BMs.

The small molecular weight natural product BMs are represented by Swainsonine [Elbein AD 1984; Bowlin et al 1989; Dennis et al 1990], an indolizidine alkaloid with a known structure and at least one known biological target [alpha-mannosidase II]. Another, quantitatively more potent BM, is represented by Cellular Activator and Differentiator [CAD] which is a small molecular weight, non-protein, BM extracted from a common soil fungus, *P. restrictum.* This molecule [or rather a mixture of 5 to 7 natural products] has been extensively characterized in terms of its biological activity but has been refractory to detailed structural characterization. It has exclusively differentiative activity in our models, and we believe it is related structurally to Colletruncoic Acid [CA], a natural product from another soil fungus (Stoessl and Stothers 1986). CA is a unique natural product, 2-hydroxymethylhexa-2,4-dienol, which has an identical side-chain component to the Hydroxy-Heptenoate Derivatives [HHDs] discussed below. Both the CA and HHD [and hypothetically the CAD] possess chiral hydroxyls in the C-3 and C-5 positions. There are two remaining questions with the CA/CAD molecules, aside from the final demonstration of the relatedness of their structures, and these are; 1] the CA as described is an unstable methyl ester, whereas the CAD is a very stable carboxylic acid [we hypothesize that the CA and CAD exist as free acids in nature]; and 2] little is known about the impact of the two chiral centers at C-3 and C-5 on the differentiative functionality of the BMs.

The third major class of BMs is represented by synthetic Hydroxy-Heptenoate Derivative [HHD] compounds [Chen et al 1987; Wareing JR, Damon RE 1988; Wareing JR 1986] . These synthetics have been characterized as BMs for two specific reasons; 1] the use of a synthetic compound is assurance that the BM component of the natural products is not some unknown sub-component of the molecule or mixture which we are studying; and 2] the question of the chirality of the C-3/C-5 hydroxyls can be addressed in detail. Experimental data indicates that the 3R/5R stereochemistry possesses the majority of the BM activity. This finding in itself confirms the element of the BM hypothesis which contends that the original BMs had to be structurally simple and therefore able to express control through chemically specific means. The functional relationship between these classes of BM is established by the use of specific model systems described below.

TABLE 3 MODELS FOR THE BIOMODULATOR CHARACTERIZATION

MODEL	*IN VIVO/VITRO*	*CHARACTERISTIC*
IMR-90 [PH I & II]	**IN VITRO**	**NORMAL DIFFERENTIATION**
IMR-90 [PH III]	**IN VITRO**	**TERMINAL DIFFERENTIATION**
LEUKOCYTES	**IN VITRO/ IN VIVO**	**DIFFERENTIATION CONTROL - HI & CMI**
TUMOR LINES	**IN VITRO**	**ABERRANT DIFFERENTIATION**
IMPLANTED AND SPONTANEOUS TUMORS	**IN VIVO**	**INTEGRATION**

In order to investigate the BM approach and its assumptions we must broaden our experimental view to include: 1] models of normal differentiation; 2] models where normal control is suspended or superceded in response to a non-aberrant biological pressure [terminal differentiation]; 3] models where essential sub-components of specific control can be studied in a sequential fashion [immune differentiation]; 4] models of aberrant growth control/differentiation [neoplastic transformation]; and finally 5] models where all of the crucial components of #1-#4 can be assessed as a functional unit. Table 3 shows some of the pertinent characteristics of these model systems.

MODEL 1 NORMAL CELLULAR DIFFERENTIATION

The IMR-90 normal fetal lung fibroblast cell line is used as a representative of normal cell growth/differentiation [Mann, Lopez-Colberg and Kelley 1987]. This cell line has stable growth/differentiation characteristics when cultured between 30% and 80% surface confluence, and between Population Doubling Level [PDL] 13 and 40. Above and below these surface confluence conditions the line shows growth/differentiation changes in phenotype analogous to low and high density growth inhibition. At approximately PDL 40 the line undergoes specific phenotypic changes in growth/differentiation pattern [cellular senescence], which we refer to as terminal differentiation [Bell et al 1978]. Thus this model is useful for studying the effects of BMs on normal growth/differentiation control, and several "normal" phenotypic alterations of this control.

In order to study the purported BM effect on differentiation we also developed assays which could be used to investigate structural/conformation changes of oligosaccharides on the cellular surface. The hypothesis is that all information must transit the cell membrane and therefore its ability to interact with the environment should have some predictive indices regarding subsequent differentiative behavior. In several studies [Mann, Lopez-Colberg and Kelley 1987; Mann, Swartz and Kelley 1987; Mann, Swartz and Holmes 1988a] we showed that cell surface oligosaccharides were modulated as a precursor to cellular phenotypic change. In our studies we showed that these oligosaccharide changes were arrangement-dependent. In other studies the cell surface oligosaccharide alterations were attributed to changes in specific epitopes [Aizawa, Mitsui and Kurimoto 1980], related to insertion and/or production anomalies. However, in the present studies, using a lectin-based competition assay which provides data for Scatchard analyses, and in turn affinity classes [K_{ij}] and binding capacities [R_{ij}] for 0 to 5 classes [DeLean, Munson and Rodbard 1978] we found that conformational re-arrangements could explain the data. Given the number of glycopeptides and their glycosylation diversity, the original hypothesis was that these analyses would yield a standard "noise" pattern, suggesting the absence of any coherent supra-organizational structure. However, a very specific pattern of one, or two [in some rare cases a three] binding site classification per sugar specificity was observed. This accounted for >85% of the total specific lectin binding on the IMR-90 model. These specific patterns have been interpreted [Mann , Swartz and Holmes 1988a and b; Mann 1988] as evidence for a supra-organization of carbohydrate-dependent moieties on the distal cellular surface. Thus the developing hypothesis is that this conformationally sensitive oligosaccharide coat acts as a filter for information from the environment. It controls the flow and/or efficiency of information interpretation. A simple initial mechanism for this control suggests that the oligosaccharides form a three-dimensional structure at the cell surface where transference of information is controlled by energy-dependent interactions [ligand/receptor binding]. Changes in these patterns have been related in a predictive manner to phenotypic cellular changes [Mann, Swartz and Kelley 1987], thus establishing a functional linkage for the analytical tool. The lectin-dependent affinity-based class changes are dependent on specific sugar moieties [nominally mannose, galactose, and glucosamine] and therefore are complex. Using the Gibb's Free Energy [GFE] calculation with all its inherent assumptions we have simplified the individual carbohydrate binding data by calculating aggregate binding values in KCals for all the individual binding events. The GFE data for the *in vitro* cell models represents a measure of the overall "organization" of the interactive surface. A principle component of this surface interaction is the specific arrangements of oligosaccharides. In other studies, discussed below, this concept of a modulatable, organized energy-dependent barrier can be related to the strongly temperature-dependent proton NMR linewidth measurements of bound water at the cell surface. These types of integrated studies permit the testing of hypotheses from several different directions. The cell surface oligosaccharide display of IMR-90s changes specifically at PDL 38-39 before any phenotypic changes have occurred [Mann, Swartz and Holmes 1987b]. A "down-regulation" of the mannose-dependent oligosaccharides is observed. No change in the total number of binding sites is seen but a significant change in the total GFE results from the loss or conversion of a high affinity mannose-dependent class. Thus the decrease in GFE is a prelude to the phenotypic changes. This phenotype change is the definition of terminal differentiation in this model. Similar changes in individual carbohydrate characteristics and GFE are observed with both low and high density growth inhibition, suggesting that the carbohydrate "regulation" is a generic differentiation-dependent phenomenon and not simply related to a terminal event. This finding is important because it places the structural events squarely in the midst of normal regulatory mechanisms.

In another series of experiments [Busse, Mann and Griffey, 1989] the proton NMR linewidth of cell surface water was measured at 300.1 MHz. These linewidth data can be related to the mobility and fraction of the water at the "bound" cellular surface. In simple terms this experimental parameter measures the association/dissociation of water with the cellular surface. We further infer from this data that there is a relationship between proton linewidth measurement and the physical/structural status of the cell surface. These measurements are related to entropy [$T\Delta S$] and therefore complement the GFE studies [$\Delta G^o = \Delta H - T\Delta S$] and provide new insights into the biochemical/physiological effects of the BM treatment. These measurements coincide with the oligosaccharide "regulation" and GFE measurements discussed above. Serial assessments of both surface oligosaccharide status and linewidth as a function of PDL yielded a direct correlation. Tert-butanol treatment of the cell surfaces was used as a control as it randomized both sets of data. Thus there appears to be a link developing between phenotype, biochemical assessment, and physical characteristics all centered around the oligosaccharide display status of the cellular surface.

MODEL II TERMINAL CELLULAR DIFFERENTIATION

Figure 1 shows data from a series of experiments designed to test the hypothesis that the BMs modulate the normal expression of IMR-90 differentiation. The experimental parameter in this case is the aggregated GFE for all oligosaccharide contributions to the cell surface [Mann, Swartz and Holmes, 1987b]. The first set of bars represents Phase II [normal growth] for the IMR-90 cells. In this case the GFE values are maximal and strongly dependent on culture conditions. This is evidenced by the last two sets of bars in Figure 1 where both high and low cell density culture conditions were tested. The GFE is reduced under these culture conditions for the Phase II cells. IMR-90 cells were treated with BMs [in this particular case PWM at 10 ug/passage] beginning at PDL32. This PDL represents a population doubling level which is before any phenotypic changes related to terminal differentiation were observed. Control cultures without the BM terminally differentiate at or about PDL 44, at this point there is a 64% reduction in the GFE ["down-regulation"] whereas the BM treated cultures terminally differentiate at approximately PDL 100. There are specific but minor changes in the individual oligosaccharide display in the cells from PDL 60 to PDL 100 but the control PDL45 display does not appear until PDL 100 in the treated cultures. Other experimental parameters include; proton NMR linewidth measurements, generation time [T_g], GFE, and surface complex [lectin/ligand] mobility. These remain normal [Phase II] with the BM treatment until PDL 100. The data in Figure 1 on the effect of high and low density culture conditions demonstrate the generic nature of the display assay. The BM treatment of low density growth inhibited cultures partially "up-regulated" their surface oligosaccharide displays, and increased their growth rates. The treatment of high density inhibited IMR-90s had little or no effect.

FIGURE 1 THE EFFECT OF BM TREATMENT ON NORMAL DIFFERENTIATION

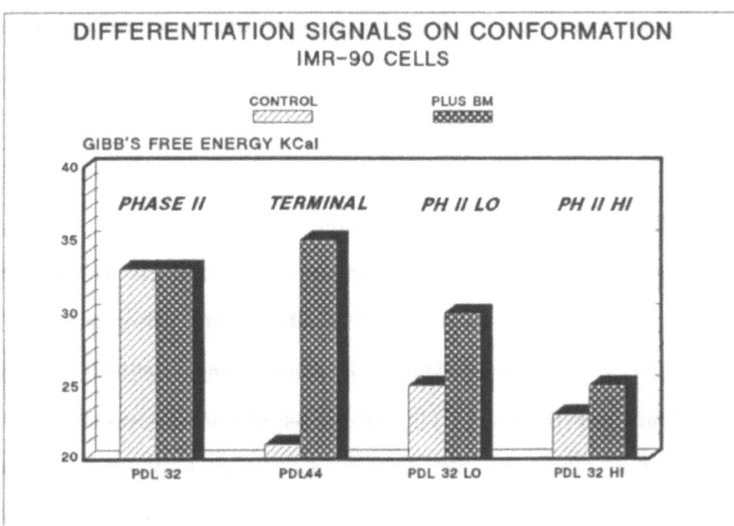

THIS COMPILATION OF DATA REPRESENTS NORMAL GROWTH CONTROL [PHASE II]; TERMINAL DIFFERENTIATION [TERMINAL] AS WELL AS PHASE II CELLS CULTURED UNDER EITHER LOW OR HIGH DENSITY CULTURE CONDITIONS, AND THE EFFECT OF BMs ON THESE PARAMETERS

The representatives of all classes of BM have been tested and in fact one of the criteria for a BM is the ability to delay terminal differentiation in a dose-dependent manner correlated to oligosaccharide and physical standards. There is no discernable effect of the BM treatment on Phase II IMR-90 cells with the possible exception of a slight decrease in the overall T_g; thus the BM effect is apparently selective for the events which ultimately progress into terminal [?inappropriate?] differentiation. The clear causal potential is shown by the fact that treatment must begin before phenotype is expressed, and that there is a clear relationship to the cell surface oligosaccharide display. A major question which is being addressed currently is whether the PDL 100 cell at terminal differentiation is the same as the PDL 45 cell at terminal differentiation. The hypothesis contends that the PDL 45 control event was the result of reversible events under differentiation control whereas the PDL 100 cell is the result of irreversible events. There is only one terminal differentiation.

MODEL III IMMUNE DIFFERENTIATION

FIGURE 2 THE BM EFFECT ON HUMORAL IMMUNITY

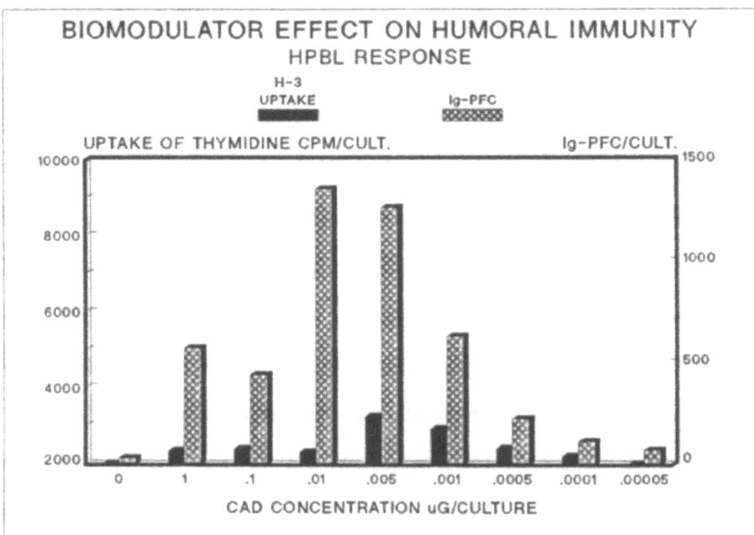

THE EFFECT OF CAD ON THE PROLIFERATIVE AND DIFFERENTIATION: HPBLs WERE TREATED WITH VARIOUS CONCENTRATIONS OF CAD AND THE INCORPORATION OF H-3 THYMIDINE WAS ASSESSED

Figure 2 shows data from a series of experiments designed to investigate the effects of BMs on lymphoid differentiation patterns. In these experiments both human peripheral blood leukocytes [HPBLs] and murine splenocytes were tested *in vitro* for the response to BMs. The plaque assay was used so that specificity, class and sub-class of individual clones could be determined. Figure 2 shows a five log dose-dependency response curve for HPBLs with CAD. The relative lack of any proliferative response can be seen. In general very little proliferative response is noted with CAD [especially when the 200,000 to 300,000 cpm values of the typical mitogen response are considered]. The optimal response is approximately 5 to 10 ng, and the optimal kinetics is approximately 72 to 96 hours after stimulation. The differentiation parameter in this case is non-antigen specific immunoglobulin [Ig-PFC] synthesis and secretion. The shape of the dose-response curve is typical of the BMs and unusual in that there is not the traditional broad optimal response plateau. This bell-shaped curve suggests a concentration-dependent control mechanism. Another unique feature of the BM effect is that, in general, the optimal doses determined by any of the *in vitro* models are directly transferable into the *in vivo* models. A 10 ug dose of BM which is optimal at the tissue culture level is also optimal in the animal models. This has the advantage that the *in vitro* models can be used to predict activity at the *in vivo* model level.

Figure 3 shows data from experiments designed to investigate the impact of the BM effect on antibody specificity. Traditional thought considers that antigen [in this case sheep red blood cells, SRBCs] is required in the *de novo* immune response to initiate the differentiation response. When HPBLs are exposed to SRBC for 96 hours a small primary [IgM-PFC] but no secondary [IgG-PFC] immune response is noted. When a BM [in this case CAD at 10 ng] is added to the culture in the absence of antigen for 24 hours followed by an antigen exposure of 72 hours an intense antigen specific IgG-PFC response is observed. Thus it appears that the BM has some control over specificity and class/sub-class switching. Similar effects have been seen with *in vivo* administration of the BMs in the murine model. There is a mild splenomeglia and a marked increase in IgG-PFCs specific to the injected antigen. The optimal kinetics is a 10 day to 21 day pre-treatment with BM [10 ng], ip on Monday,

Wednesday, and Friday, followed by antigen [small doses] 3 to 5 days before sacrifice.

FIGURE 3 THE BM EFFECT ON ANTIBODY SPECIFICITY

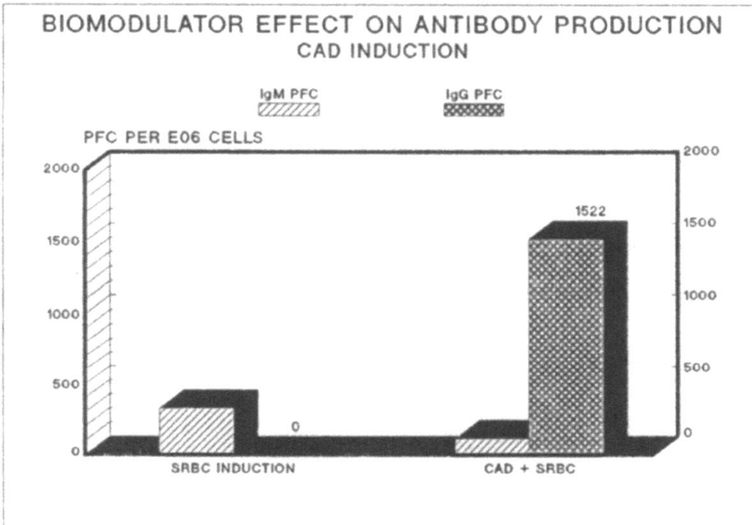

THE EFFECT OF BMS ON ANTIBODY SPECIFICITY: HPBLs, CONTROL AND TREATED WITH BM WERE EXPOSED TO ANTIGEN AND SUBSEQUENTLY ASSAYED FOR SPECIFIC ANTIBODY

Particulate, protein and haptenic antigens have been tested successfully. The spleens from these animals have been used for hybridization protocols and therefore the ability of the BMs to affect lymphoid differentiation has been immortalized. These clones will enable us to ascertain whether the BM treated clones produce higher affinity product than do their untreated counterparts.

The pre-dominant target for the BM effect on lymphoid differentiation appears to be the T-cell lineage, as anti-B-cell reagents have little effect on the outcome of the reaction until very late in the response; whereas anti-T-cell reagents are inhibitory at all stages of the response. This T-cell-dependence is also suggested by the BM data on Cell Mediated Immunity [CMI]. BM treatment affects both cytotoxic T cell [CTL] and natural killer [NK] phenotypic expression. Values of 4 to 10 fold increases in cytotoxicity were observed with treatment. The BM treatment changes the target oligosaccharide display. This is related to marked increases in effector/target conjugation potential.

Thus BMs have a demonstrated non-cell-lineage specific stimulation, in that they have similar differentiative effects on both fibroblasts and lymphoid cells, at similar concentrations and conditions. They appear to affect lymphoid differentiation at a very primitive control level effecting expression, specificity and the molecular properties of the product. The BMs use *in vitro* appears directly translatable to the *in vivo* situation with similar results.

MODEL IV NEOPLASTIC TRANSFORMATION

FIGURE 4 THE BM EFFECT ON TRANSFORMED CELL SURFACES

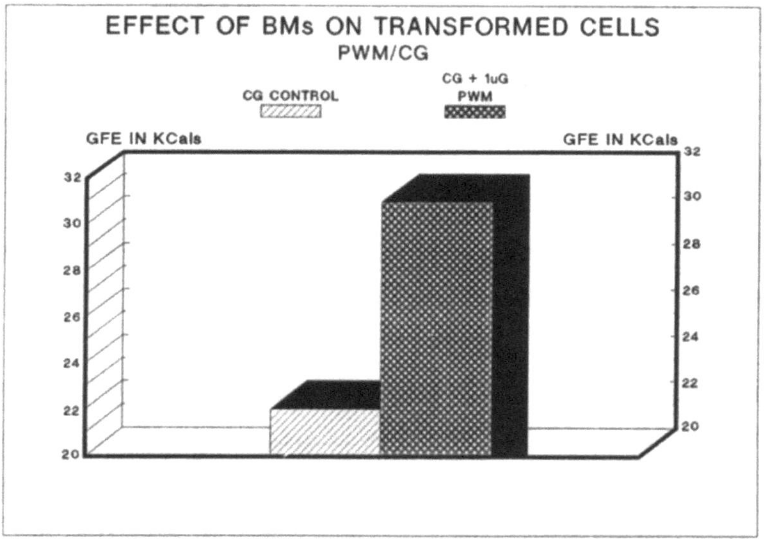

THE EFFECT OF BM TREATMENT ON THE REGULATION OF CG CELL SURFACES: CG CELLS WERE CULTURED IN THE PRESENCE AND ABSENCE OF BM AND THE GFE CALCULATED AS DESCRIBED

The canine glioma [CG] cell line developed by Salcmann is the principle model used in the study of the BMs on neoplastic cell surfaces [Mann, Swartz and Holmes, 1988b]. Additional work has been done on a number of human transformed cell lines as well as the B-16 melanoma [both parental and lung specific sub-clones]. The physical and cultural characteristics of the CGs are similar to the IMR-90s.

In this experiment the cell surface oligosaccharide display was characterized as described above. Again, there were cell-specific alterations in the sugar-dependent profiles [Mann, 1988], but for simplicity Figure 4 shows the aggregated GFE values for these analyses. The control CGs have a PDL-independent GFE value of 22 KCals, which is similar to the terminally differentiated IMR-90. The CGs show no "regulation" of their oligosaccharide displays in response to high or low cell density, culture conditions or PDL and are therefore referred to as "un-regulated". Treatment of the CGs with BM over a 10 to 21 day period induced a stable increase in the GFE. This change in GFE was accompanied by phenotypic changes in T_g, morphology, and the cells' ability to participate in immune reactions.

The BM treated CGs were 4 to 8 fold more susceptible to lysis in standard natural killer chromium release assays than their untreated controls. Biopsy samples from implanted tumor showed BM-dependent changes in their characteristics which mirrored the *in vitro* data.

FIGURE 5 THE BM EFFECT ON NMR PARAMETERS OF CGs

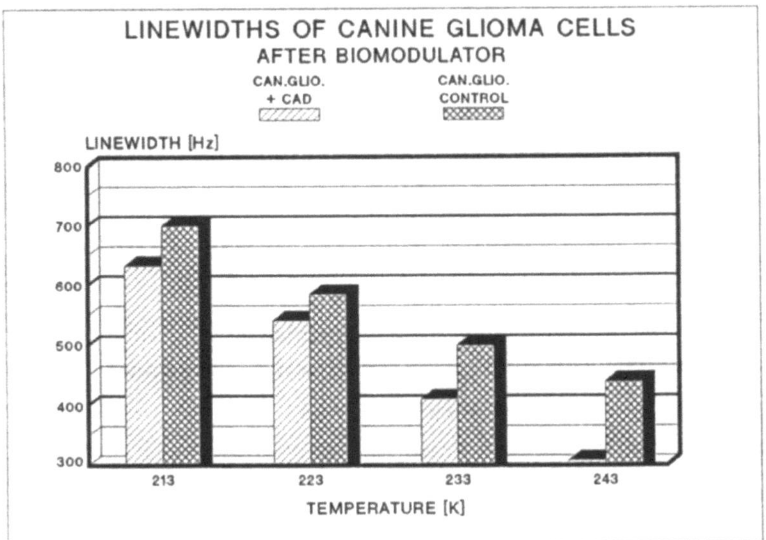

THE EFFECT OF BM ON NMR LINEWIDTH: CG CELLS WERE TREATED WITH BM AND PROTON NMR LINEWIDTH MEASUREMENTS PERFORMED AS DESCRIBED

Figure 5 shows the proton linewidth data from experiments on CGs. In these experiments the BM [CAD in this case at 10 ng] induced a significant decrease in the measured linewidths. Like the data presented for the normal cellular model [IMR-90], these data are interpreted from the perspective of the measurement of dis-organization [the strongly temperature-dependent T ΔS parameter], and integrated with the GFE data. Unlike the experimental parameters described above these line-width changes occurred with as short as a 20 minute to 2 hour BM exposure. Thus it appears as if the water-dependent changes are very rapid and followed by the phenotypic and oligosaccharide changes. These are considered as part of the immuno-regulatory facet of the response. Finally these oligosaccharide and NMR organizational parameters correlate with functional changes in the host's immune system.

FIGURE 6 THE BM EFFECT ON CELL MEDIATED IMMUNITY

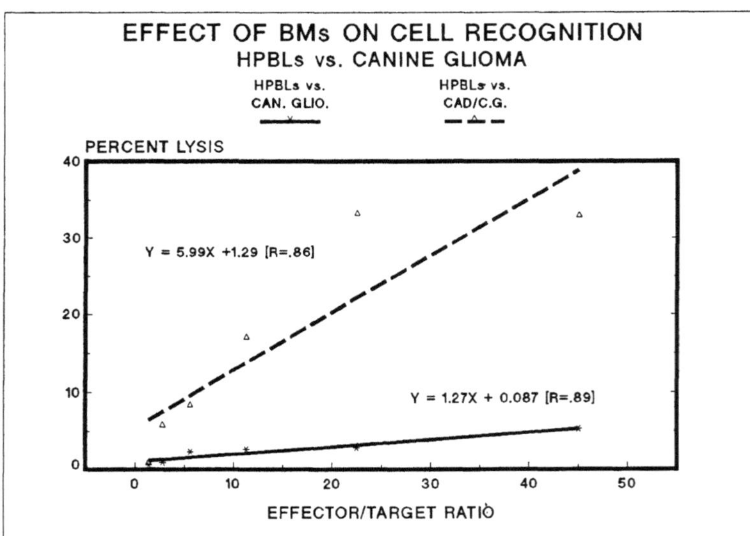

THE EFFECT OF BMs ON CMI: THE HPBLs AS NATURAL KILLER EFFECTOR CELLS WERE TESTED FOR ACTIVITY AGAINST CONTROL AND BM TREATED CGs

Figure 6 shows data on the BM effect on NK activity of HPBLs against chromium labelled control and BM treated [CAD in this case, 10 ng] CGs. The slope estimate from the linear regression of the six pairs of effector/target ratios provides an indication of the efficiency of kill. Under the BM treatment the CGs are markedly better targets for NK activity. Similar results have been observed with CTL experiments following both *in vitro* and *in vivo* treatment of either host or effectors.

Thus in summary the BMs effect the physical [NMR] and biological [phenotypic and oligosaccharide] characteristics of CGs *in vitro*. The B-16 parental line is more refractory to BM treatment, but the F_4 lung-specific clones show specific galactose anomalies in their cell surface oligosaccharide display, and are very sensitive to the BM treatment. This may suggest that their is a spectrum of differentiation patterns within any aberrant population and that clones of this population may have altered sensitivity to the BM treatment.

MODEL V IMPLANTED AND SPONTANEOUS NEOPLASIAS

Figure 7 shows data from a series of experiments designed to integrate all the findings of the previous models. In this case the CG tumour line was implanted into nude rats [RNU] so that approximately 25 days of growth could be documented before humanely sacrificing the animals. This model shows approximately 11% spontaneous regression rate. This spontaneous effect is easily observed as a slowing in the growth of the tumors after day 20 to 22 when exponential growth is normally observed. This does not effect the BM-induced changes, as they occur between day 9 and 11, but does suggest that their is an inherent element of recognition in this model. As shown in Figure 7 doses of 10 ug of PWM induce a significant regression of the implanted tumors. In this case the regression approaches 100% by day 27. In general the same dose-dependency is observed for these experiments as was found for the *in vitro* models. In fact, the *in vitro* models are used to screen potential BMs for efficacy and dose-dependency before *in vivo* testing. In the case of the syngeneic implantation model, B-16 into C57BL/6 hosts, the BMs have been shown to produce significant tumour regression but less frequent total regressions. We attribute this to the very rapid growth pattern of the parental bulk tumour. The more controlled B-16 model where implantation is followed by excision of the bulk tumour permits longer growth periods and is currently being assessed. In this model the bulk of the parental tumor will be removed as it enters into log phase growth thus providing a more realistic model. This model also is very useful for testing the BM effect on metastatic potential. We have observed that there are specific galactose-dependent oligosaccharide changes on the surfaces of lung-specific metastatic clones from the parental lines. These specific oligosaccharide changes are particularly sensitive to modulation by the BM treatment *in vitro*. This extension of the model will assess the BM's effect on *in vivo* metastatic potential.

FIGURE 7 THE BM EFFECT ON IMPLANTED TUMOUR GROWTH

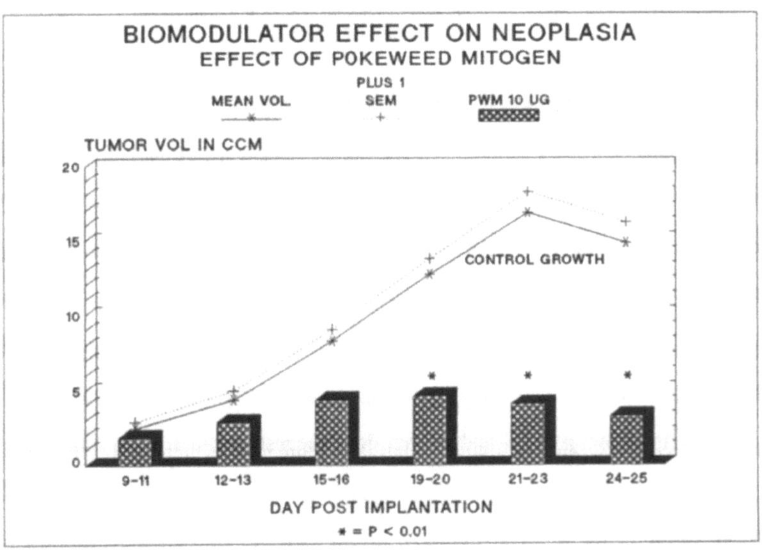

THE EFFECT OF BM ON IN VIVO TUMOUR GROWTH: CG CELLS WERE IMPLANTED INTO THE FLANK OF RNU RATS, TREATED ANIMALS WERE DOSED WITH 10 UG PWM ON M/W/F AND THE TUMOUR GROWTH MEASURED.

Extensive studies have been performed with the HHD synthetics in both *in vitro* and *in vivo* models. The 3R/5R stereochemistry is apparently crucial to the tumour regression properties. This stereo-isomer shows a dose-dependent immune stimulatory potential which is 3 logs more effective than the 3R/5S configuration.The heteronuclear structure [represented on Table 2 as R] appears to have little effect on the BM activity [two series of different Rs and all their associated stereo-isomers were tested]. In unrelated experiments it was demonstrated that the BM effect was independent of the HHDs anti-cholesterol activity.

FIGURE 8 THE BM EFFECT ON T-1 MRI

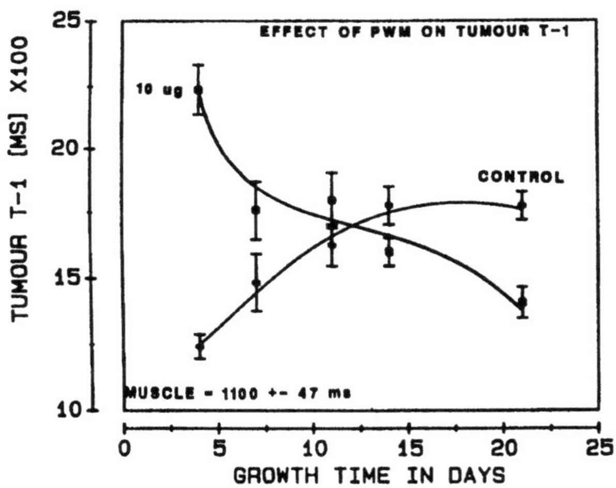

THE EFFECT OF BM ON T-1 MEASUREMENTS IN VIVO: CG CELLS WERE IMPLANTED INTO THE FLANKS OF RNU RATS, THEN T-1 MEASUREMENTS WERE MADE AT 1.5 T AND COMPARED WITH THE NON-IMPLANTED FLANK

Figure 8 shows a T-1 NMR study on the implanted CG tumour as a function of growth and BM treatment. Normal muscle [from the opposite flank of the animal] gives a T-1 value of approximately 1100 +- 47 ms. The CG tumour in the control shows little change in

T-1 until the tumour gets quite large [day 9]. Within 2 hours of the first BM treatment the T-1 of the specific tumour site goes to 2200 +- 195 ms. This T-1 change, at this point in the treatment schedule, is absolutely predictive of the eventual tumour regression by day 17-22. We believe that this T-1 change is related to the NMR water experiments performed in the other models and represents a physical ramification of host/tumor recognition. It is the physical manifestation of an alteration in the interstitial structure [perhaps oligosaccharide/ glycocalyx] of the whole tumour. This is related to the GFE estimation of ΔG and the proton NMR linewidth estimates of TΔS measurements performed on the *in vitro* models.

FIGURE 9 THE BM EFFECT ON THE ALLOGENEIC MODEL

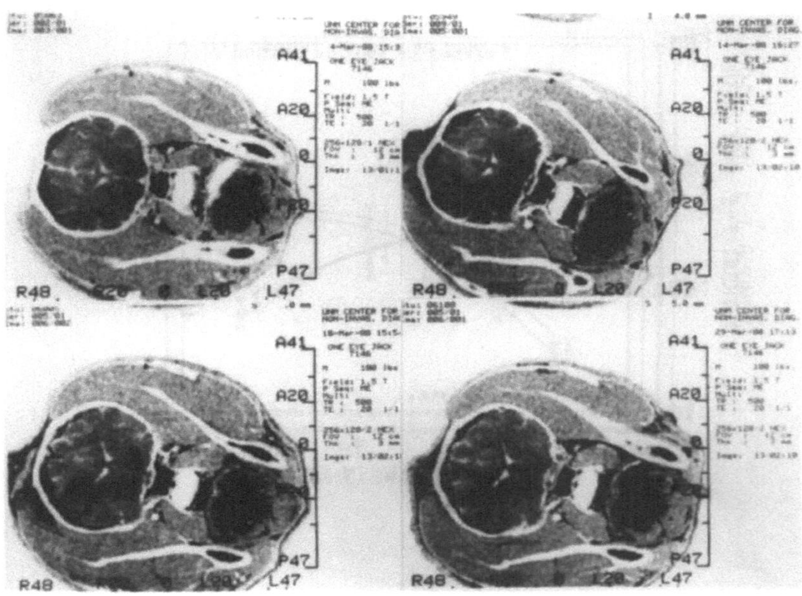

THE EFFECT OF BM ON ALLOGENEIC IMPLANTATION: CG CELLS WERE IMPLANTED INTO THE BRAINS OF OUTBRED DOGS, TREATMENT WAS INITIATED WHEN THE MRI DEMONSTRATED NEO-VASCULARIZATION OF THE IMPLANTED TUMOUR

Detailed histopathology of these tumors indicate an initial infiltration of host mast cells into the host/tumour interface [day 1 to 3]. This is followed by evidence of degranulation and massive recruitment. On day 7 to 9 there is evidence of T-cell invasion and the movement of host effector cells into the base of tumour. By day 11 to 13 there is histological evidence of a massive inflammatory response. This inflammatory response is not accompanied by the usual edema [as shown by the decrease in the T-1 values] but rather by a large infiltration of host immune cells. Histologically, the regressing tumour appears to simply "melt" away from the base with very little overt evidence of massive cell death or debris.

Figure 9 shows a four panel representation of the course of events after CG implantation into the brain of an outbred canine. The CG cells are placed surgically [sub-dural] so that approximately 20 to 25 days of growth can be observed before the control animals show the first signs of opposite-sidedness weakness, at which point they are humanely sacrificed and their brains taken for pathology. The first panel in Figure 9 shows a pre-surgery image of the dog's brain. The second panel was taken after 10 days and the first physical indication of the tumour growth is evident. The third panel was taken 4 days later and the tumour and accompanying edema is very large. A Gd-DTPA contrast study shows significant neo-vasculariza tion at this point. Treatment with the BM was initiated at this point. The fourth panel shows an image on day 25. There is little or no evidence of tumour. Of five treated animals all five showed complete tumor regression. This level of success is largely due to the development of the models discussed above. These models were used to screen and predict dosage level and frequency for optimal results. Again the same dose that was optimal in the *in vitro* models was effective in this model.

In one case the treated canines was sacrificed as soon as the MRI showed no further evidence of tumour growth. In this case the neuro-pathologist found evidence of a substantial host immune infiltration still present in the affected area. There was no evidence of tumour cells. In other cases the treated animals were left for several months after regression before sacrifice. In these cases the pathology indicated that there was the re-establishment of the normal architecture of the brain with no evidence of the tumour. The original needle track was used for orientation. In one case the treated animal is still alive 3 years after the implantation and treatment.

TABLE 4 THE BM EFFECT ON SPONTANEOUS NEOPLASIA

TUMOUR	*PROGNOSIS*	*POST BM*
MELANOMA	POOR[<6 MO]	REGRESSION AT 10 WEEKS [>9 MO]
ADENOCARC.	GUARDED [6-12 MO]	REGRESSION AT 4 WEEKS [>15 MO]
MELANOMA	GUARDED [6-12 MO]	REGRESSION AT 1 WEEK
FIBROSARCOMA	POOR [6 MO]	REGRESSION AT 10 WEEKS

Table 4 shows a summary of four canines treated with BM [in this case PWM]. These canines were diagnosed with various spontaneous tumors and treated according to the best veterinary practice. This included surgery, and the use of chemotherapy. When no other modality showed any level of success these canines were enrolled for BM treatment. The results are striking. In some cases marked improvement was observed within weeks of treatment.

In the continuing process of mechanism definition we became interested in the very early occurrence of the BM-induced aberrant tissue NMR changes. The T-1 studies clearly allowed us to "image" the changes in water in the tumour relative to the surrounding normal muscle. In essence the T-1 weighted image was a demonstration of BM-dependent imaging of aberrant tissue without the obvious use of a specific contrast agent [at least in the traditional sense].

The hypothesis which these data have generated is that the BM-induced water-dependent tissue distributional changes can be used as a unique methodology for non-specific imaging purposes. Figure 10 shows an image of an implanted B-16 tumour in a C57BL/6 mouse. The large bright spot is stomach and the smaller bright spot is tumour. This image was produced by injecting Tc-99m labelled PWM into the mouse [iv] followed one hour later by the imaging procedure [the animals were pre-treated by a 1 hour ip injection of unlabelled PWM]. As was the case with the T-1 study the radionuclide-labelled BM provided an image of the tumour mass. This appears to be a very sensitive procedure in that we have imaged very small tumors [within 5 to 7 days of implantation]. We have no explanation of the apparent stomach uptake. These data were confirmed with extensive bio-distribution studies.

FIGURE 10 BMs AS POTENTIAL IMAGING AGENTS

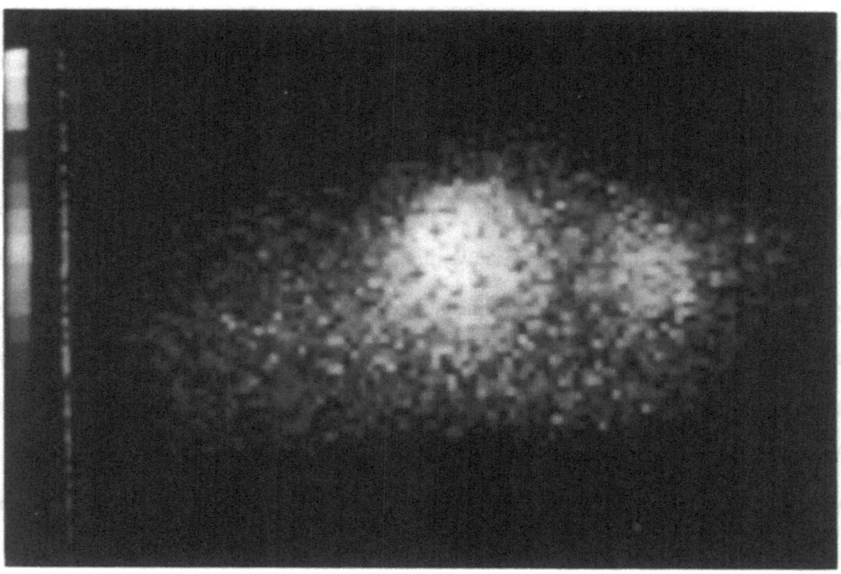

BMs AS UNIQUE IMAGING AGENTS: Tc-99M LABELLED PWM WAS INJECTED INTO A MOUSE IMPLANTED WITH B-16 TUMOUR AND IMAGED WITH A GAMMA CAMERA APPROXIMATELY ONE HOUR POST INJECTION

The image is more transient than was observed with the T-1 imaging process, suggesting that the imaged material [Tc-99m PWM] was simply delayed in its normal wash-out process. Labelled endogenous proteins, such as albumin are effective imaging agents in the presence of BM suggesting that the images are not the result of some specific tumour/agent interaction. In other studies a collaborator has evidence that BMs markedly improve the imaging potential of biologicals such as TNF. This suggests that the BM effect may be useful in the imaging area both as an imaging agent itself or as an enhancer of the specific binding properties of other imaging agents. In a continuing series of experiments it appears that the BM induces a change in the relative clearance rate within the host [DTPA studies] and this change can be used, in a transient fashion, to image the aberrant tissue. The presence of BM increases the clearance rate of DTPA by approximately 20 to 30%. BMs also alter the shape of the clearance kinetic curve, suggesting that the imaging potential of the BMs is related to the more rapid clearance of the labelled material from the normal surrounding tissue than the aberrant tissue. In turn, this hypothesis relates the GFE changes at the oligosaccharide level to the proton NMR linewidth data and to the T-1 MRI findings. It is predicted that the BM approach may be particularly useful in improving the imaging characteristics of McAbs and other biologicals which have an inherent specificity for the aberrant tissue.

In summary these data indicate the potential utility of a re-discovery of the original Burnet concept on the modulation of host as a possible approach to enhance recognition and stimulation of subsequent host defenses. We propose that the BMs interact with the biological system at a very primitive non-cell-lineage specific access point. The interaction involves a very rapid change in the TΔS-dependent dis-organization parameter of the cell surface as reflected by the "bound" water mobility. Later this change in organization is reflected by a change in the conformation of the oligosaccharide display [ΔG] of the cell surface. These changes are part of everyday normal growth control and differentiation and become aberrant in the neoplastic cell population. Treated neoplastic cells "re-adjust" their surfaces in response to the BM and as a consequence of this re-adjustment are readily recognized by the host and their existence brought under control. A physical ramification of these processes is our ability to image the purported changes in the cell surface interstitial spaces within the aberrant tissue with both MRI and radio-nuclides modalities.

REFERENCES:

Aizawa S, Mitsui Y, Kurimoto F (1980) Cell surface changes accompanying aging in human diploid fibroblasts II Two types of age related changes revealed by Concanavalin A mediated red blood cell absorption. Exp Cell Res 125:287-296

Baird LG, Kaplan AM (1980) Effects of polyanion immunomodulators on the immune system. In: Donaruma LG, Ottenbrite RM, Vogl O (eds) Anionic polymeric drugs, John Wiley, New York, p 185 - 210

Balmer CM (1990) Clinical use of biological response modifiers in cancer treatment DICP, The Annals of Pharmacotherapy 24:761-767

Bell E, Marek LF, Levinstone DS, Merill C, Sher S, Young IT, Eden M (1978) Loss of division potential in vitro: Aging or differentiation. Science 202:1158-1163

Bowlin TL, McKown BJ, Kang MS, Sunkara PS (1989) Potentiation of human lymphokine-activated killer cell activity by Swaisonine, an inhibitor of glycoprotein processing. Can Res 49:4109-4113

Burnet FM (1970) The concept of immunological surveillance. Progr Exp Tumor Res 13:1-27

Busse SC, Mann PL, Griffey RH (1989) Senescence-induced alteration in cell surface carbohydrates correlated using proton NMR spectroscopy and lectin-based affinity-binding assay. Biochimica et Biophysica Acta 984:183-187

Catane R, Schlange S, Gottlieb P, Halpern S, Treves AJ, Fuks Z, Fridkin M (1981) Toxicology and anti-tumor activity of Tuftsin in mice. Proc Am Ass Ca Res 22:371

Chen KM, Hardtmann GE, Lee GT, Linder J, Wattanasin S, Kapa PK (1987) Process for preparation of antiatherosclerotic 7-substituted-3R,5S-dihydroxy-6-heptenoates. CA108(15):131038q

Chirigos MA (1984) The introduction. In: Fenichel RL Chirigos MA (ed) Immune modulation agents and their mechanisms, vol 25 M Dekker, New York Basel, p3

Dennis JW, Koch K, Yousefi S, VanderElst I (1990) Growth inhibition of human melanoma tumourxenografts in athymic nude mice by Swainsonine. Can Res 50:1867-1872

Elbein AD (1984) Inhibitors of the biosynthesis and processing of N-linked olgosaccharides. CRC Critical Reviews in Biochemistry 16:21-49

Gmelig-Meyling F, UytdeHaag AGCM, Ballieux RE (1977) Human B-cell activation in vitro: T-cell depednet pokeweed mitogen induced differentiation of blood B-lymphocytes. Cell Immunol 33:156-169

Hooks JJ, Detrick B (1985) Immunoregulatory functions of interferon. In: Biological response modifiers Torrence PF (ed) Acad Press, New York p57 - 75

Isaacs A, Lindenmann J (1957) Virus interference I The interferons. Proc R Soc London Ser B 147:258-267

Janossy G, Gomez de la Concha E, Snajgr MJ, Waxdal MJ, Platts-Mills TAE (1977) T-cell regulation of immunoglobulin synthesis and proliferation in pokeweed (Pa-1) stimulated human lymphocyte cultures Scand J Immunol 6:109-123

Low YLK, Goldstein AL (1984) Thymosin, peptide moieties and related agents. In: Fenichel RL Chirigos MA (ed) Immune modulation agents and their mechanisms, vol 25 M Dekker, New York Basel, p 135-162

Mann P, (1983) In vitro differentiation of human peripheral blood leukocytes: Considerations for human monoclonal antibodies In: Burchiel SW, Rhodes BA (eds) Radioimmunoimaging and radioimmunotherapy, Elsevier Sci Publ, New York, p 121-141

Mann PL, Lopez-Colberg I, Kelley RO (1987) Cell surface oligosaccharide modulation during differentiation I Modulation of lectin binding. Mech Ageing Devel 38:207-217

Mann PL, Swartz CM, Kelley RO (1987) Cell surface oligosaccharide modulation during differentiation II Membrane mobility of oligosaccharide lectin conjugates. Mech Ageing Devel 38:219-230

Mann PL, Swartz CM, Holmes DT (1988a) Cell surface oligosaccharide modulation during differentiation III Lectin affinity class distributions. Mech Ageing Devel 44:1-16

Mann PL, Swartz CM, Holmes DT (1988b) Cell surface oligosaccharide modulation during differentiation IV Normal and transformed cell growth control. Mech Ageing Devel 44:17-33

Mann PL (1988) Membrane oligosaccharides: Structure and function during differentiation. Inter Rev Cyto 112:67-95

Mathe G, Amiel JL, Schwartzenberg L, Schneider M, Cattan A, Schlumberger JR, Hayat M, de Vassal F (1969) Active immunotherapy for acute lymphoblastic leukaemia. Lancet 1:697-699

Qintans J, Lefkovits I (1974) Clones of antibody-forming cells in pokeweed mitogen stimulated microcultures. II estimation of the frequency of precursor cells and average clone size. Eur J Immunol 4:617-621

Stoessl A, Stothers JB (1986) Carbon-13 magnetic resonance studies. Colletruncoic acid methyl ester, a unqie meroterpenoid from Colletotrichum truncatum. Z Naturforsch 41:877-880

Tuttle RL, Cantrell J (1981) *C. Parvum* - determinants of biologic activity In: Hersh EM, Chirigos MA, Mastrangelo MJ (eds) Augmenting agents in cancer therapy, Raven Press, New York, p 53-69

Vilcek J, De Maeyer E (1984) Interferon. In: Interferons and the immune system, Finter NB(ed) Elsevier Sci Publ, Amsterdam pp 268

Von Leyden VE, Blumenthal F (1902) Attempts to immunize humans by innoculation of their own cancer. Dtsch Med Wochenschr 28:637-638

Wareing JR (1986) Imidazole analogs of mevalonolactone and derivatives thereof as cholesterol synthesis inhibitors. CA107(7):59029x

Wareing JR, Damon RE (1986) Preparation of (heterocyclyalkenyl)mevalonates as hypolipemics and antiatherosclerotic agents. CA108(13):112221c

Immunotherapeutic effects of purified mistletoe lectin (ML-I) on murine large cell lymphoma

S. S. Joshi*, K. C. Komanduri*, S. Gabius+, and H.-J. Gabius#

*Department of Anatomy, University of Nebraska Medical Center,

Omaha, NE 68198-6395, U.S.A.,

The immune system in the tumor bearing host, particularly those with advanced disease, is deficient in eliciting specific or non-specific immune response against an immunogen including its own tumor associated antigens. This decreased immune response is thought to play an important role in tumor progression-metastasis which results in the death of tumor bearing host (Lovett et al., 1982; Joshi et al., 1991; Messbarger and Joshi, 1991). Therefore, an effective in vivo immunomodulator/ immunostimulator would increase the host's immune defense mechanism which, in turn, would help in decreasing tumor growth and their metastasis. Among the different types of modulators, specific binding of carbohydrate moieties to cellular glycoconjugates such as endogenous lectin is increasingly gaining attention (Leiner et al., 1986; Olden and Parent, 1987; Gabius, 1988a; 1988b). Some plant lectins/products have beneficial effects on modulating the immune system (Hajto et al., 1986). The proprietary mistletoe extract, Iscador, has an immunomodulatory action in humans and in rabbits (Nienhaus and Leroi, 1970; Nienhaus et al., 1970; Rentea et al., 1981; Kwaja et al., 1981; Blokshma et al., 1982; Hajto et al., 1986; 1989; 1990; Gabius et al., 1990a; 1990b). Except for the last two reports by Hajto et al. (1989; 1990), the rest of the studies on mistletoe were based on crude mistletoe extract. The purified mistletoe lectin ML-I, is a ß-galactoside specific lectin with an approximate molecular mass of 63 kD. ML-I consists of two

+Department of Hematology and Oncology, University Clinic, Gottingen, Germany,
#Department of Chemistry, Max-Planck Institute for Experimental Medicine, Gottingen, Germany.

subunits with different functional domains, the toxic subunit with an apparent molecular weight of 29 kD, termed A, and the carbohydrate-binding (ß-galactoside specific) subunit with an apparent molecular weight of 34 kD, termed B. At a very low dose, the purified lectin, ML-I, has demonstrated antitumor/immunostimulatory effects in rabbits and in humans (Hajto et al., 1989; 1990). Preliminary studies by Gabius et al. (1990a; 1990b; Gabius and Gabius, 1990; Gabius, 1990a) have shown that the ML-I administration in vivo did not cause any toxic effects in normal subjects.

In this report, we describe an antitumor/immunostimulatory effect of ML-I against a murine large cell lymphoma.

MATERIALS AND METHODS

Cell Culture: For these experiments, RAW117-H10 lymphoma cells were maintained in 60 mm x 10 mm Petri dishes (Falcon 1007, Falcon Plastics, Oxnard, CA) containing RPMI 1640 medium with 10% fetal calf serum (FCS) (Flow Laboratories, McLean, VA), penicillin (100 units/ml) and streptomycin (100 μg/ml) (Flow Laboratories).

In Vivo ML-I Treatment: A group of 12 normal Balb/C mice were injected with ML-I intraperitoneally (10ng/Kg) on alternate days up to day 9 after initiation of the treatment. The mistletoe lectin was purified as described by Gabius (1990b) from the crude plant extract. The immune status of the ML-I treated and control animals were determined on day 6 and 12 using mitogen assay, natural killer (NK) cell mediated cytotoxicity assay and flow cytometric analysis of the spleen cells for the frequencies of various immune cells. Normal untreated Balb/c mice were used as controls in these experiments.

In Vitro Mitogen Response: Spleen cells from ML-I treated and untreated Balb/c mice (5×10^4) were added with Concanavalin-A (10 μg ml) in a 96 well microtiter plate. As a control, 5×10^4 normal Balb/c splenocytes were cultured without mitogen. These co-cultures were pulsed with 1 μCi of ^{3}H-thymidine at 52 hours to evaluate the synthesis of DNA. The cells were then incubated for an additional 18 hours and harvested on glass fiber paper discs using a cell harvester. The filter paper discs were dried and counted in a liquid scintillation counter (LKB-1882) by putting each disc in scintillation solution.

Natural Killer (NK) Cell Cytotoxicity Assay: Spleen cells from control and ML-I treated mice were prepared and resuspended in RF10 medium, and used as NK effector cells. YAC-1 and/or RAW117-H10 target cells (5×10^6) were incubated with 100 μCi of ^{51}-Chromium (as sodium chromate) for 45 minutes at 37°C. The labelled cells were washed twice with RF10 medium and adjusted to a concentration of 2.5×10^5 cells/ml. The NK cytotoxicity assay was performed using varying effector to target cell ratios (E:T), i.e. 100:1, 50:1, 25:1, 12.5:1, as we have described previously (Joshi et al., 1987a). The supernatants were harvested using a Titertech supernatant harvester. ^{51}-Chromium release was measured using a LKB gamma counter and the percentage of target cells lysed was calculated.

Flow Cytometric Analysis: One million spleen cells from control and ML-I treated mice were placed in 96 well microtiter plates. Antibody to asialo GM-1 (10 μg in 0.1 ml) (Wako Chemical, Dallas, TX) was added to these cells. The cells were incubated on ice for 45 minutes and washed three times with phosphate buffered saline (PBS) containing 2% fetal calf serum (FCS). The cells were then stained with fluoresceinated antirabbit immunoglobulins and positive cells were quantitated using an Ortho 50H cytofluorograph equipped with a Cytomation Cicero Data Acquisition Controlled Analyses System (Denver, CO) as we have described previously (Joshi et al., 1987a). In addition, the normal Balb/c spleen cells were analyzed for their reactivity with ML-I using flow cytometric techniques. In these experiments various concentrations (10ng, 25ng and 50ng/ml) of biotinylated ML-I were treated with 1×10^5 cells for 30 minutes, the cells were then washed and stained with avidin-FITC. The positive cells were analyzed as described above.

In Vivo Studies: The antitumor effects of ML-I was tested by injecting the RAW117-H10 lymphoma bearing syngeneic Balb/c mice with ML-I (10ng/Kg) intraperitoneally on days 1, 3, 5, 9 and 12 after tumor transplantation. The animals were observed daily and the survival period was noted. When the animals became morbid, the animals were euthanized and the target organ liver and spleen, lungs were removed and fixed in 10% buffered formalin and evaluated for the presence of metastatic tumor nodules using standard histological techniques (Joshi et al 1987b).

Statistical Analysis: The statistical significance of the results obtained was determined using the Student's t-test.

RESULTS AND DISCUSSION

a. Immunostimulatory Effects of Mistletoe Lectin:

In order to study the effects of mistletoe lectin on host immune response, we injected purified mistletoe lectin, ML-I, at a concentration of 10 ng/kg intraperitoneally into 16-20 week old normal Balb/c mice on alternate days. The purified mistletoe lectin used in this study has been tested for endotoxin contamination and was free of effective levels of endotoxins. On days 6 and 12, we sacrificed these mice and the immune status was measured using in vitro mitogen assay, natural killer cell mediated cytotoxicity assay, and flow cytometric analysis of spleen cells for the frequencies of various immune cells using antibodies to the markers for T cells, B cells and NK cells. We also weighed the spleens and thymuses of these animals. In all these studies, untreated age matched animals were used as controls.

There was an increase in the mitogen response of the splenic T cells to the mitogen, Concanavalin-A, on control animals (P<0.01) (Table 1).

TABLE 1

EFFECTS OF MISTLETOE EXTRACT ON MITOGEN RESPONSE ON SPLEEN CELLS

	Counts Per Minute (Mean + S.E.) Day 6	Day 12
Control	40,418 ± 2,504	50,655 ± 2,344
Mistletoe Extract	57,506 ± 506	121,416 ± 22,117**

The natural killer cell mediated cytotoxicity levels of the spleen cells was also enhanced on days 6 and 12 after treatment with purified ML-I. Figure 1 demonstrates the NK cell mediated cytotoxicity levels of the spleen cells on day 12 after ML-I treatment. Our results with flow cytometric analysis of spleen cells revealed that there was an increase in asialo GM1 positive cells (NK cells) on days 6 and 12 when compared to control animal spleen cells (Table 2). However, we did not see such an increase with T and B cell populations.

** Significantly different from control (P<0.01).

TABLE 2

FLOW CYTOMETRIC ANALYSIS OF ASIALO GM1 POSITIVE SPLEEN CELLS FROM CONTROL AND MISTLETOE LECTIN TREATED ANIMALS+

Percent Positive Cells (Mean + S.E.)		
	Mistletoe Treated	
Control	Day 6	Day 12
20.5 + 0.2	23.4 ± 1.2	31.3 ± 2.9**

**Significantly different from control ($P<0.05$).
+Intraperitoneal injection of ML-I at a dose of 10 ng/kg.

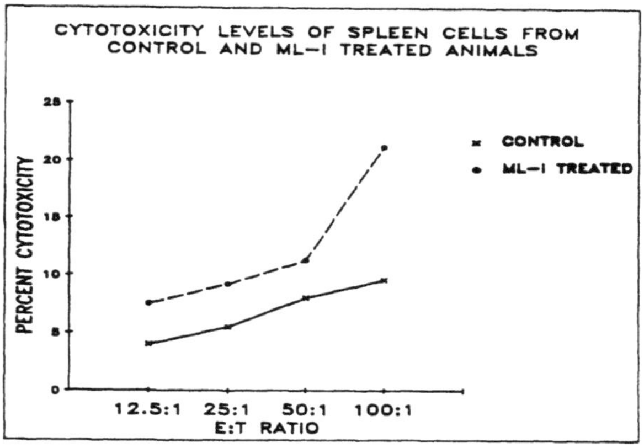

[Figure 1]

There was a slight increase in spleen weight of treated animals and a significant increase in thymus weight in mistletoe lectin treated animals (Table 3).

TABLE 3
EFFECTS OF MISTLETOE EXTRACT ON SPLEEN AND THYMUS WEIGHT* IN MICE

	Spleen Weight (mg)		Thymus Weight (mg)	
	Day 6	Day 12	Day 6	Day 12
Control	88.0 ± 5.5	104.5 ± 11.8	16.7 ± 0.6	12.9 ± 1.1
Mistletoe Extract	95.6 ± 4.5	128.2 ± 10.5	17.3 ± 0.2	23.0 ± 0.2**

* - Mean ± S.E.
**- Significantly different from control ($P<0.01$).

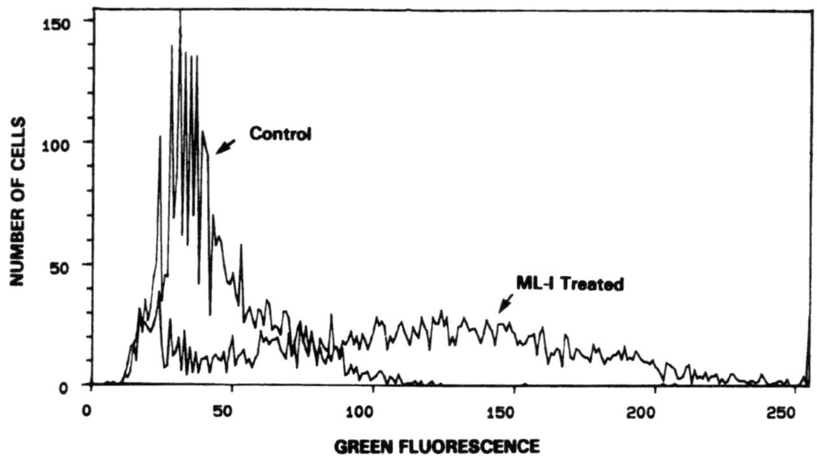

[Figure 2]

We also studied the reactivity of the ML-I with normal Balb/c spleen cells directly using flow cytometry techniques and we have found that the ML-I reacts with spleen cells. Figure 2 shows the reactivity of ML-I at a concentration of 25ng/ml.

b.) Antitumor/Immunostimulatory Effects of Mistletoe Lectin, ML-I

Since our preliminary studies showed immunostimulation by mistletoe lectin, we also studied the antitumor effects of mistletoe lectin using a murine lymphoma model system (Brunson and Nicolson, 1978; Brunson and Joshi, 1982; Joshi et al., 1987a; 1987b). Syngeneic Balb/c mice bearing liver metastatic RAW117-H10 lymphoma were intraperitoneally injected with purified mistletoe lectin (10 ng/kg) on alternate days following intravenous inoculation of 5,000 lymphoma cells. Survival of the control and treated animals was noted. In these experiments, 10 animals were used for control and treated groups. As shown in Figure 3, the tumor-bearing animals were treated with ML-I survived longer than untreated control animals. These results demonstrate an anti-tumor effect of ML-I against murine lymphoma. The ML-I treatment also reduced the numbers, and size of the tumor nodules in the target organ liver. Figure 4 demonstrate the reduced tumor burden in the liver of the treated animals (Figure 4A) when compared to that of the untreated tumor bearing mice (Figure 4B).

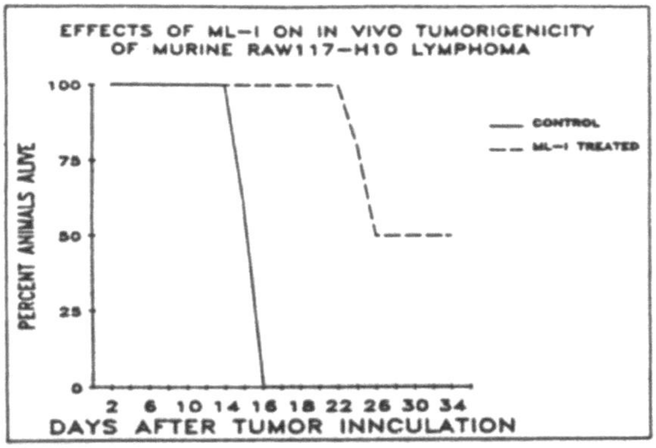

[Figure 3]

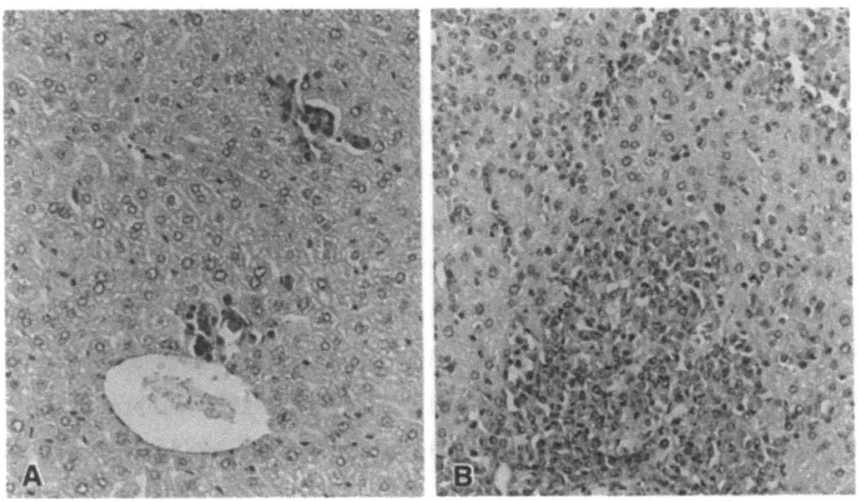

[Figure 4]

Our studies demonstrated an immunostimulatory/antitumor effects of purified mistletoe lectin ML-I on lymphoma bearing syngeneic Balb/c mice. Unlike other recently reported studies by Hajto et al (1989, 1990), we have used multiple intraperitoneal route of administrations of ML-I into these mice. Our analysis of the immune status of these animals indicated that the ML-I significantly enhanced mitogen response and NK cell mediated cytotoxicity of the syngeneic Balb/c mice. Furthermore, the ML-I treatment also enhanced the number of asialo GM1 positive (NK) cells in the spleens of the treated animals. However, we did not see any significant increase in other immune effector cells populations in the spleen of the treated animals. As reported earlier (Nienhaus et al, 1970) for the crude mistletoe extract, the ML-I treatment also increased the thymus weight in these animals. It appears that these enhancements in nonspecific immune response, reduced the in vivo tumorigenicity of highly malignant/metastatic RAW117-H10 lymphoma (Figures 3 and 4).

The higher and multiple doses of ML-I injections in these studies, did not appear to show any hematologic/immunologic toxic effects in these animals as measured by the normal spleen cellularity and functions. Because the in vivo effective concentration of ML-I is at the level of a nanograms or less per kilogram body weight, the toxic subunit of ML-I does not seem to have any unwanted side effects at this low concentration. Whereas the ß-galactoside specific lectin subunit appears to have a significant immunostimulatory/ antitumor effect at the low dose levels.

At present we do not know the exact mechanism of ML-I induced antitumor/ immunostimulatory effects. Hajto et al (1990) have shown that the ML-I induce the stimulation of cytokines such as TNF-α, IL-1 and IL-6. These cytokines or other cytokines might have enhanced the antitumor immune response against metastatic lymphoma. In a previous study we have shown that the RAW117-H10 lymphoma cells induce host immunosuppression in vivo and administration of IL-2 (intraperitoneally on alternate days enhanced the in vivo immune response and consequently decreased the in vivo tumorigenicity of RAW117-H10 cells. Based on our results and other reports, mistletoe lectin has a potential antitumor/immunostimulatory effects. Therefore, a systematic study designed to address the mechanism of action of ML-I is essential to understand the effects of ML-I in vivo and for its future use in clinical trials. In addition, we also need to evaluate the in vivo effects of 34kD ß-galactoside specific lectin component of the ML-I and the effects of the other toxic subunit of ML-I. These studies are currently being performed in our laboratories using the same experimental model with different dose levels and routes of administration of ML-I. The preliminary results of these studies also indicate the therapeutic usefulness of the purified mistletoe lectin.

ACKNOWLEDGEMENT

The authors would like to thank Ms. Roberta Anderson for an excellent job of preparing this manuscript.

REFERENCES.

Blokshma N, Schmiermann P, de Reuver M, van Dijk H. (1982) Viscum preparation. Planta medica 46: 221-227.

Brunson KW and Nicolson GL (1978) Selection and biologic properties of malignant variants of a murine lymphosarcoma. JNCI 61: 1499-1503.

Brunson KW and Joshi SS. (1982) Properties of malignant lymphosarcoma variant cell lines associated with enhanced liver metastasis. In: Tumor Progression and Markers, Kugler Pub., Netherlands (K. Lapis, Ed.), pp. 137-143.

Gabius H-J. (1987) Endogenous lectins in tumors and the immune system. Cancer Invest 5: 39-46.

Gabius H-J. (1988a) Mammalian lectins: their structure and their glycobiological and glycoclinical roles. ISI Atlas of Science: Biochemistry 1: 210-214.

Gabius H-J. (1988b) Tumor lectinology: At the intersection of carbohydrate chemistry, biochemistry, cell biology and oncology. Angew Chem. Int. Ed. 27: 1267-1279.

Gabius H-J, Hostanska K, Trittin A, Gabius S, Hajto T. (1990a) Klinischer Nachweis von immunmodulation durch lektine im rahmen der misteltherapie. Therapentikon 4: 38-45.

Gabius H-J and Gabius S. (1990) Tumorlektinologie: Status und perspektiven klinischer anwendung. Naturwissenschaften 77: 505-514.

Gabius S, Hajto T, Gabius H-J. (1990b) Erstellung von Richtlinien fur eine standardisierte Misteltherapie durch harmonisches Zusammenspiel von biochemischer, immunologischer und klinischer Forschung. Med Organica 14: 121-124.

Gabius H-J. (1990a) Lektin aus mistelextrakt als potenter immunmodulator. Dt. Arztebl. 87: B-2628.

Gabius H-J. (1990b) Influence of type of linkage and spaces on the interaction of ß-galactoside-binding proteins with immobilized affinity ligands. Anal. Biochem. 189: 91-94.

Hajto T. (1986) Immunomodulatory effects of Iscador: A viscum album preparation. Oncology 43 (Suppl 1): 51-65.

Hajto T, Hostanska K, Gabius H-J. (1989) Modulatory potency of the ß-galactoside specific lectin from mistletoe extract (Iscador) on the host defense system in vivo in rabbits and patients. Cancer Res. 49: 4803-4808.

Hajto T, Hostanska K, Frei K, Rordorf C, Gabius H-J. (1990) Increased secretion of tumor necrosis factor-α, interleukin-1 and interleukin-6 by human mononuclear cells exposed to ß-galactoside-specific lectin from clinically applied mistletoe extract. Cancer Res. 50: 3322-3327.

Joshi SS, Tilden PA, Jackson JD, Sharp JG, Brunson KW. (1987a) Cell surface properties associated with malignancy of metastatic large cell lymphoma cells. Cancer Res. 47: 3551-3557.

Joshi SS, Sharp JG, Brunson KW. (1987b) Differential growth characteristics of metastatic variant RAW117 murine lymphosarcoma cells. Oncology 44: 180-185.

Joshi SS, O'Connor SJ, Weisenburger DD, Sharp JG, Gharpure HM, Brunson KW. (1991) Enhanced antiproliferative activity by metastatic RAW117 lymphoma cells. Clin. Exp. Metastasis (in press).

Joshi SS, Messbarger LJ, Hao W. (1990) Differential sensitivity of metastatic variant lymphoma cells to natural immunity and effects of rIL-2 on their tumorigenicity in vivo. Proc. Am. Assn. Cancer Res. 28: 275.

Kwaja TA, Diaz CB, Papoion T, Pentecost S. (1981) Studies on cytotoxic and immunologic effect of viscum album (mistletoe). Proc. Am. Assn. Cancer Res. 22: 253.

Kwaja TA. (1983) Experimental basis for the use of Iscador in cancer treatment. 13th Int. Congr. Chemother., Vienna.

Liener IE, Sharon N, Goldstein IJ (eds.). (1986) The lectins: properties, functions and applications in biology and medicine. New York: Academic Press.

Messbarger LJ, Joshi SS. (1991) Effects of IL-2 on in vivo immune response and tumorigenicity. Tumor Biol. (submitted).

Nienhaus J, Stoll M, Vester F. (1970) Thymus stimulation and cancer prophylaxis by viscum proteins. Experientia 26: 523-525.

Nienhaus J, Leroi R. (1970) Tumorhemmung and thymus stimulation durch mistletoe prepaparate. Elemente Naturwissenschaft 13: 45-54.

Rentea R, Lyon E, Humter R. (1981) Biologic properties of Iscador: a viscum album preparation. Lab. Invest. 44: 43-48.

The rationale of mitogenic lectin therapy for cancer

B. M. Wimer

JBMW Immunotherapeutics
1609 Catron Ave SE
Albuquerque, NM 87123
USA

The momentous observation of Hungerford, et al. (1959) that phytohemagglutinin (PHA) is a potent lymphocyte mitogen challenged the plant lectin's long-established pre-eminence as a hemagglutinin. The ensuing deluge of articles on PHA mitogenicity enabled an extensive literature survey by Naspitz and Richter in 1968. The reviewers recognized an analogy with the mitogenesis of alloactivation, and their awareness that pokeweed mitogen (PWM) generated similar effects predicted that other plant lectins would be found mitogenic. Other important perceptions on the mitogenic activation included its predominant T lymphocyte distribution; its association with enhanced cytokine production; its nonspecific polyclonal nature as contrasted to antigen-specific activation; its capacity to provide protection against irradiation and cytotoxic drugs; and its tendency to inhibit antibody responses temporarily.

The clinical trials initiated by Humble (1964) for the PHA treatment of aplastic anemias did much to inspire this review. The aplastic anemia study was based on the mistaken assumption that PHA would prove mitogenic for hematopoietic stem cells because of their morphologic similarity to lymphocytes. It may have been unwarranted alarm about the toxicity of PHA in small animals that curtailed the aplastic anemia investigation, even though mildly favorable clinical results had been realized without the occurrence of serious side effects. The hope of Naspitz and Richter that their elucidation of PHA mitogenesis would help validate this therapy and lead as well to other decisive immunotherapeutic applications did not materialize.

Prohibitive toxicities that might have been avoided by the use of properly-dosed nonerythroagglutinating preparations severely compromised many of the subsequent animal studies concerning PHA applications in cancer therapy and allograft transplantation. Most of the clinical trials involving the use of PHA as an adoptive cell activator for cancer treatment were destined to fail because of the disparity in giving meager numbers of activated effector cells to destroy massive tumor burdens. Although a nonerythroagglutinating form of PHA was used in the intravenous trials, dosage inadequacies and other misapplications compromised the responses.

Their consistent omission from classifications of biotherapeutic agents best indicated the continuing failure to appreciate the immunotherapeutic potential of the mitogenic lectins. Most notable was the classification of Lotz and Rosenberg (1988), whose research had involved PHA and concanavalin A (Con A). The Rosenberg group restricted the role of PHA to activation and expansion of peripheral blood mononuclear cells (PBMC) in culture for the treatment of advanced cancers (Mazumder, et al. 1984). However, they quickly recognized the need for concomitant systemic administration of interleukin-2 (IL-2) when it was similarly applied in lymphokine-activated killer (LAK) cell therapy. The unimpressive performance of PHA as an autologous PBMC activator erroneously branded it as an ineffective immunostimulant even though LAK cell therapy without concomitant systemic IL-2 administration fared no better (Lotz and Rosenberg 1988). The concept that the mitogenic lectins can best be applied as PBMC activators rather than as directly administered mitogenic stimulants persists in the current attempt to develop an effective cancer treatment using the extracorporeal circulation of blood through PWM immobilized in a minicolumn packed with synthetic polymer beads (Numa, et al. 1990).

The detailed audits of old investigations to explain the failures in prior research with mitogenic lectins are likely too mundane to command much attention in this era of sophisticated biotechnology. Nevertheless, a conviction that the proper application of these agents has much to offer in the treatment of cancer and related disorders prompted the publication of a series of articles that attempted first to define a clinician's view of the attributes of an ideal immunomodulator (Wimer 1989); second to document the capacity of the L4 isolectin of PHA to at least partially fulfill many of the requirements (Wimer 1990a,1990b); and finally to justify a broad array of potential applications of L4 in human and veterinary medicine (Wimer 1990c).

The fact that the champion of mitogenic lectins continues to be a lonely figure is well illustrated by this presentation's being the only discussion of the subject in this volume on lectins and cancer. Certain evidences now suggest, however, that this status limitation could be changing, the most obvious being the granting of a patent for concanavalin A dimers as immunoregulatory agents (Morris 1989). A more pertinent, albeit subtle, example is a presentation by the chairman of this symposium concerning the application of mistletoe extracts for cancer therapy (Gabius 1991). The action of the ß-galactoside-specific lectin of mistletoe to increase antitumor cellular and cytokine responses has been elucidated by in vitro and clinical studies. Whether or not the weak mitogenicity of mistletoe lectin has any relevance to these intriguing observations, this research clearly favors the concept that, as with the mitogenic lectins, specific lectin-carbohydrate interactions can provide the keys to favorable therapeutic immunomodulations.

An important element in proposing the L4 isolectin of PHA as a model immunomodulator is that other mitogenic lectins should offer modulation differences that might be used in alternative, alternating, or supplemental fashion (Wimer 1990b). Which lymphocyte subsets are activated by a lectin will be determined by its membrane binding characteristics. Next to PHA, concanavalin A (Con A), derived from the jack bean *Canavalia ensiformis*, and pokeweed mitogen (PWM), extracted from the plant *Phytolacca americana,* are currently the most useful and widely researched mitogenic lectins. These three lectins are selected as the mitogenic prototypes because of their distinctive types of membrane binding. Since PHA is the mitogenic lectin most completely studied and the only one that has been given clinically, the PHA data will serve as the model for the application of other mitogenic lectins.

Binding Characteristics of the Key Mitogenic Lectins and Their Significance

PHA, Con A, and PWM are all thought to bind with the T cell receptor (TCR) intimately associated with CD3. The nature of the lymphocyte subset responses to the various mitogenic lectins suggests that each may also interact uniquely with other CD receptors. The response to PHA may be relatively balanced between T helper (Th) lymphocytes and T cytotoxic/suppressor (Tcs) lymphocytes, although biased toward the latter. Con A activation more clearly favors CD8+ Tcs lymphocytes and PWM the CD4+ Th lymphocytes, the latter subset also being an inducer for CD8+ Tc lymphocytes. The activations are further complicated by evidence that one of the five Con A isolectins (Pa-1) is directly mitogenic for B cells (Yokayama, et al. 1976), although the lymphocyte receptor involved has not been identified. That Con A may additionally interact with the CD2 receptor is suggested by the lectin's singular capacity to activate thymocytes, which distinctly lack the other CD receptors. The proposed sites of interaction would not necessarily be sharply defined, and they might vary with lectin dosage and other factors. In any event, the general identification of membrane receptor sites with which each mitogenic lectin interacts should prove helpful to its therapeutic applications.

Therapeutic Attributes of Mitogenic Stimulation

Contradistinctive to the small percentage of monospecific T cells activated by a single antigen, the polyclonal activation by a mitogenic lectin will include all lymphocyte subsets sufficiently involved in its binding to the membrane. In contrast to the approximate 85 to 15 ratio between T and B cells activated by PHA and Con A, PWM-induced blasts are equally of T and B cell origin (Stobo 1980). PWM is regarded as a B cell mitogen, although T cells play an essential role in many B cell activations. To what degree the B cell stimulation is direct, rather than a consequence of Th lymphocyte activation, remains to be determined. Nevertheless, nonspecific T cell activation is the primary event in mitogenic stimulation that leads further to activation of B cells, NK cells, and monocytes in conjunction with the production of multiple lymphokines and cytokines.

Activated cellular effectors can initiate destructive actions against microorganisms as well as malignant cells, but it must be recognized that the accompanying destruction of vital tissues can be detrimental if excessive, a risk that is greatest when the response is directed against intracellular viruses. Concerns have also been generated by in vitro studies indicating that mitogenic activations can stimulate HIV proliferation, but these must be reconciled against other reports that Con A and PHA inhibit HIV in culture, possibly by blocking the viral glycoproteins (Robinson, et al. 1987). That the stimulation of growth factor production occurs with lectin activation of lymphocytes is indicated by the myeloproliferative responses to PHA therapy in aplastic anemias (see Wimer 1990c). PHA has also been shown to enhance recovery from the damaging effects of irradiation and cytotoxic drugs. The usefulness of mitogenic lectins as vaccine adjuvant is suggested by the attraction of mononuclear cells to sites of injection and the amplification of target cell antigenicities. These effects could also be of importance in intralesional therapy by enhancing the activities of tumor infiltrating lymphocytes (TIL). The use of mitogenic lectins in vitro to initiate the activation of adoptive PBMC before their administration and then intravenously in the recipient to inhibit rejection and GVH responses should be restricted to application with partially-matched cells to patients previously unresponsive to systemic administration but not at high risk for serious GVH reactions. The methods employed should be consistent with those described below for PHA-L4.

The concept that lectin mitogenesis can also be immunosuppressive becomes more credible with the recognition that activation will make T cells unresponsive to antigen encounter. The stimulation of suppressor cell function and blocking actions on CD3, CD4, and CD8 receptors may also be important. Thus, maintaining the lectin-saturation of all T cell receptors would likely provide the most complete possible suppression by these agents. PHA has also been shown to augment the suppressive effects of other agents in allograft transplantations. The advantages of using lectins for therapeutic suppression include the concomitant antitumor and antimicrobial protection provided by both myeloproliferative stimulation and total activation of effector systems (see Wimer 1990b). Other major advantages of lectin mitogenesis are that endogenous responses are incited, and these responses are neatly packaged in mononuclear cells for delivery to their sites of action.

Phytohemagglutinin

Phytohemagglutinin (PHA) is the paradigm of mitogenic lectins, in its crude form inducing erythroagglutinating and leukoagglutinating effects as well as mitogenicity by its interaction with D-GalNAc. Intravascular hemagglutination associated with the use of the cruder extracts have been shown to contribute substantially to the toxicity, particularly when given in excessively high doses to small animals. As reviewed by Wimer (1990a), Yachnin and colleagues and Weber and colleagues concluded that PHA glycoproteins comprise a class of five isolectins that are isomeric tetramers composed of varying proportions of two different subunits, one lymphocyte-reactive (L) and the other erythrocyte-reactive (E), designated as follows: L4 (purely mitogenic-leukoagglutinating), L3E1, L2E2, L1E3, and E4 (purely erythroagglutinating). This concept was substantiated initially by Bachur et al. and more recently by Glad and Borrebaeck (1984) as well as by Fleischmann et al. (1985). Erythroagglutination can be eliminated by utilizing the L4 isolectin, which is leukoagglutinating only at higher concentrations, and minimally or virtually nonagglutinating at the maximally mitogenic level of 5 μg/mL in serum or 2.5 μg/mL in serum-free media. PHA-L4 is currently not produced for clinical application but is available commercially for laboratory use in a well-standardized form that is stable from E-Y Laboratories (San Mateo, California) and Vector Labs (Burlingame, California) as well as other sources. Burroughs Wellcome in Beckenham, England and Research Triangle Park, North Carolina have infrequently released purified mitogenic PHA-P for clinical trials. This product, which is supplied as Purified Phytohemagglutinin for special laboratory procedures, is produced by erythroagglutinin absorption of the Hurn PHA preparation; it may be may be roughly equivalent to PHA-L4 in mitogenicity.

The available methods of isolating the L4 isolectin by extraction and chromatography methods are not conducive to mass production and costly enough to be a deterrent to its large scale evaluation as an immunomodulator. Fleischmann, et al. (1985) have described a fast-working single-step procedure for isolation and resolution of PHA isolectins by affinity chromatography that may be cost-reducing. The recombinant DNA methodology for synthesis of nonerythroagglutinating PHA-L4 in *Escherichia coli* (Eco PHA-L) was reported by Hoffman and Donaldson (1987) and subsequently patented (Hoffman 1989), but the method has not been utilized and therefore remains untested as a mass-production source of PHA-L4 at reasonable cost. The separation of the mitogenic and leukoagglutinating properties of PHA has not been shown to be feasible. Hence, simply maintaining concentration of L4 at maximally mitogenic levels and below the threshold for leukoagglutination is currently the one reliable means of segregating the two.

Studies (see Wimer 1990a) have confirmed that PHA binds rapidly to lymphocytes with the largest amount of the lectin bound at 20° C being twice as much as at 37° C. Binding is completed within one hour at 37° C, but also nearly complete within 90 minutes at 20° C. After one hour in culture using standard concentration of cells and PHA, the stimulation can no longer be reversed by repeated washing of the cells. This suggests that once PHA leukoagglutinin becomes irreversibly bound, the fate of the cell, (i.e., whether stimulated or agglutinated) would depend on the initial level of exposure to the lectin. A subsequent drop in serum level would presumably not alter the effect, but it is not certain that increasing the PHA concentration would not change the effect from stimulation to agglutination. The point is that failure to achieve maximum mitogenicity may just as likely result from PHA-overdose as from underdose.

This effect of PHA means that adoptive cells would become fully activated in vivo after they have been administered immediately after incubation with lectin for 60-90 minutes at room temperature, although their proliferative expansion would require a concomitant source of IL-2. While some IL-2 should be produced by the donor's activated Th lymphocytes, this source would not likely be sufficient to provide for adequate proliferation of the adoptive cells. Systemically administered PHA-L4 should assure adequate production of IL-2 by host lymphocytes. In this manner the isolectin can be used to initiate adoptive cell activation in vitro and then given systemically to support continued in vivo activation and proliferation of both the transfused adoptive and host lymphocytes.

An important pharmacologic aspect is the relatively narrow range of PHA concentration for maximum mitogenicity to take place, although lesser degrees of mitogenicity will occur within a broader range (see Wimer 1990a). The in vitro mitogenicity of purely mitogenic PHA in whole serum peaked at 5 μg/mL, with a range of 0.2 μg to 1.0 mg/mL, while mitogenicity of the L4 isolectin crested at 2.2 μg/mL with a range of 0.1 to 150 μg/mL in serum-free cultures. These results are consistent with the finding that a 50% serum concentration inhibits L4 binding by over 50%.

PHA isolectins have been shown (Glad and Borrebaeck 1985) to differ in their affinities toward unrelated serum glycoproteins (E4 high, L4 low). Thus, isolectins rich in E-subunits bind preferentially to serum glycoproteins and are barely available to binding with lymphocytes. E4 exhibits affinity for 14 serum glycoproteins and L4 for 9 of these. Isolectins rich in L-subunits with only weak affinity toward serum glycoproteins can stimulate lymphocytes at low concentrations even in the presence of serum. E4 was shown to possess strong mitogenic activity that was, however, masked by the interference of serum components. The mitogenicity of E4 increases 200 fold in the absence of other glycoproteins, while that of L4 increases 25 fold. In normal human serum, peak mitogenicity of L4 was found to occur at a 5 μg/mL concentration with a range of 0.2 to 100 μg/mL. Hence, a reasonable estimate of the optimal L4 level in serum for maximum mitogenicity in vitro would be 5 μg/mL, with a range of 0.2 to 300 μg/mL. Based on the assumption that L4 will diffuse throughout the extracellular fluid space, the estimated loading dose of PHA-L4 needed to induce maximum mitogenicity in a 70-kg-person would be 1 mg/kg body weight (see Wimer 1990a).

The intradermal injection of PHA attracts mononuclear cells to the site leading to their local activation and release of cytokines that produces peak erythema and induration at 24 hours. The reactivity is nonspecific to the mitogen and independent of prior sensitization. The PHA skin test has been used as a test of immune competence (Zuckerman and LoBuglio 1977), but the inconsistent correlation with the in vitro PHA tests has raised doubts as to its accuracy for that purpose. The inhibition of the response to intradermal PHA likely to occur from increased levels of unrelated glycoproteins in serum and tissue fluids might explain this discrepancy. These glycoprotein increments could be expected in various clinical settings such as general anesthesia, major surgery, hyperthermia, infection, advanced malignancy, extensive tissue necrosis, and GVH reactions, in which the standard PHA skin test responses are prone to be negative (Wimer - unpublished observations). The tendency for negative skin test responses to become positive with increases of the intradermal doses suggests a correlation with glycoprotein levels independent of immunocompetence. Hence, the skin test might serve to indicate the need for PHA dosage adjustments, with the therapeutic escalations required being proportional to the increase in skin test dose over normal needed to elicit a positive reaction.

PHA has been injected into various animal species and earlier reports suggested relatively low toxicity. Varying degrees of morbidity and rates of mortality have been observed, however, depending on the preparation given (PHA-P > PHA-M), its erythroagglutinating activity (high > low), dosage (higher > lower), route of administration (intravenous > intraperitoneal), rate of administration (rapid > slow or divided dose), small size of the species (mice > rabbits), and maturity of the animal (newborn > mature). Although considerable data are available in various reports scattered throughout the literature, the interpretations are impaired by the heterogeneity with respect to these factors. The adverse effects of erythroagglutinating PHA may be vascular-circulatory in character, some investigations having provided evidences that they are related to intravascular erythroagglutination. Hence, the greatest morbidity and lethality have associated with rapid intravenous injections of potent erythroagglutinating preparations given to immature small animals (see Wimer 1990a).

If intravascular erythroagglutination accounted entirely for the lethal effects of PHA, a purified mitogenic preparation should be not be deadly. On the contrary, the LD50 of L4 in mice has been found to lie between 4 and 8 mg/kg body weight with intravenous administration. Macroscopic autopsy findings have included marked splenomegaly and hemorrhages in the lungs, intestines, and liver. The high range of the L4 doses and the early-onset of lethal reactions are consistent with the assumption that intravascular leukoagglutination rather than mitogenicity would have been causative (see Wimer 1990a).

The use of Wellcome purified PHA-P for the treatment of aplastic anemias provided the broadest experience with intravenous administration in humans, the side-effects including mild chills to severe rigor, sensations of heat to frank fever, and generalized aching or muscle and joint pains (see Wimer 1990b). Superficial thrombophlebitis was noted in two patients. Humble reported a tendency to pains in the bones of the thorax and legs occurring 8 to 10 days after the first injections, although this was not mentioned by others. Hypersensitivity reactions were not reported. The administration of PHA-activated marrow cells or leukocytes was also accompanied by a tendency to chills and fever. Heat, swelling, and pain in the tumor nodules occurred with the treatment of a patient with metastatic melanoma by Lewis et al. (1971).

In a phase I clinical trial in 8 cancer patients of autologous lymphocytes activated by PHA in 4-day culture before infusion, Mazumder et al. (1984) observed toxicity to include fever and chills in 8 of 8, headache in 5 of 8, transient nausea and vomiting in 3 of 8, and a mean decrease of 30% in pulmonary diffusion capacity that reversed within 24 hours in 6 of 6 patients tested. Six of 8 patients required one unit of RBC transfusion because of losses associated with leukaphereses. The nonerythroagglutinating form of PHA from Burroughs Wellcome was given in this and most other human trials, but PHA-L4 has not been evaluated in clinical applications.

It is not likely that these side-effects were induced by lectin-induced hemagglutination, but the use of optimal mitogenic doses of the L4 isolectin rather than excessive doses of the cruder preparations would exclude concerns about agglutination effects. A more likely explanation for the chills, fever, and malaise would be the production of pyrogenic lymphokines, particularly interleukin-1 (IL-1) and tumor necrosis factor or cachectin. The growth factors, particularly GM-CSF and G-CSF, have been reported to cause bone ache from expansion of the marrow cavity, and this would best explain the bone pain described by Humble (1964) with PHA treatment of aplastic anemias. Protracted PHA administration might give rise to some cytokine side-effects, but these should be minimized by the inherent on-site delivery system. Acceptable toxicity and absence of healing-inhibitory activities along with the stimulating effects of L4 on myeloproliferation and immune responses should make it readily applicable as an adjuvant for tumor surgery, as well as to the severely debilitated patient. There are no data on the tolerance for PHA in pregnancy, but L4 should be much better endured than chemotherapy or radiation, at least by the mother. A greater concern to be assessed would be the possibility of damage to the fetus such as teratogenicity.

The possibility of PHA-induced leukemoid responses was shown in the report of Askoy, et al. (1966) concerning the treatment of red cell aplasia in a 46-year-old man with PHA given as 18 intravenous injections of 50 mg each. For three months, no response was noted, but the leukocytes then rose from 8,200 to 40,000/mm^3 with 77% atypical lymphocytes. The lymphocytic proliferation lasted six months, but recovery of normal erythropoiesis ensued. The nature of this occurrence can readily be understood on the basis of the intense lymphoproliferative effects of PHA injection demonstrated experimentally, as discussed by Wimer (1990a).

The extensive tumor destruction in vital tissues theoretically possible with effective PHA therapy would demand due caution to avoid serious complications such as brain hemorrhage or bowel perforation. The theoretical hazards of treating certain infections are exemplified by extensive hepatic necrosis that might occur from the immunotherapeutic destruction of infected cells in the acute phase of viral hepatitis and the predicted proliferative stimulation of human immunodeficiency virus (HIV) from the activation of infected T4 lymphocytes. Hence, PHA therapy would be contraindicated during phases of infectious diseases known to be adversely affected by immunostimulation.

The risk of serious GVH reactions from giving even minute numbers of competent lymphocytes into immune incompetent individuals has been well documented (see Wimer 1990a). While care must be exercised to avoid transfusions in patients at high risk, the proper application of L4 for immunosuppression and selection of partially matched donors should reduce the risks of PBMC administration.

Experimental evidences of sensitization to PHA, both with respect to hypersensitivity manifestations and antibody inhibition of mitogenic effects, are contradictory (see Wimer 1990a). The inconsistencies might be explained by quantitative differences related to the duration of exposure to sensitizing factors. Since agglutinins and impurities in the preparation might be involved, these should be eliminated as much as possible. Actually, hypersensitivity responses have not interfered in most experimental investigations, and in one study, sensitization to PHA enhanced rather than blocked the mitogenic effect on autochthonous lymphocytes. Sensitization was reported in the clinical treatment of aplastic anemia with intravenous PHA that resulted in production of antibody with a capacity to inhibit the mitogenic action of PHA in vitro, but no significant effect was evident in vivo. The tendency for PHA to substantially suppress antibody responses when given before or concomitantly with antigen exposure suggests that sensitization would be unlikely with the applications anticipated. Sensitization is not induced by repeated application of minute doses in the PHA skin test, the responses to which are highly reproducible.

Concanavalin A

Composed of four identical subunits, concanavalin A is a tetrameric glycoprotein that becomes dimeric at a pH < 5.6 or upon succinylation or acetylation. The dimeric form may be a less potent mitogen but it offers the important advantage of being nonagglutinating. The carbohydrate specificity is for alpha-Mannose, alpha-Glucose, and alpha Glc-Nac. Con A reacts with bacterial and animal cells and can distinguish between normal and tumor cells. The perception that the lectin favors suppressor activity is consistent with the assumption that it may also favor CD8 binding. Membrane binding that includes CD2 would predict the occurrence of immature leukemoid responses to full doses of the lectin because of thymocyte activation, but published experimental data are lacking Con A is the least expensive of the three lectins to produce, but some variability in composition of the products obtained from different sources presumably relates to variations in impurities. Con A and succinylated Con A are available in partially purified forms for laboratory purposes from lectin suppliers such as E-Y Laboratories and Vector Labs, but recombinant methods of production have not been reported. The mitogenicity of Con A peaks at a concentration of 4 μg/mL (Morris, et al. 1987) in the absence of serum compared to 2.5 μg/mL for PHA-L4, making the comparative dose 1.6 times higher. Maximum mitogenicity of succinylated Con A occurs at 12.5 μg/mL in the absence of serum suggesting a dosage requirement 5 times higher than L4. These dose comparisons are valid for serum-free systems, but the comparative affinities of Con A and succinylated Con A for interfering serum glycoproteins have not been assessed.

Pokeweed Mitogen

The mitogen is obtained as crude saline extracts of pokeweed roots purified by heat treatment and protein precipitation. The product is made up of 5 isolectins designated Pa-1, Pa-2,

Pa-3, Pa-4, and Pa-5 with minor differences in mitogenic and hemagglutinating properties (Waxdal 1974). The major mitogenic effects of all are seen at concentrations well below those required for agglutination. While all five isolectins are mitogenic for T cells, only the Pa-1 isolectin is also mitogenic for B cells (Yokoyama 1976). Preparations of PWM contain batch-variable mixtures of all five of isolectins The lectin yields and proportions of the different isolectins also vary with the season of harvesting the root. PWM has a carbohydrate specificity for oligomers of N-Acetylglucosamine and N-Acetyllactosamine. A comparison between the in vitro mitogenicities of certain purified forms of PHA and PWM based on the amounts of the substances necessary for the stimulation of 3 x 10^6 lymphocytes was found to be 1 µg for PHA with 80-90% transformations versus 0.1 µg for PWM with 50-60% activations (Borjeson, et al. 1966). PWM was shown in subsequent articles to stimulate a qualitatively distinct transformation with the expression of biochemical, histochemical, and fine structural features resembling early antibody-forming cells.

The validity of this estimated mitogenicity ratio between PHA and PWM may be limited to the specific preparations studied, but the potency of PWM is indicated by a previous report of remarkable responses in two laboratory workers to accidental inoculations of the extract (Barker, et al. 1965). One occurred through the conjunctiva and the other through a subcutaneous puncture wound. No local or systemic toxic effects were observed, but large immature basophilic lymphocytes and typical plasma cells appeared in the peripheral blood shortly after the exposures and these cells persisted in blood smears and buffy coat preparations for the eight weeks prior to the report. A few of the cells were interpreted as immature plasma cells. Complete hematologic assessments were found to be normal in all other respects. No further data have been published concerning experimental or clinical responses or toxicities. Unpublished observations apparently unique to PWM exposure concern the occasional occurrences of type I hypersensitivities in laboratory personnel involved in production of the lectin (Stanton 1991). The fact that this has sometimes been seen on first exposure should raise the suspicion that the culprit might be an ingredient other than the lectin involved in the processing. PWM is available in a partially purified form from E-Y Laboratories, but a recombinant method of production has not been reported.

Discussion

The binding differences among the selected mitogenic prototypes offer the possibility of expanding the potential of therapeutic modulations. These differences are defined well enough now for their exploitation, although further clarification would be desirable. The element of agglutination should ideally be eliminated from any mitogenic lectin to be used for therapy to exclude any toxicity or sensitization that might be related. Production procedures should also eradicate contaminants that might contribute to sensitization risks, and the focus should be on the output of well-standardized materials. Economical mass production methods would be essential to facilitating evaluations and applications in larger animals and humans.

Whether other glycoproteins in serum and tissue fluids interfere with mitogenic responses to Con A and PWM should be ascertained, as this would influence the utility and interpretation of skin tests that might be developed with these agents. An absence of interference would validate the respective skin tests as measures of immune competence, but its presence would indicate the need to apply them as guides to lectin dosages.

Except for the inadequacy of the sources of supply, PHA-L4 may be ready for clinical trials that should best commence with veterinary applications, especially for cancer treatment in dogs. Favorable responses without serious side effects could provide the key to other uses and justify human trials. There is a void in published data on toxicities and hypersensitivities of Con A and PWM in small and large animal applications that must be remedied before any human trials can be considered.

In spite of the vast therapeutic potential perceived for these agents, major obstacles stand in the way of their ever reaching the market place. The mitogenic lectins are complex molecules that are difficult and costly to produce in an ideal form necessary for their being reliably assessed and utilized. Realistic sensitization concerns have to be addressed. Periods of prolonged administration of these powerful mitogenic stimulants are likely to exaggerate the recognized side-effects and generate additional ones. Certain evidences reported with PHA and PWM exposures assure the occurrence of lasting leukemoid reactions severe enough to be frightening even though their benignancy would be quite certain. The versatility of applications possible with these agents itself discourages their investigation because of the resources required to establish safety and efficacy. Obtaining patent protection of substances that are naturally-occurring is a concern that is not be lessened by the large number of competing lectins theoretically available. Finally, the preoccupation of the investigational establishment with the development of biotherapeutic agents that are justifiably seen as vastly more exciting from technical and economic standpoints is a strong deterrent. Ironically, the therapeutic benefits realized for some applications of biotechnical marvels such as monoclonal antibodies, interferons, IL-2, GM-CSF, and G-CSF could be relatively minor aspects of those theoretically possible with effective mitogenic stimulation. Despite the magnitude of the problems to be faced, the reluctance to at least explore the therapeutic potential of the mitogenic lectins seems unwarranted, considering the devastating nature of cancer and the other disorders involved.

References

Askoy M, Erdem S, Dincol K (1966) Phytohaemagglutinin in refractory anaemia. Lancet 2:1464-1465

Barker BE, Farnes P, Fanger H (1965) Mitogenic activity in Phytolacca americana (pokeweed). Lancet 1:170

Börjeson J, Reisfeld R, Chessin LN, Welsh PD, Douglas SD (1966) Studies on human peripheral blood lymphocytes in vitro. I. Biological and physiochemical properties of the pokeweed mitogen. J Exp Med 124:859-872

Fleischmann G, Mauder I, Illert W, Rudiger H (1985) A one-step procedure for isolation and resolution of *Phaseolus vulgaris* isolectins by affinity chromatography. Biol Chem Hoppe-Seyler 366:1029-1032

Glad C, Borrebaeck CAK (1984) Affinity of phytohemagglutinin (PHA) isolectins for serum proteins and regulation of the lectin-induced lymphocyte transformation. J Immunol 133:2126-2132

Hoffman LM, Donaldson DD (1987) Synthesis of mitogenic phytohemagglutinin-L in Escherichia coli. Biotechnology 5:157-160

Hoffman LM (1989) Method and composition for producing lectin in microorganisms. United States Patent 4,870,015

Humble JG (1964) The treatment of aplastic anaemia with phytohaemagglutinin. Lancet 1:1345-1349

Hungerford DA, Donnelly AJ, Nowell PC, Beck S (1959) The chromosome constitution of a human phenotypic intersex. Am J Human Genet 11:215-236

Lewis MG, Humble JG, Lee ES, Phillips TM (1971) The effects of intravenous phytohaemagglutinin in a patient with disseminated malignant melanoma. A clinical and immunological study. Rev Europ Etudes Clin Biol 16:924 -927

Lotz MT, Rosenberg SA (1988) The immunologic treatment of cancer. CA 38:68-94

Mazumder A, Eberlein TJ, Grimm EA, Wilson DJ, Keenan AM, Aamoldt R, Rosenberg SA (1984) Phase I study of adoptive immunotherapy of human cancer with lectin activated autologous mononuclear cells. Cancer 53:896-905

Morris RE, Hoyt G, DeValeria P (1987) A comparison of the in vitro and in vivo immunologic activities of concanavalin A and its succinylated derivative. Transplantation Proc 19:626-627

Morris RE (1989) Concanavalin A dimers as therapeutic agents. United States Patent 4,889,842

Naspitz CK, Richter M (1968) The action of phytohemagglutinin *in vivo* and *in vitro*, a review. Prog Allergy 12:1-85

Numa K, Tani T, Kodama M (1990) Trial of anticancer immunotherapy with immobilized pokeweed mitogen: immunotherapy by extracorporeal circulation. Cancer Immunol Immunother 32:125-130

Robinson WE, Montefiori DC, Mitchell WM (1987) Evidence that mannosyl residues are involved in Human Immunodeficiency Virus Type 1 (HIV-1) pathogenesis. Aids Res Human Retroviruses 3:265-282

Stanton G (1991) E-Y Laboratories - Personal communication

Waxdal MJ(1974) Isolation, characterization, and biological activities of five mitogens from pokeweed. Biochem 13:3671-3677

Wimer BM (1989) The ideal biological response modifier. Molecular Biotherapy 1:311-317

Wimer BM (1990) Characteristics of PHA-L4 the mitogenic isolectin of phytohemagglutinin as an ideal biological response modifier. Molecular Biotherapy 2:4-17

Wimer BM (1990) Therapeutic activities of PHA-L4 the mitogenic isolectin of phytohemagglutinin. Molecular Biotherapy 2:74-90

Wimer BM (1990) Potential therapeutic applications of PHA-L4 the mitogenic isolectin of phytohemagglutinin. Molecular Biotherapy 2:196-200

Yokoyama K, Yano O, Terao T, Osawa T (1976) Purification and biological activities of Pokeweed (*Phytolacca americana*) mitogens. Biochim Biophys Acta 427:443-452

Zuckerman KS, LoBuglio AF (1977) Phytohemagglutinin skin tests in cancer patients. Cancer 39:2355-2361

New perspectives in the treatment of acute myeloid leukemia by hematopoietic growth factors

W. Hiddemann, B. Wörmann, M. Zühlsdorf, C. Reuter, E. Schleyer, C. Busemann, M. Kiehl, H. Garritsen, M. Königsmann, A. Boeckmann, and T. Büchner

Department of Internal Medicine
University of Münster
Albert-Schweitzer-Str. 33
4400 Münster
FRG

In the treatment of acute myeloid leukemia (AML) major improvements have been achieved in the late 1960ies and 1970ies when anthracyclines and cytosine arabinoside (AraC) were introduced and applied in combination. Since then progress has substantially slowed down although more effective supportive measures led to a reduction in early death rate and provided the basis for intensified initial therapy such as "double induction" developed by the German AML-Cooperative Group (AMLCG) (1,2). A current update of this approach indicates not only a high remission rate but also strongly suggests a considerable increase in long term survival (3). Still, the major problems in AML treatment remain early deaths during the initial phase of therapy, the primary resistance of leukemic cells against the applied cytostatic drugs and the high rate of leukemic relapses after successful first line treatment which is reduced but not prevented by prolonged myelosuppressive maintenance therapy (4).

A new perspective to at least partially overcome these problems has recently arisen from the discovery of hematopoietic growth factors such as GM-CSF and G-CSF, and other cytokines including IL 2 and their introduction into clinical studies.

1. Reduction of early deaths by GM-CSF following induction therapy

Inspite of intensive supportive care early deaths during the phase of induction therapy still occurs in 17 to 32% of patients at all ages and is even in the range of 27 - 52% in patients older than 60 years of age (2). The majority of lethal events is due to uncontrollable infections related to the severity and duration of treatment associated neutropenia. Thus, shortening the phase of severe neutropenia by the use of GM-CSF or G-CSF might lead to a reduction of the early death rate.

In fact, in first clinical studies an acceleration of neutrophil recovery was observed when applying GM-CSF after intensive chemotherapy and autologous bone marrow transplantion in patients with solid tumors and malignant lymphomas (5-8).

In AML, however, a similar approach is hampered by the possible risk to stimulate the re-growth of leukemic cells which were shown to be responsive to both G- and GM-CSF in in vitro studies (9-11). Hence, the risk to facilitate the re-growth of leukemic blasts must be jugded against the clinical benefits of shortening the critical period of neutropenia, especially in patients with a high probability of dying during the induction therapy from uncontrollable infections. Therefore, the first clinical phase II study initiated at the Department of Internal Medicine of the University of Münster was restricted to a well defined subgroup of AML patients carrying a risk for early deaths of more than 30% (12). This included patients above the age of 65 with de novo AML as well as patients with early first relapses or second and subsequent leukemic recurrences. Cytostatic therapy consisted of TAD 9 for induction treatment and S-HAM for relapsed disease (12). GM-CSF was not started before day 3 after completion of cytostatic therapy and was only begun if a bone marrow evaluation at this time revealed no residual leukemic blasts. GM-CSF was administered by continuous intravenous infusion at a dose of 250 $\mu g/m^2/d$ until a neutrophil count of $2000/mm^3$ was reached.

Between September 1987 and December 1989, 36 patients entered the study and 30 received GM-CSF. Fifty-six cases treated by the identical chemotherapy at similar indications served as control. In the GM-CSF treated group neutrophil recovery was significantly shorter as compared to the control group with medians of 10 versus 16 days ($p = 0.009$) after treatment with TAD-9 and 15 versus 24 days ($p = 0.043$) after S-HAM. Recovery of platelets was not significantly different. From the 30 patients receiving GM-CSF 18 (60%) attained a complete remission which substantially exceeds the corresponding response in the control group being 18 of 56 (32%) only ($p = 0.09$). Also, the frequency of early deaths was significantly reduced by GM-CSF to 14% versus 39% in the control group ($p = 0.009$).

2. Enhancement of AraC cytotoxicity on clonogenic leukemic blasts by GM-CSF pretreatment.

In addition to death from infection the primary resistance of leukemic cells against cytostatic drugs represents the second major cause for failure of induction therapy. Besides biochemical and genetic mechanisms of drug resistance, a low proportion of proliferating cells may limit the antileukemic activity, especially of cell cycle specific agents such as AraC and may cause a cell kinetic resistance (13). The stimulatory effect of hematopoietic growth factors on clonogenic leukemic cells offers in a new approach to overcome this restriction. By in vitro investigations several groups could demonstrate an increase in the colony formation of leukemic cells and of cells in S-phase after pretreatment with GM-CSF, G-CSF, and IL 3 which resulted in a higher cytotoxic effect of subsequently applied AraC (14-18). In order to further explore this strategy the effects of a 48 hr administration of GM-CSF (100 U/ml) or IL 3 (100 U/ml) on the proliferative activity of leukemic cells and on the intracellular metabolism and cytotoxic efficacy of a subsequent 12 hr application of AraC at doses of 0.1 - 100 μM were evaluated on bone marrow cells from 17 patients with AML. After GM-CSF or IL 3, an 1.2 to 2.4 fold increase in S-phase cells was observed in 9 of 14 (GM-CSF) and 7 of 11 (IL 3) cases. ^{3}H-AraC incorporation into the DNA was enhanced by 1.33 to 18.3 fold over respective controls in 14 of 17 patients. These effects translated into a 2.2 to 229.0 fold increase in the cytotoxic activity of AraC against clonogenic leukemic cells after GM-CSF or IL 3 pretreatment (19). Hence, GM-CSF and IL 3 enhance the intracellular metabolism of AraC and its incorporation into the DNA of leukemic cells leading to a higher antileukemic activity of AraC on CFU-L.

Based on these data, first clinical studies have been initiated already to explore the approach of pretherapeutic priming by GM-CSF before and during induction therapy for AML (20).

3. Inhibition of leukemic blast colony formation by IL 2 stimulated autologous cytotoxic lymphocytes.

Recent results from in vitro investigations and indirect evidence deduced from allogeneic bone marrow transplantation strongly suggest an antileukemic effect of autologous cytotoxic lymphocytes which may substantially contribute to the control of residual leukemic cells in complete remission after initial cytoreductive chemotherapy. In order to further explore this approach and to examine a potential stimulatory effect of IL 2 on the generation of anti-leukemic effector cells mononuclear cells were collected from 10 AML patients in complete remission and stimulated with IL 2 in vitro for 2 to 3 days. Leukemic blasts from the same patients collected at initial diagnosis served as target cells and were mixed with isolated effector cells at a ratio of 1:5 in a colony assay for 10 - 14 days. Successful colony formation was obtained in 4 samples. Leukemic blast colony formation was inhibited by CD16 positive cells in 4 of the 4 cases and by gamma/delta T-cells in 3 of 4 samples. These preliminary data strongly suggest an anti-leukemic effect of IL 2 stimulated cytotoxic lymphocytes against autologous leukemic blasts.

Based on these data and on supporting reports from other groups (22-25) the potential anti-leukemic effect of postremission treatment with IL 2 is currently explored in a clinical phase II study.

These examples illustrate the potential impact of hematopoietic growth factors on the treatment of acute leukemias and the new perspectives that may arise from their clinical application.

1. Gale RP, Foon KA (1987) Therapy of acute myelogenous leukemia. Semin Hematol 2224: 40-54
2. Büchner T, Hiddemann W (1990) Treatment strategies in acute myeloid leukemia - First line chemotherapy (AML) Blut 60: 61-67
3. Büchner T, Hiddemann W, Löffler H, Maschmeyer G, Aul C, Nowrousian M, Straif K,Hossfeld D, Heinecke A (1991) Improved cure rate by very early intensifica tion combined with prolonged maintenance chemotherapy in patients with acute myeloid leukemia (AML): Data from AMLCG. Sem Hematol: in press
4. Büchner T, Urbanitz D, Hiddemann W et al. (1985) Intensified induction and consolidation with or without maintenance chemotherapy for acute myeloid leukemia (AML): Two multicenter studies of the German AML Cooperative Group. J Clin Oncol 3: 1583-1589
5. Bronchud MH, Scarffe JH, Thatcher N, Crowther D, Souza LM, Alton NK, Testa NG, Dexter TM (1987) Phase I/II study of recombinant human granulocyte colony-stimulating factor in patients receiving intensive chemotherapy for small cell lung cancer. Br J Cancer 56: 809-813
6. Antman KS, Griffin JD, Elias A, Socinski MA, Ryan L, Cannistra SA, Oette D, Whitley M, Frei III E, Schnipper LE (1988) Effect of recombinant human granulocyte macrophage colony-stimulating factor on chemotherapy-induced myelosuppression. N Engl J Med 319: 593-598
7. Brandt SJ, Peters WP, Atwater SK, Kurtzberg J, Borowith MJ, Jones RJ,Shpall EJ, Bast RC, Gilbert CJ, Oette D (1988) Effect of recombinant human granulocyte-macrophage colony-stimulating factor on hematopoietic reconstitution after high-dose chemotherapy and autologous bone marrow transplantation. N Engl J Med 318: 869-876
8. Herrmann F, Schulz G, Lindemann A, Meyenburg W, Oster W, Krumwieh D, Mertelsmann R (1989) Hematopoietic responses in patients with advanced malignancy treated with recombinant human granulocyte-macrophage colony-stimulating factor. J Clin Oncol 7: 159-169
9. Vellenga E, Young DC, Wagner K, Wiper D, Ostapovicz D, Griffin JD (1987) The effects of GM-CSF and G-CSF in promoting growth of clonogenic cells in acute myeloblastic leukemia. Blood 69: 1771-1776
10. Asano Y, Shibuya T, Okamura S, Yamaga S, Otsuka T, Niho Y (1987) Effect of human recombinant granulocyte/macrophage colony stimulating factor and native granulocyte colony-stimulating factor on clonogenic leukemic blast cells. Cancer Res 47: 5647-5648
11. Cannistra SA, Groshek P, Griffin JD (1989) Granulocyte-macrophage colony-stimulating factor enhances the cytotoxic effects of cytosine arabinoside in acute myeloblastic leukemia and in the myeloid blast crisis phase of chronic myeloid leukemia. Leukemia 3: 328-334
12. Büchner T, Hiddemann W, Koenigsmann M, Zühlsdorf M, Wörmann M, Boeck mann A, Aguion-Freire E, Innig G, Maschmeyer G, Ludwig WD, Sauerland MC, Heinecke A, Schulz G (1991) Recombinant human granulocyte-macrophage colony-stimulating factor following chemotherapy in patients with acute myeloid leukemia at higher age or after relapse. Blood: in press

13. Raza A, Maheshwari Y, Mandava N, Yasin Z, Perven S, Mayers G, Browman G, Goldberg J, Gottlieb A, Grunwald H, Larson , Miller K, Vogler R, Preisler HD (1987) Cell cycle and drug sensitivity studies of leukemic cells that appear relevant in determining response to chemotherapy in acute nonlymphocytic leukemia. Sem Oncol 14 (Suppl 1): 217-225
14. Lista P, Porcu P, Avanzi GC, Pegoraro L (1988) Interleukin 3 enhances the cytotoxic activity of 1-ß-D-arabinofuranosylcytosine (araC) on acute myeloblastic leukaemia (AML) cells. Br J Haematol 69: 121-127
15. Miyauchi J, Kelleher CA, Wang C, Minkin S, McCulloch EA (1989) Growth factors influence the senisitivity of leukemic stem cells to cytosine arabinoside in culture. Blood 73: 1272-1278
16. Cannistra SA, DiCarlo J, Groshek P, Kanakura Y, Berg D, Mayer RJ, Griffin JD (1991) Simultaneus administration of granulcyte macrophage colony stimulating factor and cytosine arabinoside for the treatment of relapsed acute myeloid leukemia. Leukemia 5: 230-238
17. Andreeff M, Tafuri A, Hegewisch-Becker S (1990) Colony stimulating factors (rh G-CSF, rh GM-CSF, rh IL 3, and BCGF) recruit myeloblastic and lymphoblastic leukemic cells and enhance the cytotoxic effects of cytosine-arabinoside. In: Büchner T, Schellong G, Hiddemann W, Ritter J (eds) Acute Leukemias-Prognostic Factors and Treatment Strategies II, Springer Berlin, Heidelberg: 747-762
18. Brach M, Klein H, Platzer E, Mertelsmann, Herrmann F (1990) Effect of inter leukin 3 on cytosine arabinoside-mediated cytotoxicity of leukemic myeloblasts. Exp Hematol 18: 748-753
19. Hiddemann W, Kiehl M, Zühlsdorf M, Busemann C, Schleyer E, Wörmann B, Büchner T. (1991) Granulocyte-macrophage colony stimulating factor (GM-CSF) and interleukin 3 (IL 3) enhance the incorporation of cytosine arabinoside into the DNA of leukemic blasts and the cytotoxic effect on clonogenic cells from patients with acute myeloid leukemia (AML). Proc Am Ass Cancer Res, in press
20. Bettelheim P, Valent P, Andreeff M, Tafuri A, Haimi J, Gorischek C, Muhm M, Sillaber Ch, Haas O, Vieder L, Maurer D, Schulz G, Speiser W, Geissler K, Kier P, Hinterberger W, Lechner K (1991) Recombinant human granulocyte-macrophage colony-stimulating factor in combination with standard induction chemotherapy in de novo acute myeloid leukemia. Blood 77: 700-711
21. Pelzer P, Garritsen HSP, Wörmann B, Doornbos R, Büchter U, Büchner T, de Groot B, Greve J, Hiddemann W (1990) Inhibition of colony formation of leukemic blasts in acute myeloid leukemia by interleukin-2 stimulated autologous lymphocyte subsets. Blut 61: 176
22. Lotzová E, Savary CA, Herberman RB (1987) Induction of NK cell activity against fresh human leukemia in culture with interleukin 2. J Immunol 138: 2718-2727
23. Adler A, Chervenick PA, Whiteside Tl, Lotzová E, Herberman RB (1988) Inter leukin 2 induction of lymphokine-activated killer (LAK) activity in the peripheral blood and bone marrow of acute leukemia patients. I. Feasibility of LAK generation in adult patients with active disease and in remission. Blood 71: 709-716
24. Teichmann JV, Ludwig WD, Seibt-Jung H, Thiel E (1989) Induction of lymphokine-activated killer cells against human leukemic cells in vitro. Blut 59: 21-24

25. Slavin S, Eckerstein A, Hardan I, Ben Shahar M OR , Naparstek E, Weiss L (1990) Toward improvement of therapeutic strategies in leukemia by amplification of the immune responses against leukemia. In: Acute Leukemias: Prognostic Factors and Treatment Strategies II, Büchner T, Schellong G, Hiddemann W, Ritter J (eds) Springer Verlag Berlin Heidelberg New York: 36-40

Enzyme replacement therapy with macrophage-targeted human placental glucocerebrosidase in Gaucher disease

R. O. Brady and N. W. Barton

Developmental and Metabolic Neurology Branch
National Institute of Neurological Disorders and Stroke
National Institutes of Health, Bethesda, Maryland 20892, USA

Signs and Symptoms of Gaucher Disease

As part of the requirement for completion of his medical education, the French medical student Philippe C. E. Gaucher wrote a thesis in 1882, in which he described a patient with an enlarged spleen that he believed was a primary epithelioma of that organ (Gaucher, 1882). Descriptions of patients with similar presentations began to appear sporadically, and eventually the eponym "Gaucher disease" was applied to designate these individuals. The manifestations of Gaucher disease vary greatly, and three major clinical phenotypes are currently distinguished. The most prevalent is Type 1 Gaucher disease that comprises patients with involvement of systemic organs and the skeleton but the central nervous is spared The most frequent manifestations are splenomegaly, hepatomegaly, anemia, thrombocytopenia, and leukopenia. The abdomen is protuberant because of gross enlargement of the spleen and liver. The lungs may also be involved. Lymph nodes, pericardium, and the kidneys are less frequently affected. However, the skeleton is virtually always damaged. Acute episodes of bone pain and deformities of the femur are characteristic of the disorder. Other bones may also be involved, and fractures of the hip and vertebrae are not uncommon. Type 1 Gaucher disease is the most common hereditary lipid storage disorder, and it is the most prevalent genetic disease in persons of Ashkenazic Jewish ancestry where the carrier frequency may be as high as 1 in 10 persons.

The second phenotype is called Type 2 Gaucher disease. These patients exhibit hepatosplenomegaly very early in life. The central nervous system is extensively involved, and progressive neurological damage is the principal cause of death that generally occurs at about 9 months.

The third phenotype is known as Type 3 Gaucher disease. Two subtypes have been described in this category (Filling-Katz et al., 1990; Yu et al., 1990). Patients in both of these groups exhibit hepatosplenomegaly and bone involvement. Patients in the first subtype develop a progressive neurological syndrome in adolescence or early adulthood that includes cognitive decline, myoclonic seizures and horizontal gaze palsy. Usually their systemic disease is not severe. Patients in the second group exhibit horizontal gaze palsy as the sole neurological sign during childhood. The subsequent clinical course is marked by aggressive systemic disease with death in childhood or adolescence from hepatic or pulmonary complications.

Pathology

Enlarged histiocytic cells engorged with lipid that are called Gaucher cells are present in the spleen, liver, bone marrow and other organs of patients with all forms of Gaucher disease. These cells arise from precursors of the reticuloendothelial (monocyte/macrophage) system in the bone marrow. Accumulated glucocerebroside appears in the form of twisted tubular structures by electron microscopy. In the liver, the Kupffer cells are principally involved in the storage process. The liver may become cirrhotic, and occasionally patients die of acute hepatic failure. This process seems to be accelerated in some individuals by splenectomy. In the past, this operation was frequently performed to alleviate the anemia and thrombocytopenia caused by splenic hypertrophy and hyperfunction.

Gaucher cells also are found in the bone marrow. In some patients, the entire marrow space may become completely infiltrated with Gaucher cells. The bones become demineralized and weakened. Gaucher cells may also occur in alveolar capillaries and lymphatic vessels of the lungs. Many children with Gaucher disease are delayed in growth. The immune system is frequently affected, and monoclonal and polyclonal gammopathies occur in patients with Gaucher disease. Serum acid phosphatase activity is increased, and angiotensin-converting enzyme activity is elevated. Glucocerebroside is elevated in the blood as much as ten-fold over normal values.

Enzymatic Defect

It was reported in 1924 that the principal accumulating material in patients with Gaucher disease was cerebroside (Lieb, 1924). The specificity of the carbohydrate moiety was not determined, but was presumed to be galactose. A decade later, it was declared that the hexose of the accumulating cerebroside in Gaucher disease was glucose, not galactose (Aghion, 1934). Metabolic experiments carried out in slices of spleen tissue obtained from patients at splenectomy with ^{14}C-glucose and ^{14}C-galactose showed that there was no abnormality of carbohydrate metabolism in Gaucher disease (Trams and Brady, 1960). These experiments also revealed that there was no acceleration in the rate of glucocerebroside formation in this condition. This information led to the hypothesis that a defect in glucocerebroside catabolism was the cause of Gaucher disease (Trams and Brady, 1960).

In order to explore this possibility, radioactive glucocerebroside was chemically synthesized with ^{14}C in the glucose or in the fatty acid portion of the molecule. Metabolic experiments revealed that all mammalian tissues contain an enzyme called glucocerebrosidase that catalyzes the hydrolytic cleavage of glucose from glucocerebroside [Fig. 1] (Brady *et al.*, 1965a).

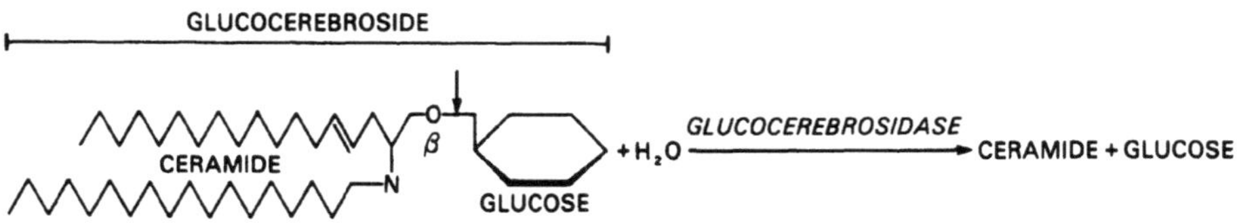

Figure 1. Structure of glucocerebroside that accumulates in Gaucher disease and site of hydrolytic cleavage catalyzed by glucocerebrosidase (vertical arrow).

It was discovered soon thereafter that glucocerebrosidase activity is significantly reduced from normal in tissues obtained from patients with Gaucher disease (Brady *et al.*, 1965b; Brady *et al.*, 1966). This information led to the development of reliable diagnostic and genetic counseling procedures that are widely used at the present time (Kampine *et al.*, 1966; Brady *et al.*, 1971; Schneider *et al.*, 1972). Most glucocerebrosidase activity is localized in lysosomes indicating that Gaucher disease is a lysosomal disorder (Weinreb *et al.*, 1968). The gene for glucocerebrosidase is located on human chromosome 1 at band q21 (Barneveld *et al.*, 1983). A number of deleterious mutations in this gene have been identified in patients with Gaucher disease (Eyal *et al.*, 1990). Much of the glucocerebroside that accumulates in the liver and spleen appears to arise from sphingoglycolipids in the membranes of senescent red and white blood cells (Kattlove *et al.*, 1969). Some of the glucocerebroside in the brain of patients with Type 2 and Type 3 Gaucher disease may be derived from these cells; however, glucocerebroside arising from the turnover of gangliosides in the brain is probably also a significant source of the accumulating lipid (Svennerholm *et al.*, 1982).

THERAPY

Background

In the past, total or partial splenectomy has been performed in numerous patients with Gaucher disease in order to relieve their anemia, thrombocytopenia, and leukopenia. Although these manifestations generally improve following this procedure, the benefit may be only temporary and other serious difficulties may arise. In some patients, the pathologic process in the liver and skeleton appears to be accelerated by splenectomy (Rose *et al.*, 1982; Ashkenazi *et al.*, 1986).

Four specific possibilities for treatment of patients with Gaucher disease have been considered. Organ transplantation including kidney, liver, and spleen allografts has been attempted. However, this approach has been uniformly unbeneficial. When successful, bone-marrow transplantation can produce a permanent cure. However, there is a high degree of morbidity and mortality associated with this time-consuming and difficult procedure. Moreover, because bone marrow replacement can be a cure for the disorder if it is successful, it is anticipated that gene replacement will be tried in patients with Gaucher disease. The rationale for this approach lies in the fact that the lipid-storing macrophages arise from stem cells in the bone marrow. A number of fundamental studies along this line are underway at present. Indeed, the enzymatic defect has been corrected by *in vitro* transfer of the human glucocerebrosidase gene into progenitor cells derived from patients with this disorder (Fink *et al.*, 1990). However, it is difficult to predict how soon clinical trials of gene replacement will be undertaken in patients with Gaucher disease or what clinical effects will result from this approach.

Enzyme Replacement

Soon after the enzymatic defect was discovered in Gaucher disease, replacement of the attenuated enzyme was suggested as a therapeutic strategy (Brady, 1966). In order to reduce the chance of sensitizing recipients to the exogenous enzyme, human placental tissue was selected as the source of glucocerebrosidase. Small quantities of glucocerebrosidase were isolated (Pentchev *et al.*, 1973) and injected intravenously into a boy with Type 3 Gaucher disease and into adult female with Type 1 Gaucher disease . Significant reduction of hepatic glucocerebroside was observed in the liver of both patients (Brady *et al.*, 1974). In addition, glucocerebroside associated with erythrocytes in the circulation declined from their previous three-fold elevation to normal values by 72 hours. The level of circulating glucocerebroside remained low for a number of months (Pentchev *et al.*, 1975). It became obvious that large quantities of glucocerebrosidase would be necessary in order to expect clinical benefit. We therefore developed a novel enzyme isolation procedure that took advantage of the exceptionally strong hydrophobic properties of the enzyme (Furbish *et al.*, 1977). It was discovered that glucocerebrosidase prepared by the newer method was not efficiently delivered to cells of the reticuloendothelial system such as the Kupffer cells in the liver (Furbish *et al.*, 1978). We became aware that glucocerebrosidase was a glycoprotein and that the majority of its oligosaccharide chains were of the complex (sialic acid terminal) type [Fig. 2]. Stahl and coworkers reported that macrophages contain a

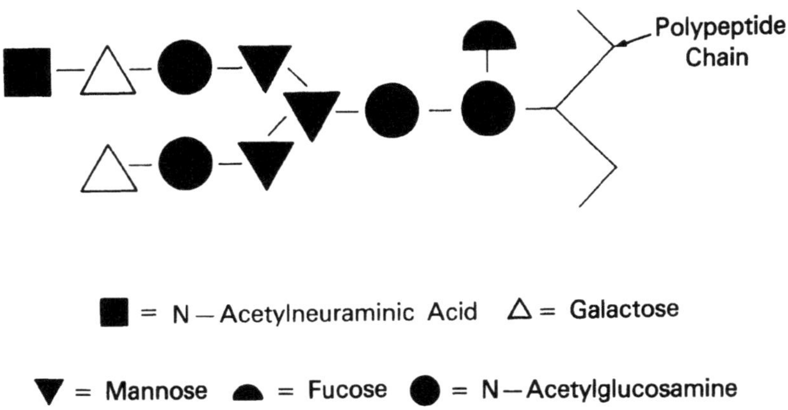

Figure 2. Carbohydrate sequence of three of the four oligosaccharide chains of human placental glucocerebrosidase (Takasaki *et al.*, 1984).

lectin that interacts with mannose-terminal glycoconjugates (Stahl *et al.*, 1984). We therefore examined several procedures to enrich purified placental glucocerebrosidase with mannose (Brady and Furbish, 1982). The most efficient approach was through sequential deglycosylation yielding a mannose-terminal enzyme [Fig. 3] (Furbish *et al.*, 1984). This modification led to greatly increased delivery of glucocerebrosidase to Kupffer cells in rats (Table I).

When weekly injections of 190 I.U. of this macrophage-targeted enzyme preparation were administered to a young male with Type 1 Gaucher disease, there was a dramatic improvement in the hemoglobin level [Fig. 4] . In addition, his platelet count increased. The size of his liver and spleen decreased, and there was improvement of his

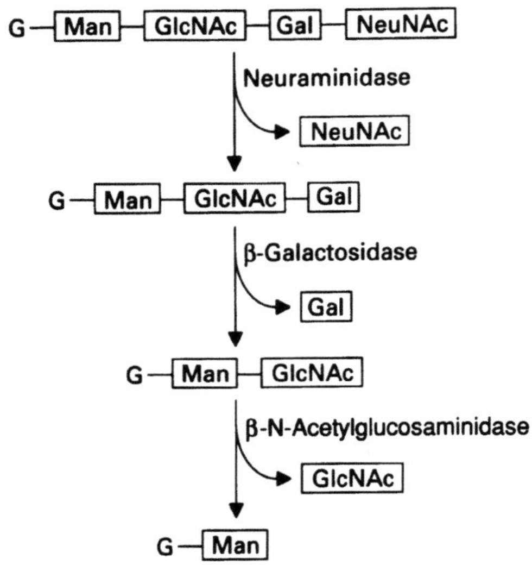

Figure 3. Enzymatic modification of human placental glucocerebrosidase (G). NeuNAc = N-acetylneuraminic acid; Gal = galactose; GlcNAc = N-acetylglucosamine; Man = mannose

Table I. Cellular distribution of mannose-terminal human placental glucocerebrosidase in rat liver.

Cell type	- Enzyme	+ Enzyme	Enrichment
	Enzymatic activity		
	(nanomoles/mg protein/hr)		(fold)
Hepatocytes	98	123	1.3
Kupffer cells	39	2216	57.

The values represent the mean of two separate experiments. Each rat received 50 IU of mannose-terminal glucocerebrosidase in the tail vein. One hour later, the animals were euthanized and the livers were removed. Cells were separated according to the procedure of Steer and Clarenburg (1979).

skeleton (Barton *et al.*, 1990; Brady and Barton, 1991). Figure 5 is a photograph of this patient at four years of age before receiving macrophage-targeted glucocerebrosidase. Figure 6 is a photograph of this patient at age 10 following weekly infusions of the enzyme.

We then carried out a comprehensive dose-response study in 23 patients with Gaucher disease. On the basis of information obtained in this investigation, we developed a protocol to investigate the therapeutic efficacy of macrophage-targeted human placental glucocerebrosidase in 12 non-splenectomized patients with Type 1 Gaucher disease. There were 8 children and 4 adults in this trial with gender equally divided. Intravenous infusions of 60 I.U. of macrophage-targeted human placental glucocerebrosidase per kg of

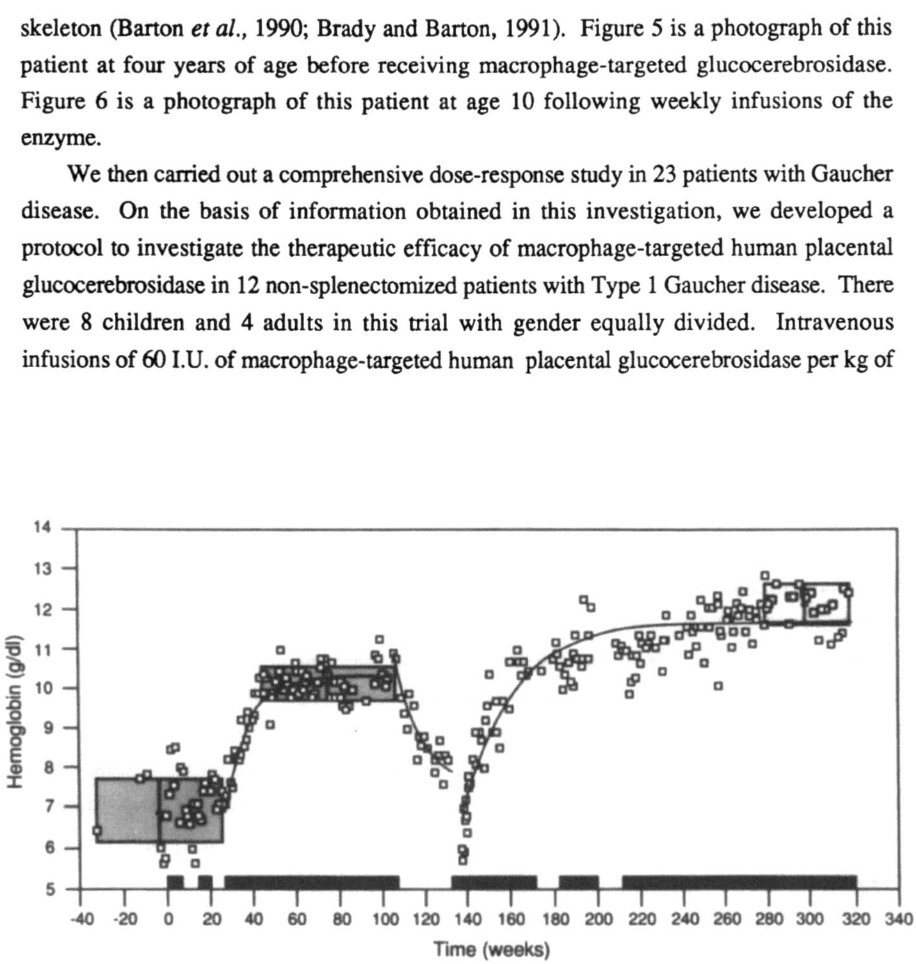

Figure 4. Increase in the hemoglobin level in a young male with Gaucher disease following the intravenous administration of macrophage-targeted human placental glucocerebrosidase. The black bars on the abscissa indicate the periods when the enzyme was administered.

body weight were administered over a one-hour period. After six months of bi-weekly infusions of this enzyme preparation, the hemoglobin concentration had increased significantly and the splenic volume had decreased in all patients (Barton *et al.*, 1990; Brady *et al.*, 1990). An increase in platelet count was observed in 7 of 10

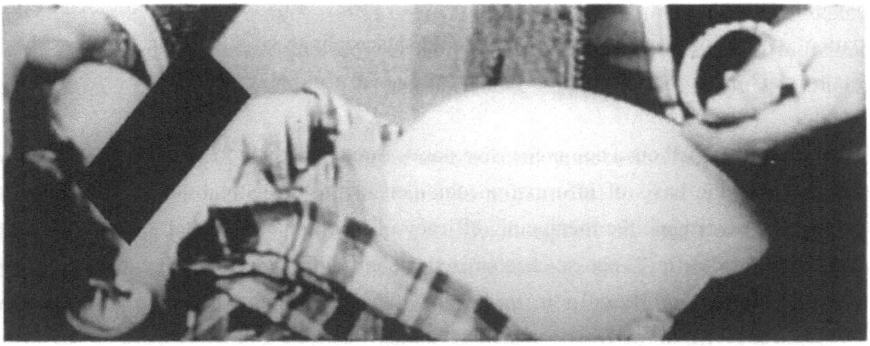

Figure 5. Photograph of the patient whose hemoglobin response is shown in Figure 3 at age four before enzyme replacement therapy. Note the abnormal contour of his abdomen caused by enlargement of the spleen and liver.

thrombocytopenic patients. Hepatic volume was significantly reduced in five of the recipients, and early signs of skeletal improvement were observed in three recipients during the trial (Barton *et al.*, 1991). No untoward reactions were encountered nor did any patient become sensitized to the macrophage-targeted enzyme (Murray *et al.*, 1991). Since enzyme replacement with macrophage-targeted human placental glucocerebrosidase produces a clear reversal of the signs and symptoms of Type 1 Gaucher disease and it has a remarkable safety profile, it is the treatment of choice for this form of the disease at this time. Enzyme replacement therapy is becoming widely used to treat patients with Type 1 Gaucher disease.

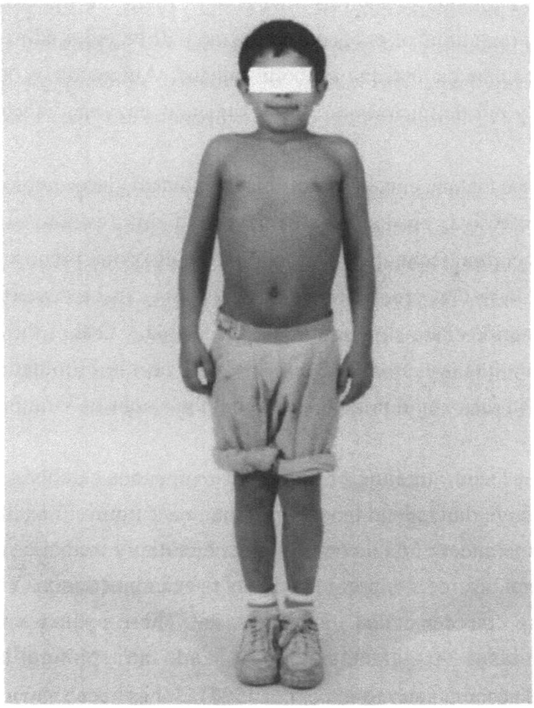

Figure 6. The same patient at age 10 after enzyme replacement therapy with macrophage-targeted glucocerebrosidase showing normalization of abdominal contour.

Future Prospects

It seems appropriate to consider enzyme replacement as a successful model for the use of lectins to target a wide range of therapeutic agents for. In a major sense, we have carried out an up-front debulking of the pathologic stores of glucocerebroside. This reduction is followed by a decrease in the size of the enlarged spleen and liver in these patients. The Gaucher cells in the bone marrow were replaced and the normal hematopoietic cell population was re-established (Parker *et al.*, 1991). We believe that once the stored glucocerebroside has been removed by comparatively large doses of the exogenous enzyme it will be possible to reduce the dose of enzyme dramatically. Conceivably, as little as one-fifth or one-tenth of the amount initially required to affect clinical improvement will be sufficient to maintain patients in a healthy state. This supposition is supported by the fact that even though patients have reduced glucocerebrosidase activity in their tissues, only a comparatively small amount of the glycolipid that arises from erythro- and leukocytophagy accumulates in these individuals daily (Kattlove *et al.*, 1969). It is therefore expected that comparatively small quantities of exogenous enzyme will be required to prevent further accumulation and keep the patients in metabolic balance. Alternatively, one could simply reduce the frequency of administration of the exogenous enzyme, which would also be advantagous.

One can anticipate that the clincial effects of recombinant glucocerebrosidase will soon be examined. If effective, production of enzyme in this fashion will have several advantages over the current technique of isolation from placental tissue and carbohydrate remodeling. It is likely that precise bioreactor systems can be developed to produce whatever oligosaccharide chain structure might be desired. One of the most important advantages of recombinantly produced enzyme will be the elimination of potential biohazards inherent in the natural product and avoid the extensive viral inactivation steps that are now used.

We believe that lectin targeting of numerous exogenous catalysts will become an extraordinarily effective therapeutic modality in the near future. In addition to Gaucher disease, successful treatment of a number of other hereditary metabolic storage disorders appears likely. Several disorders appear to be likely to examined soon. One of these is the sphingolipid storage disorder called Fabry disease. These patients lack the enzyme ceramidetrihexosidase (α-galactosidase A) and the sphingolipid known as ceramidetrihexoside accumulates (Brady *et al.*, 1967). It has been shown that infusion of human placental ceramidetrihexosidase reduces the accumulation of ceramidetrihexoside in the blood (Brady *et al.*, 1973). It is expected that clinical efficacy trials of placental or recombinant ceramidetrihexosidase will soon be carried out. In like manner, one can anticipate that enzyme replacement will be attempted in glycogen storage disorders such as Pompe's disease. The latter condition is caused by a lack of α-glucosidase. Here one would expect that the terminal oligosaccharide of the exogenous α-glucosidase should be mannose-6-phosphate since a lectin for this phosphorylated sugar is present on the surface of muscle cells which are extensively involved in this disorder.

Finally, the brain is a site of extensive pathology in a large number of hereditary metabolic diseases such as Type 2 and Type 3 Gaucher disease, Type A Niemann-Pick disease, Tay-Sachs disease, the Hurler syndrome (Mucopolysaccharidosis 1H) and many other conditions (Scriver *et al.*, 1989). Enzyme replacement for such disorders is a difficult at this time. Unless something is done to alter the blood-brain barrier, little or no intravenously injected enzyme reaches the substance of the brain. A number of experiments have been carried out in an effort to temporarily alter the blood-brain barrier, usually by the intracarotid injection of hyperosmolar solutions of mannitol (Barranger *et al.*, 1979). More effective methods for enzyme delivery to the brain is required in order to encourage examinations of enzyme replacement for metabolic disorders that involve the central nervous system. Here also, we believe that enzymes must be targeted to the specific types of cells within the nervous system that are involved in the storage process. Mannose-terminal oligosaccharide side chains may very well be required for the delivery of exogenous enzymes to neurons. The lessons learned during the development of successful enzyme replacement therapy for Gaucher disease appear to be applicable to many hereditary enzyme deficiency disorders of humans.

REFERENCES

Aghion H (1934) La maladie de Gaucher dans l'enfance. Thèse, Paris
Ashkenazi A, Zaizov R, Matoth Y (1986) Effect of splenectomy on destructive bone changes in children with chronic (Type 1) Gaucher disease. Eur J Pediatr 145:138-141.
Barneveld RA, Keijzer W, Tegelaers *et al.* (1983) Assignment of the gene coding for human β-glucocerebrosidase to the region q21-q23 of chromosome 1 using monoclonal antibodies. Hum Genet 64:227-231
Barranger JA, Rapoport SI, Fredericks WR *et al. (1979)* Modification of the blood-brain barrier: increased concentration and fate of enzymes entering the brain. Proc Natl Acad Sci USA 76:481-485
Barton NW, Furbish FS, Murray GJ, Garfield M, Brady RO (1990) Therapeutic response to intravenous infusions of glucocerebrosidase in a patient with Gaucher disease. Proc Natl Acad Sci USA 87:1913-1916
Barton NW, Brady RO, Doppelt SH et al (1990) Clinical effectiveness of enzyme replacement in Gaucher's disease. Clin Res 38:457A
Barton NW, Brady RO, Dambrosia JM et al (1991) Replacement therapy for inherited enyzme deficiency: effectiveness of macrophage-targeted glucocerebrosidase in Gaucher disease. N Engl J Med, in press
Brady RO, Kanfer J, Shapiro D (1965a) The metabolism of glucocerebrosides. I. Purification and properties of a glucocerebroside-cleaving enzyme from spleen tissue. J Biol Chem 240:39-42
Brady RO, Kanfer JN, Shapiro D (1965b) Metabolism of glucocerebrosides: evidence of an enzymatic deficiency in Gaucher 's disease. Biochem Biophys Res Commun 18:221-225
Brady RO, Kanfer JN, Bradley RM, Shapiro D (1966) Demonstration of a deficiency of glucocerebroside-cleaving enzyme in Gaucher's disease. J Clin Invest 45:1112-1115.
Brady RO: The sphingolipidoses (1966) N Engl Med 275:312-318
Brady RO, Gal AE, Bradley RM, Martensson E, Wasrshaw AL, Laster L (1967) Enzymatic defect in Fabry's disease: ceramidetrihexosidase deficiency. N Engl J Med 276:1163-1167
Brady RO, Johnson WG, Uhlendorf BW (1971) Identification of heterozygous carriers of lipid storage diseases. Am J Med 51:423-431
Brady RO. Tallman JF, Johnson WG *et al.* (1973) Replacement therapy for inherited enzyme deficiency: use of purified ceramidetrihexosidase in Fabry's disease. N Engl J Med 289:9-14
Brady RO, Pentchev PG, Gal AE, Hibbert SR, Dekaban AS (1974) Replacement therapy for inherited enyzme deficiency: use of purified glucocerebrosidase in Gaucher's disease. N Engl J Med 291:989-993
Brady RO, Furbish FS (1982) Enzyme replacement therapy: specific targeting of exogenous enzymes to storage cells. In: Martonosi AN (ed) Membranes and Transport, Vol 2. Plenum, New York, p587
Brady RO. Barton NW, Doppelt SH, Mankin HJ (1990) Efficacy of enzyme replacement in Type 1 Gaucher disease. Am J Hum Genet 47:A151
Brady RO, Barton, NW (1991) Enzyme replacement therapy for Type 1 Gaucher disease. In: Desnick RJ (ed) Treatment of genetic diseases, Churchill-Livingston, New York, in press.
Eyal N, Wilder S, Horowitz M (1990) Prevalent and rare mutations among Gaucher patients. Gene 96:277-283
Filling-Katz MR, Barton NW, Katz NNK (1990) Gaucher's disease. In: Gold DH, Weingeist TA, eds. The Eye in Systemic Disease. Philadelphia: JB Lippincott, p 365
Fink JK, Correll PH, Perry LK, Brady RO, Karlsson S (1990) Correction of glucocerebrosidase deficiency following retroviral-mediated gene transfer into

hematopoietic progenitor cells from patients with Gaucher disease. Proc Natl Acad Sci USA 87:2334-2338

Furbish FS, Blair HE, Shiloach J, Pentchev PG, Brady RO (1977) Enzyme replacement therapy in Gaucher's disease: large-scale purification of glucocerebrosidase suitable for human administration. Proc Natl Acad Sci USA 74:3560-3563

Furbish FS, Steer CJ, Barranger JA, Jones EA, Brady RO (1978) The uptake of native and desialylated glucocerebrosidase by rat hepatocytes and Kupffer cells. Biochem Biophys Res Commun 149:1047-1053

Furbish FS, Oliver KL, Zirzow GC, Brady RO, Barranger JA (1984) Interaction of human placental glucocerebrosidase with hepatic lectins. In: Barranger JA, Brady RO (eds) Molecular basis of lysosomal storage disorders, Academic Press, Orlando, p 219

Gaucher PCE (1882) De l'épithélioma primitif de la rate. Thèse de Paris

Kampine JP, Brady RO, Kanfer JN, Feld M, Shapiro D (1967) The diagnosis of Gaucher's disease and Niemann-Pick disease using small samples of venous blood. Science 155:86-88

Kattlove HE, Williams JC, Gaynor E, Spivack M, Bradley RM, Brady RO (1969) Gaucher cells in chronic myelocytic leukemia: an acquired abnormality. Blood 33:379-390

Lieb (1924) Cerebrosidespeicherung bei Splenomegalie Typus Gaucher. Ztschr Physiol Chem 140: 305-313

Murray GJ, Howard KD, Richards SM, Barton NW, Brady RO (1991) Gaucher's disease: lack of antibody response in 12 patients following repeated intravenous infusions of macrophage-targeted glucocerebrosidase. J Immunol Methods, in press

Parker RI, Barton NW, REad EJ, Brady RO (1991) Hematologic improvement in a patient with Gaucher's disease on long-term enzyme replacement therapy: evidence for decreased splenic sequestration and improved red blood cell survival. Am J Hematol, manuscript submitted

Pentchev PG, Brady RO, Hibbert SR, Gal AE, Shapiro D (1973) Isolation and characterization of glucocerebrosidase from human placental dissue. J Biol Chem 248:5256-5261

Pentchev PG, Brady RO, Gal AE, Hibbert SR (1975) Replacement therapy for inherited enyzme deficiency: sustained clearance of accumulated glucocerebroside in Gaucher's disease following infusion of purified glucocerebrosidase. J Mol Med 1:73-78

Rose JS, Grabowski GA, Barnett SH, Desnick RJ (1982) Accelerated skeletal deterioration after splenectomy in Gaucher Type 1 disease. AJR [Am J Roentgenology] 139:1202-1204.

Schneider EL, Ellis WG, Brady RO, McCulloch JR, Epstein CJ (1972) Infantile (Type II) Gaucher's disease: in utero diagnosis and fetal pathology. J Pediatr 81:1134-1139

Scriver CR, Beaudet AL, Sly WS, Valle D (1989) The metabolic basis of inherited disease, 6th edn. McGraw-Hill, New York

Stahl PD, Wileman TE, Shepard VL (1984) The mannose-recognition pathway-implications for lysosome physiology. In: Barranger JA, Brady RO (eds) Molecular basis of lysosomal storage disorders. Academic Press, Orlando, p 209

Steer CJ, Clarenburg R (1979) Unique distribution of glycoprotein receptors on parenchymal and sinusoidal cells of rat liver. J Biol Chem 254:4457-4461

Svennerholm L, Håkansson G, Månsson JE, Nilsson O (1982) Chemical differentiation of the Gaucher subtypes. In: Desnick RJ, Gatt S, Grabowski GA (eds) Gaucher disease: a century of delineation and research, Alan R Liss, New York, p 231

Takasaki S, Murray GJ, Furbish FS, Brady RO, Barranger JA, Kobata A (1984) Structure of N-asparagine-linked oligosaccharide units of human placental β-glucocerebrosidase. J Biol Chem 259:10112-10117.

Trams EG, Brady RO (1960) Cerebroside synthesis in Gaucher's disease. J Clin Invest 39:1546-1560.

Weinreb NJ, Brady RO, Tappel AL (1968) The lysosomal localization of sphingolipid hydrolases. Biochim Biophys Acta 159:141-146.

Yu K-T, Merrick HFW, Verderese C et al (1990) Horizontal supranuclear gaze palsy: a marker for severe systemic involvement in Type III Gaucher's disease. Neurology 40:S357.

Preparation of neoglycoprotein-bearing liposomes and their interaction with cells and tissues

N. Yamazaki, S. Kojima*, S. Gabius**, and H.-J. Gabius***

Industrial Products Research Institute
Agency of Industrial Science and Technology
Higashi 1-1-4, Tsukuba-city,
Ibaraki 305, Japan

Carbohydrate-binding proteins such as endogenous lectins are involved in various biological processes of the cells, including cell-to-cell recognition, communication and adhesion (Barondes, 1981; Gabius, 1988; Rademacher et al., 1988; Sharon and Lis, 1989). To evaluate the importance of carbohydrate-protein interactions for intercellular recognition, model systems which are based on an artificial lipid membrane, like liposomes, as a model of the cell membrane, can provide appropriate tools. Studies on carbohydrate-binding proteins using liposome-based systems might be not only of basic glycobiological importance, but also of clinical benefit. Cell-biological recognition processes can be mimicked to direct therapeutic drugs to diverse cell types, by utilizing neoglycoprotein-conjugated liposomes that convey the transport signal and the carbohydrate ligands on the membrane surface. The design of novel drug-targeting carriers involves routine methods of preparation of liposome conjugates (Yamazaki, 1987; Yamazaki, 1989), as well as assay-methods for the evaluation of in vitro and in vivo characteristics (Gabius et al., 1989; Gabius

*Research Institute for Biosciences, Science University of Tokyo, 2669 Yamazaki, Noda-shi 278, Japan, **Medizinische Universitatsklinik, Abteilung Hamatologie-Onkologie, Robert-Koch-Str. 40, D-3400 Gottingen, FRG, and *** Max-Planck-Institut fur experimentelle Medizin, Abteilung Chemie, Hermann-Rein-Str. 3, D-3400 Gottingen, FRG et al., 1990a; Kojima and Gabius, 1988; Kojima et al., 1990). We here report a procedure for the preparation of a new type of drug-targeting device which is based on neoglycoprotein/liposome conjugates and which exploits cellular functions of carbohydrate-binding proteins, and also assay-results for the evaluation of their interaction with cells and tissues.

PREPARATION OF LIPOSOMES, AND COUPLING OF NEOGLYCOPROTEINS TO LIPOSOMES

Fig.1 summarizes a procedure for the preparation of a type of neoglycoprotein/liposome conjugates. Liposomes were prepared by a standard method involving controlled detergent dialysis (Zumbuehl and Weder, 1981). The coupling of neoglycoproteins to liposomes was carried out according to a standard

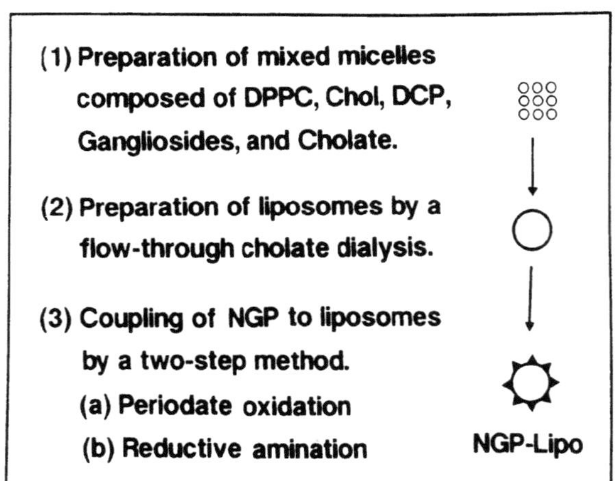

Fig. 1. Procedure for NGP/liposome conjugation. NGP, neoglycoproteins; NGP-Lipo, neoglycoprotein-coupled liposomes.

two-step procedure (Heath et al., 1981), including some modifications as described in detail elsewhere (Yamazaki, 1987; Yamazaki, 1989).

Covalent linking of proteins to liposomes by inclusion of glycolipid or ganglioside, followed by periodate oxidation and reductive amination, was reported many years ago (Urdal and Hakomori, 1980; Heath et al., 1980 and 1981) when avidin or antibody was linked to liposomes respectively. According to the detailed studies by Heath et al.(1981), periodate treatment at pH 8.4 oxidized a large proportion of the external glycolipid of liposomes without oxidizing the internal glycolipid. This method provides a useful basis for covalent coupling of biologically active proteins to the membrane surface of liposomes. We reported recently (Yamazaki, 1987; Yamazaki, 1989) a procedure for the coupling of lectins to liposomes by applying this method, which resulted in coupled protein/total lipid ratios between 0.03 and 0.3 g/g.

Table I. Neoglycoprotein coupling to liposomes

Type[a] of neoglycoprotein coupled	Protein/lipid ratio in the products (g/g)
Man-BSA	0.45
Mel-BSA	0.08
Fuc-BSA	0.05
GalNAc-BSA	0.02
Sialic acid-BSA	0.02
BSA	0.05

[a]Abbreviations: BSA, bovine serum albumin; Man, α -Mannose; Mel, Melibiose; Fuc, α -Fucose; GalNAc, α-N-acetylgalactosamine.

We have now attempted coupling of five types of neoglycoproteins including non-glycosylated bovine serum albumin which was used as a control (Table I). Under conditions of the coupling procedure, coupled protein/total lipid ratios in the final products were between 0.02 and 0.08 g/g, except liposomes bearing mannosylated neoglycoproteins when coupled protein was 0.45 g/g. We do not have a clear answer to the reason of this variability, but one possible explanation can be found in the different accessibility of various types of neoglycoproteins toward the membrane surface of liposomes. Heath _et al_.(1981) reported that the coupled protein/lipid ratio was related to the protein concentration of the reaction mixture. We need to find out optimal coupling conditions for each type of neoglycoproteins in order to achieve its high conjugation.

ANALYSIS OF THE PURITY AND THE STABILITY OF NEOGLYCOPROTEIN-COUPLED LIPOSOMES, AND IDENTIFICATION OF THE BOUND PROTEINS

Liposome preparations obtained had a mean diameter of about 100 nm, as judged by dynamic laser light scattering measurement. Characterization of all neoglycoprotein-coupled liposome preparations was carried out by using high-performance gel-permeation chromatography, as illustrated in Figs.2 and 3. Figs. show only 280 nm-absorbancy of peaks, but separation were also monitored by 90^{0} light scattering at 633 nm, which detected liposome-content of the peaks. The stability of the fucosylated neoglycoprotein-coupled liposome preparation is shown in Fig. 2. This preparation was stable at 37^{0} C for 24 hr. It is noteworthy that all these preparations had been kept at 7^{0} C over four months without changing their property. This high stability allows their use as drug targeting devices. Bound proteins of the BSA-coupled liposomes were almost completely dissolved by Triton-treatment (Fig.3), which enabled identification of the bound proteins.

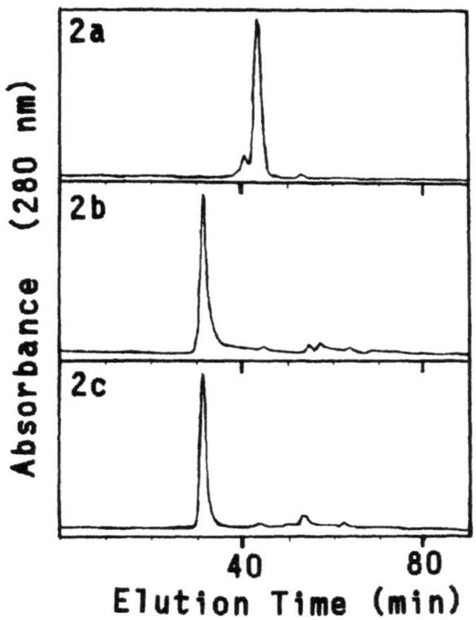

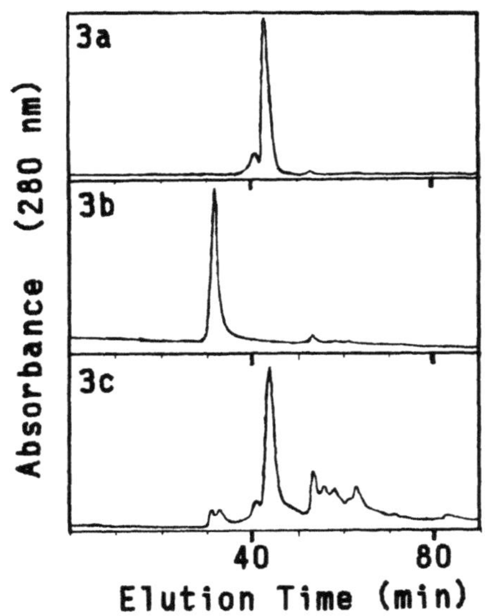

Fig. 2. Gel-permeation chromatography on HPLC-columns. Elution profiles of (2a) fucosylated BSA, (2b) fucosylated BSA-coupled liposomes, and (2c) fucosylated BSA-coupled liposomes after incubation at 37 °C for 24 hr.

Fig. 3. Gel-permeation chromatography on HPLC-columns. Elution profiles of (3a) BSA, (3b) BSA-coupled liposomes, and (3c) BSA-coupled liposomes after Triton-treatment at 37 °C for 3 hr.

AFFINITY OF NEOGLYCOPROTEIN-COUPLED LIPOSOMES TOWARD CELL-SURFACE SUGAR RECEPTORS OF HUMAN COLON ADENOCARCINOMA CELLS

Affinity of the liposome conjugates described above toward colon carcinoma cells was determined by using an inhibition assay employing homologous neoglycoenzyme. This assay is based on the principle that neoglycoprotein-coupled liposomes or neoglycoproteins, as inhibitors being tested, compete with neoglycoenzymes, as probes detecting sugar receptors, in binding specific cell-surface sugar receptors on cultured cells, as described in detail recently (Gabius *et al.*, 1989; Gabius *et al.*, 1990a). Therefore, the percentage of inhibition of neoglycoenzyme binding to cells relates to the affinity of the inhibitors toward the same type of cell-surface sugar receptors.

The model system presented here had been already shown to be feasible for targeted drug delivery in a system of colon carcinoma cells and neoglycoprotein-toxin conjugates (Gabius *et al.*, 1987). Presence of cell-surface sugar receptors on this colon carcinoma cell line has been ascertained by comparative analysis with a panel of neoglycoenzymes (Gabius *et al.*, 1990b), additionally emphasizing their modulation of expression by differentiation-inducing agents (Table II).

Table II. Modulation of expression of binding sites ($\times 10^3$) with specificity to carbohydrates on native human colon adenocarcinoma cells by treatment with chemical agents[a]

chemical treatment	Type of glycosylated-β-galactosidase of E. coli			
	Man-β-gal	Mel-β-gal	Fuc-β-gal	GalNAc-β-gal
without treatment	7.0	15.1	9.3	16.2
1 mM sodium butyrate	15.2	9.4	17.1	15.4
5 μM retinoic acid[b]	3.1	4.0	9.5	8.9
50 ng/ml TPA[c]	13.3	6.3	3.6	4.6

[a]Abbreviations: Man-β-gal, Mel-β-gal, Fuc-β-gal and GalNAc-β-gal are mannosylated-, melibiosylated-, fucosylated- and N-acetylgalactosaminylated-β-galactosidase, respectively. [b]all-trans retinoic acid; [c]12-o-tetradecanoylphorbol 13-acetate.

Table III. Inhibition of neoglycoenzyme binding to human colon adenocarcinoma cells

Inhibitor			Type[a] of Neoglycoenzyme (4 pmol/50 μl)	Inhibition[b] (%)
Type[a] of neoglycoprotein-coupled liposome and neoglycoprotein	Protein content (μg/50μl)	lipid content (μg/50μl)		
Man-BSA-Lipo	0.46	1.02	Man-β-gal	45
Man-BSA-Lipo	3.68	8.12	Man-β-gal	66
Man-BSA	3.73	0	Man-β-gal	c
Mel-BSA-Lipo	0.07	0.83	Mel-β-gal	
Mel-BSA-Lipo	0.56	6.64	Mel-β-gal	13
Mel-BSA	3.98	0	Mel-β-gal	
Lipo	0	8.44	Mel-β-gal	
Fuc-BSA-Lipo	0.08	1.50	Fuc-β-gal	
Fuc-BSA-Lipo	0.62	11.96	Fuc-β-gal	
Fuc-BSA	3.61	0	Fuc-β-gal	

[a]Abbreviations: Man-BSA-Lipo, mannosylated BSA-coupled liposome; Mel-BSA-Lipo, melibiosylated BSA-coupled liposome; Fuc-BSA-Lipo, fucosylated BSA-coupled liposome; Man-BSA, Mel-BSA and Fuc-BSA are mannosylated, melibiosylated and fucosylated neoglycoprotein, respectively; Lipo, protein-free oxidized liposome as a control; neoglycoenzymes are similar to those in Table II. [b]Data represent the average of duplicate determinations. [c]Blanks indicate that no inhibition was detected.

Results of the inhibition assays on untreated cells are shown in Table III. Mannosylated BSA-coupled liposomes and melibiosylated BSA-coupled liposomes were potent inhibitors, and neither protein-free oxidized liposomes nor free neoglycoproteins at the applied concentrations affected the interaction of the neoglycoenzymes with the cells, as shown in their effect on melibiosylated-β-galactosidase binding. The inhibitory potency correlated strictly with the amount of immobilized neoglycoproteins on membrane surface of liposomes. Liposomes with 0.46 μg mannosylated neoglycoprotein, corresponding to 7 pmol of the protein, reduced the cell binding of 4 pmol mannosylated-β-galactosidase by nearly 50 %, indicating that the I_{50} (the ligand concentration that inhibits 50 % of the neoglycoenzyme binding) of the mannosylated BSA-coupled liposome seems to be as low as 140 nM. It is noteworthy that such high affinity could be acquired through immobilization of the neoglycoproteins on membrane surface of liposomes. Free neoglycoproteins need to be applied at relatively high concentrations to inhibit the neoglycoenzyme binding to the cells, the dissociation constant of which can be as low as 27 nM (Gabius et al., 1990b). Decrease of molecular weight of the glycosylated carrier leads to notable increases in the dissociation constant (Gabius et al., 1990a). These results substantiated the high affinity-binding of neoglycoprotein-conjugated liposomes to the carcinoma cells via cell-surface sugar receptors.

TISSUE DISTRIBUTION OF RADIOIODINATED NEOGLYCOPROTEIN-COUPLED LIPOSOMES

To quantify tumor uptake in the more complex in vivo situation and to assess binding to individual tissues we determined the biodistribution of the ^{125}I-neoglycoprotein-coupled liposomes after injection into tumor-bearing mice (Table IV). Radioiodination was performed by application of chloramine-T, and ^{125}I-neoglycoprotein-coupled liposome injections were given seven days after Ehrlich ascites tumor cells had been inoculated into the left rear leg of male mice. The biodistribution was determined, as described for neoglycoproteins (Kojima and Gabius, 1988), 24 hr after injection of liposomes, bearing various types of ^{125}-neoglycoproteins, into the tail vein of Ehrlich solid tumor-bearing mice. The individual response on the level of organ content showed differences among different types of neoglycoprotein-conjugated liposomes. These data, together with previous works (Kojima and Gabius, 1988; Kojima et al., 1990), constitute the basis

Table IV. Biodistribution of ^{125}I-neoglycoprotein-coupled liposomes in Ehrlich-solid-tumor-bearing mice after 24 hr

Tissue	Type[a] of ^{125}I-neoglycoprotein-coupled liposomes				
	Man-BSA-Lipo	Mel-BSA-Lipo	Fuc-BSA-Lipo	GalNAc-BSA-Lipo	Sialic acid-BSA-Lipo
Blood	0.51±0.04[b]	0.12±0.03	0.09±0.01	0.08±0.01	0.20±0.02
Liver	0.57±0.01	1.17±0.07	1.56±0.12	1.53±0.17	3.32±0.28
Kidneys	0.24±0.01	0.18±0.02	0.24±0.03	0.14±0.02	0.27±0.01
Spleen	0.65±0.09	1.18±0.24	2.04±0.29	2.12±0.36	5.06±0.56
Heart	0.12±0.01	0.05±0.01	0.08±0.01	0.06±0.01	0.19±0.01
Lungs	0.16±0.01	0.11±0.02	0.11±0.01	0.13±0.02	0.29±0.01
Pancreas	0.07±0.01	0.04±0.01	0.04±0.00	0.04±0.01	0.13±0.03
Muscle	0.06±0.02	0.03±0.01	0.02±0.00	0.02±0.01	0.04±0.01
Brain	0.01±0.00	0.01±0.00	0.01±0.00	0.01±0.00	0.02±0.00
Tumor	0.21±0.03	0.14±0.05	0.09±0.02	0.09±0.01	0.31±0.05

[a]Abbreviations are similar to those in Table I and III.
[b]Each value represents % of the injected dose/g of tissue (%ID/g tissue), and is the mean binding ± SD for four animals.

for further refinements of the carbohydrate ligands of suitable neoglycoprotein-bearing liposomes to allow their potentially rational application as drug targeting devices.

SUMMARY

Five types of neoglycoprotein-coupled liposomes were prepared according to the following procedure; (1) preparation of mixed micelles composed of dipalmitoylphosphatidylcholine, Chol, dicetylphosphate, gangliosides and cholate, (2) preparation of liposomes by a flow-through cholate dialysis, and (3) coupling of neoglycoproteins to liposomes by a two-step method consisting of periodate oxidation and reductive amination. An inhibition assay of neoglycoenzyme binding to human adenocarcinoma cells indicated the high affinity-binding of neoglycoprotein-bearing liposomes to the cells. A tissue distribution assay using Ehrlich solid tumor-bearing mice showed individual response of diverse tissues toward different types of applied neoglycoprotein-bearing liposomes.

REFERENCES

Barondes SH (1981) Lectins: their multiple endogenous cellular functions. Ann Rev Biochem 50:207-231

Gabius H-J (1988) Mammalian lectins: their structure and their glycobiological and glycoclinical roles. ISI Atlas of Science: Biochemistry 1:210-214

Gabius H-J, Engelhardt R, Hellmann T, Midoux P, Monsigny M, Nagel GA, Vehmeyer K (1987) Characterization of membrane lectins in human colon carcinoma cells by flow cytofluorometry, drug targeting and affinity chromatography. Anticancer Res 7:109-112

Gabius S, Hellmann K-P, Hellmann T, Brinck U, Gabius H-J(1989) Neoglycoenzyme: a versatile tool for lectin detection in solid-phase assays and glycohistochemistry. Anal Biochem 182:447-451

Gabius S, Schirrmacher V, Franz H, Joshi SS, Gabius H-J(1990a) Analysis of cell-surface sugar receptor expression by neoglycoenzyme binding and adhesion to plastic-immobilized neoglycoproteins for related weakly and strongly metastatic cell lines of murine tumor model systems. Int J Cancer 46: 500-507

Gabius S, Yamazaki N, Hanewacker W, Gabius H-J (1990b) Regulation of distribution, amount and ligand affinity of sugar receptors in human colon carcinoma cells by treatment with sodium butyrate, retinoic acid and phorbol ester. Anticancer Res 10:1005-1012

Heath TD, Fraley RT, Papahadjopoulos D (1980) Antibody targeting of liposomes: cell specificity obtained by conjugation of $F(ab')_2$ to vesicle surface. Science 210:539-541

Heath TD, Macher BA, Papahadjopoulos D (1981) Covalent attachment of immunoglobulins to liposomes via glycosphingolipids. Biochim Biophys Acta 640:66-81

Kojima S, Gabius H-J(1988) Biodistribution of neoglycoproteins in mice bearing solid Ehrlich tumor. J Cancer Res Clin Oncol 114:468-472

Kojima S, Ishido M, Kubota K, Kubodera A, Hellmann T, Kohnke-Godt B, Wosgien B, Gabius H-J (1990) Tissue distribution of radioiodinated neoglycoproteins and mammalian lectins. Biol Chem Hoppe-Seyler 371:331-338

Rademacher TM, Parekh RB, Dwek RA (1988) Glycobiology. Ann Rev Biochem 57:785-838
Sharon N, Lis H (1989) Lectins as cell recognition molecules. Science 246:227-234
Urdal DL, Hakomori S(1980) Tumor-associated ganglio-N-triosylceramide: target for antibody-dependent, avidin-mediated drug killing of tumor cells. J Biol Chem 255:10509-10516
Yamazaki N (1987) Multivalent interaction between polysaccharide fragments and lectin-conjugated lipid vesicles. An affinity system to probe the structural features of carbohydrates. In: Zlatkis A (ed) Advances in Chromatography 1986, Part II. Elsevier, Amsterdam, pp 371-380
Yamazaki N (1989) Analysis of the carbohydrate-binding specificity of lectin-conjugated lipid vesicles which interact with polysaccharide fragments. J Membrane Sci 41:249-267
Zumbuehl O, Weder HG (1981) Liposomes of controllable size in the range of 40 to 180 nm by defined dialysis of lipid/detergent mixed micelles. Biochim Biophys Acta 640:252-262

Membrane recognition of steroid-glycoprotein complexes: A model for steroid delivery to the target cells

G. V. Avvakumov

Institute of Bioorganic Chemistry
Byelorussian Academy of Sciences
5/2 Zhodinskaya
220067 Minsk
U.S.S.R.

INTRODUCTION

Until recently, steroid-binding glycoproteins of human and animal plasma were believed to have the only physiological function, namely, protection of steroids against chemical and enzymatic attacks in the blood and rapid clearance from the circulation (Westphal, 1986). Discovery of the specific interactions of human transcortin (corticosteroid-binding globulin, CBG) (Strel'chyonok and Avvakumov, 1983) and sex-hormone-binding globulin (SHBG) (Strel'chyonok et al., 1984b) with the plasma membranes of steroid-responsive tissues indicated that physiological functions of the steroid-binding glycoproteins might be more diverse. Further investigation of the glycoprotein-membrane interactions (for review, see Hammond, 1990; Rosner, 1990; Strel'chyonok and Avvakumov, 1990) gave evidence for direct involvement of these glycoproteins in steroid targeting and steroid-mediated signalling. The exact mechanisms of these processes, properties of the corresponding membrane receptors and effector systems as well as the entire spectrum of biological effects produced by the interactions of steroid-glycoprotein complexes with cell membranes are still to be elucidated. However, the data obtained up to date allow some general conclusions to be done.

MEMBRANE BINDING OF STEROID-BINDING GLYCOPROTEINS

Table 1 summarizes the data on the CBG and SHBG interactions with cell membranes obtained in this laboratory.

Table 1. Specific binding of serum binding glycoprotein-hormone complexes to the plasma membranes of human steroid target tissues.

Complex	Tissue	K_d[a], mol/l	B_{max}[a], fmol per mg membrane protein
CBG-cortisol	Liver	$1 \cdot 10^{-6}$	10
SHBG-estradiol	Decidual endometrium	$3 \cdot 10^{-12}$	10
CBG-progesterone	Decidual endometrium	$1 \cdot 10^{-10}$	5
CBG-cortisol	Syncytiotrophoblast, type I sites	$2 \cdot 10^{-10}$	150
Pregnancy-associated CBG variant-cortisol	Syncytiotrophoblast, type II sites	$3 \cdot 10^{-12}$	3
SHBG-testosterone	Syncytiotrophoblast	$2 \cdot 10^{-12}$	10

[a] *Rounded mean values are shown for better comparison.*

The first indication to the existence of specialized binding sites for steroid-binding glycoproteins in cell membranes was obtained in a study of the interaction of asialo-CBG with human liver plasma membrane (Strel'chyonok and Avvakumov, 1983). We have found that, along with well-known asialoglycoprotein receptor, asialo-CBG bound to another type of binding sites with a Kd of approx. 10^{-10} M. Intact CBG competed effectively with asialo-CBG for these sites, the calculated Kd for the native glycoprotein (see Table 1) being close to the physiological blood level of CBG. These findings were further supported by a direct demonstration of the membrane binding of intact ^{125}I-CBG (Strel'chyonok and Avvakumov, 1990). Also, rat hepatic CBG receptor distinct from the asialoglycoprotein receptor and displaying an affinity for CBG characterized by a Kd value of about 10^{-6} M has recently been reported (Maitra et al., 1990). CBG receptors of this type were also found in various rat tissues (Khun, 1988) and human prostate (Hryb et al., 1986).

Comparison of the Kd values and physiological CBG concentration indicates the possibility of a two-way regulation of the receptor occupancy: at the level of the organism as a whole via alteration of the CBG concentration in the blood and at the level of the target tissue through variation of the exposed receptor density.

In the case of reproductive tissues such as endometrium and trophoblast, situation is quite different. CBG and SHBG binding to the plasma membranes of these tissues is characterized by K_d values ranged from 10^{-10} to 10^{-12} M (Table 1) which is much lower than the values of normal blood levels of these steroid-binding glycoproteins (10^{-6}-10^{-7} M). So, the receptor occupancy which determines receptor-mediated signalling is controlled only by the tissue itself via alteration of membrane receptor density. Hence, CBG and SHBG can supply, via membrane receptor systems, target cells with steroid hormones necessary for the normal development of feto-placental unit irrespectively of alterations in the hormone and glycoprotein levels in the maternal circulation.

Steroid dependence of the glycoprotein-membrane interactions

Conformational transitions of the steroid-binding glycoprotein molecules, accompanying formation of the complexes with steroids, depend on the nature of the complexed steroid (Westphal, 1986). This provides for a molecular basis for differential recognition of various steroid-glycoprotein complexes by the membrane binding sites. However, this phenomenon has been found to be tissue-specific. Thus, both CBG devoid of steroid and CBG complexed with either cortisol or progesterone bound to the plasma membrane of human liver with approximately equal affinity. In the case of high affinity binding to the reproductive tissue membranes (see Table 1), steroid dependence of the binding reactions is pronounced. Plasma membranes of human decidual endometrium display higher affinity for CBG complexed with progesterone than with cortisol and do not bind CBG complexed with testosterone or devoid of steroid (Avvakumov et al., 1989). Two types of the specific binding sites for CBG and its pregnancy-associated molecular variant display different steroid dependence of the binding. However, both types of the binding sites show no affinity for the CBG variants devoid of steroid (Strel'chyonok and Avvakumov, to be published). Pronounced steroid dependence is characteristic of the interactions of SHBG with the plasma membranes of both endometrium and placenta (see Table 1): only specific binding of either SHBG-estrogen or SHBG-androgen complexes was observed in the former and in the latter cases, respectively (Strel'chyonok et al., 1984b, Krupenko et al., 1990).

It should be noted that, apart from differential conformational transitions of the CBG and SHBG molecules accompanying their complexing with different steroids, steroid dependence of the membrane binding can be explained by the existence of membrane recognition systems for the steroid-glycoprotein complexes. In human endometrium membrane, such a recognition system consists of the two principal components, namely, binding site which recognizes SHBG, having the conformation that is characteristic of SHBG-estrogen complex, and estrogen membrane receptor (Strel'chyonok et al., 1984b). Each component of an SHBG-estrogen complex interacts with the corresponding component of the membrane recognition system. Existence of the recognition systems for the steroid-glycoprotein complexes could explain tissue specificity of steroid dependence of the SHBG and CBG membrane binding.

Involvement of the glycoprotein carbohydrates in the glycoprotein-membrane interaction

Evidence for the involvement of the CBG and SHBG carbohydrates in the glycoprotein interactions with cell membranes were obtained in our laboratory using naturally occurring molecular variants of human CBG as well as enzymically desialylated glycoproteins. As mentioned above, we have found (Strel'chyonok et al., 1984a) that along with "normal" CBG, one molecule of which contains three biantennary and two triantennary N-linked sugar chains of the N-acetyllactosamine type, postpartum human serum contains a CBG variant, pregnancy-associated variant or pCBG, all five sugar chains of which are triantennary oligosac-charides of the same type. The plasma membranes of human liver display about 4.5-fold lesser affinity for pCBG than for "normal" CBG (Vashkevich et al., 1984). Similarly, type I binding sites of the human trophoblast plasma membrane show about 15-fold greater affinity for "normal" CBG than for pCBG. In contrast to this, type II binding sites of the same membrane have an affinity for pCBG which is about 30 times higher than that for "normal" CBG (Avvakumov and Strel'chyonok, 1988). Since the polypeptide components of the CBG variants are similar (Avvakumov and Strel'chyonok, 1987), the only reason for their differential membrane binding is the difference of their carbohydrate structures. Moreover, tighter binding of pCBG to the type II binding sites indicates that the CBG carbohydrates are directly involved in the glycoprotein interaction with cell membrane. (An alternative explanation, which could be valid in the case of the liver and type I trophoblast binding sites, is that more branched sugar chains of pCBG shield some amino acid residues important for the membrane recognition of CBG).

It has been also demonstrated that glycoprotein desialylation influences the CBG and SHBG membrane binding. Again, the observed effects were tissue-specific. While the affinity of the liver membrane for asialo-CBG has been found to be significantly higher than that for the intact glycoprotein (see above), SHBG desialylation resulted in a complete loss of its ability to specifically bind to the endometrium membrane (Avvakumov et al., 1988).

In a search for monosaccharide residues involved in the formation of determinants for the CBG recognition by cell membranes, we have studied inhibition of the ^{125}I-CBG binding to the plasma membrane of human syncytiotrophoblast by various (neo)glycoproteins (synthesized by Dr. H.-J. Gabius and co-workers at the Max Planck Institute for Experimental Medicine, Goettingen, F.R.G.). Results of this study (Table 2) show that some of the (neo)glycoproteins tested, especially NeuAc-BSA and Man-BSA, competed with ^{125}I-CBG for the membrane binding sites. But, even introduces at very high concentrations, they could not completely inhibit the CBG binding. This means that, unlike classical lectins, CBG membrane receptor simultaneously binds several monosaccharide residues of various types.

Table 2. Inhibition, expressed as $(1\text{-}B/B_0)\cdot 100\%$, of the specific ^{125}I-CBG binding to human syncytiotrophoblast membrane by unlabeled CBG and (neo)glycoproteins introduced into the analytical samples at various concentrations.

Competitor	Concentration of competitor, mol/l			
	$2\cdot 10^{-9}$	$2\cdot 10^{-8}$	$2\cdot 10^{-7}$	$2\cdot 10^{-5}$
Unlabeled CBG	43	92	100	100
NeuAc-BSA	25	49	47	50
Man-BSA	23	40	40	39
Lac-BSA	10	23	20	25
(Glucuronic acid)-BSA	3	10	14	18
GlcNAc-BSA	2	7	7	10
Gal-BSA	0	5	8	9
Fuc-BSA	0	3	5	5

[a] *Mean values from three independent experiments*

Taken together, the above findings suggest that both the polypeptide and carbohydrate components of the steroid-binding glycoproteins are involved in the formation of determinants for the membrane recognition of the steroid-glycoprotein complexes. Conformational transitions of the glycoprotein molecule,induced by complexing particular steroid, seem to give rise to alteration of the spatial orientation of some parts of the polypeptide chain as well as carbohydrate chains located on the surface of the polypeptide globule. Multi-point interaction of the steroid-glycoprotein complex with the cell membrane requires several monosaccharide residues, presumably belonging to different sugar chains, to form a three-dimensional surface which is complementary to that of the membrane binding site. Importance of the integrity of the glycoprotein carbohydrate chains, namely, the presence of sialic acid residues could be explained by alteration of the overall conformation of the carbohydrate chains after the removal of these negatively charged, bulky residues. Such an explanation is in line with only partial inhibition of the ^{125}I-CBG binding to the tropho-blast membrane by a great excess of NeuAc-BSA (see Table 2).

PUTATIVE MECHANISMS AND PHYSIOLOGICAL SIGNIFICANCE OF THE STEROID-GLYCOPROTEIN COMPLEX INTERACTION WITH CELL MEMBRANES

There is general agreement that the main pathway through which steroids realize their hormonal activity involves modulation of gene expression via the interaction of cytoplasmic receptor-steroid complexes with acceptor sites of chromatin. So, penetration of steroids inside the responsive cells is a prerequisite for the manifestation of their biological effects.

Physiological significance of the steroid-glycoprotein complex interaction with cell membranes is that steroid-binding glycoproteins of the plasma are thus involved in targeting steroids. This means that, in spite of a wide-spread concept, the principal hormonally active fraction is represented by steroids circulating as complexes with the specific binding glycoproteins (accounting to 30-90% of the total steroid blood level) and not by "free" steroids (1-10% of the total blood level). Furthermore, unlike "free" steroids that penetrate plasma membranes of every tissue, steroids complexed with the binding glycoproteins are directly transported to the target cells.

Interaction of the steroid-glycoprotein complexes with cell membranes can follow several putative mechanisms.

1. Reception of the glycoprotein independent on the complexed steroid. This seems to be the case when CBG interacts with the human liver plasma membrane (see above). Even in this instance, retardation of CBG in the tissue will increase local concentrations of the CBG-bound steroids. Rapid dissociation of the CBG-steroid complex under physiological conditions will obviously result in the enhanced delivery of the steroid to this particular tissue. This effect will be even more pronounced if dissociation of the steroid-glycoprotein complexis facilitated by the glycoprotein-receptor interac tionidge, 1987) or proteolytic modification of the glycoprotein (Hammond et al., 1990).

2. Internalization of the steroid-glycoprotein complex. Receptor-mediated endocytosis is known to be a general mechanism of the uptake of various low-molecular-weight ligands with the involvement of the serum binding proteins. Similar mechanism appears to take place when CBG-steroid complexes interact with the plasma membrane of human syncytiotrophoblast. Using a microvesicular

fraction of this membrane, we have found (Strel'chyonok and Avvakumov, to be published) that both of the components of CBG-cortisol complex, introduced into the incubation medium, accumulated in the intravesicular space.

3. "Shuttle" mechanism. As mentioned above, binding sites for the steroid-binding glycoprotein and membrane steroid receptor can form a unified recognition system for the steroid-glycoprotein complex. In this case, membrane binding of the steroid-binding glycoprotein, having a particular conformation induced by complexing the steroid, is followed by the transmission of the steroid to its membrane receptor. The steroid is, then, transferred into the cell via established mechanism (Szego and Pietras, 1981). Devoid of steroid, the steroid-binding glycoprotein changes its conformation, loses the affinity for the membrane and re-enters the blood wherein it can bind another steroid molecule and transport it to the responsive cell. Thus, the steroid-binding glycoprotein function as a transport "shuttle" delivering steroids (as well as drugs which it can complex with) to the responsive cells.

An example of such a mechanism is represented by the interaction of cortisol complex of the pregnancy-associated CBG variant with the syncytiotrophoblast. Unlike "normal" CBG, this variant does not penetrate the plasma membrane. However, it does not reduce the rate of cortisol accumulation in the intravesicular space of the microvesicular preparation of the membrane (Strel'chyonok and Avvakumov, to be published). This clearly demonstrates that the glycoprotein carbohydrates are also involved in the processes of the transmembrane transfer of the components of the steroid-glycoprotein complexes. In our opinion, "shuttle" mechanism also takes place when SHBG-estrogen, SHBG-danazol and CBG-progesterone interact with the plasma membrane of human decidual endometrium (see Table 1).

4. Modulation of the second messenger systems. The involvement of the steroid-binding glycoproteins in the steroid interaction with cell membranes makes possible certain parallels in receptor-mediated signal propargation for steroids and protein hormones that do not penetrate cell membranes. Thus, Rosner and co-workers reported activation of the adenylate cyclase system by the action of SHBG or CBG on cell membranes (Rosner, 1990). We have observed similar effects induced by the action of CBG-cortisol complex on human trophoblasts (Strel'chyonok and Avvakumov, to be published).

CONCLUSION

The involvement of steroid-binding glycoproteins of plasma in the steroid interaction with cell membranes has dramatically changed traditional views on the mechanisms of steroid hormone action as well as on the biological functions of these glycoproteins and, in particular, their carbohydrate moieties. Carbohydrate-protein interactions play an important role in the steroid targeting and signal propargation. Various types of monosaccharide residues as well as peptide arrays are likely to be involved in the formation of determinants for the membrane recognition of the steroid-glycoprotein complexes.

REFERENCES

Avvakumov GV, Krupenko SA, Strel'chyonok OA (1989) Study of the transcortin binding to human endometrium plasma membrane. Biochim Biophys Acta 984:143-150

Avvakumov GV, Strel'chyonok OA (1987) Properties and serum levels of pregnancy-associated variant of human transcortin. Biochim Biophys Acta 925:1-6

Avvakumov GV, Strel'chyonok OA (1988) Evidence for the involvement of the transcortin carbohydrate moiety in the glycoprotein interaction with the plasma membrane of human placental syncytiotrophoblast. Biochim Biophys Acta 938:1-6

Avvakumov GV, Zhuk NI, Strel'chyonok OA (1988) Biological function of the carbohydrate component of the human sex steroid-binding globulin. Biochemistry USSR 53:726-729

Hammond GL (1990) Molecular properties of corticosteroid binding globulin and the sex-steroid binding proteins. Endocrine Rev 11:67-79

Hammond GL, Smith CL, Patterson NAM, Sibbald WJ (1990) A role for corticosteroid-binding globulin in delivery of cortisol to activated neutrophils. J Clin Endocrinol Metab 71:34-39

Hryb DJ, Khan MS, Romas NA, Rosner W (1986) Specific binding of human corticosteroid-binding globulin to cell membranes. Proc Natn Acad Sci USA 83:323-3256

Krupenko NI, Avvakumov GV, Strel'chyonok OA (1990) Binding of human sex hormone-binding globulin-androgen complexes to the placental syncytiotrophoblast membrane. Biochem Biophys Res Commun 171:1279-1283

Khun RW (1988) Corticosteroid binding globulin interactions with target cells and plasma membranes. Ann NY Acad Sci 538: 146-158

Maitra US, Khan MS, Zhang XH, Rosner W (1990) The rat hepatic CBG receptor: distinction from the asialoglycoprotein receptor. Endocrinology 127:278-284

Pardridge WM (1987) Plasma protein-mediated transport of steroid and thyroid hormones. Am J Physiol 252:E157-E164

Rosner W (1990) The functions of corticosteroid-binding globulin and sex hormone-binding globulin: recent advances. Endocrine Rev 11:80-91

Strel'chyonok OA, Avvakumov GV (1983) Evidence for the presence of specific binding sites for transcortin in human liver plasma membranes. Biochim Biophys Acta 755:514-517

Strel'chyonok OA, Avvakumov GV (1990) Specific steroid-binding glycoproteins of human blood plasma: novel data on their structure and function. J Steroid Biochem 35:519-534

Strel'chyonok OA, Avvakumov GV (to be published) Interactions of human CBG with cell membranes. J Steroid Biochem

Strel'chyonok OA, Avvakumov GV, Akhrem AA (1984a) Pregnancy-associated molecular variants of human serum transcortin and thyroxine-binding globulin. Carbohydr Res 134:133-140

Strel'chyonok OA, Avvakumov GV, Survilo LI (1984b) A recognition system for sex-hormone-binding protein-estradiol complex in human decidual endometrium plasma membranes. Biochim Biophys Acta 802:459-466

Szego CM, Pietras RJ (1981) Membrane recognition and effector sites in steroid hormone action. In: Litwack G (ed) Biochemical actions of hormones, vol VIII. Academic Press, New York, p 304

Vashkevich II, Matveentseva IV, Avvakumov GV, Strel'chyonok OA (1984) Comparative study of human transcortin isolated from normal donor serum and retroplacental serum. Biochemistry USSR 49:1977-1985

Westphal U (1986) Steroid-protein interactions, part II. Springer, Berlin Heidelberg New York

Tumor cell-bone marrow stromal cell interaction: Neoglycoproteins as tools in leukemia research

S. Gabius, S. S. Joshi*, J. G. Sharp*, and H.-J. Gabius**

Abteilung Hämatologie-Onkologie
Medizinische Universitätsklinik
Robert-Koch-Str. 40
D-3400 Göttingen
FRG

Introduction

Multifarious molecular mechanisms can contribute to establish the intriguing accuracy of cell adhesion. They also govern pattern formation in pathological conditions, e.g. tumor dissemination. In addition to the establishment of intercellular contacts individual types of molecular interactions will convey information beyond the positioning of the cell that influences the cellular behavior. To decipher the code system, which is operative in adhesion and subsequent biosignalling, it is a rational approach to deliberately select a certain class of molecules with potential for specific interactions, to assess their presence on the cell surface in a model system with physiological relevance and to infer their contribution to adhesion and, subsequently, to regulatory processes with impact on growth modulation and/or differentiation. The elucidation of such reaction pathways can clinically be translated into improved disease management, when the cellular system under investigation has appropriate significance. Obviously, this idea is one intention to put bone marrow stromal cell - tumor cell recognition under scrutiny.

*Department of Anatomy, University of Nebraska Medical Center, 600 South 42nd Street, Omaha, NE 68198-6395, USA
**Max-Planck-Institut für experimentelle Medizin, Abteilung Chemie, Hermann-Rein-Str. 3, D-3400 Göttingen, FRG

Hematological and glycobiological concepts

Since the elaboration of conditions for long-term bone marrow cultures has facilitated to reproducibly gain access to adherent stromal cell layers that are the closest known laboratory equivalents to the in vivo prototype (Dexter et al., 1977; Gartner and Kaplan, 1980; Hocking and Golde, 1980; Coulombel et al., 1983), its interaction with progenitor cells as well as tumor cells can be studied. A representative illustration of such a bone marrow culture is given in Fig. 1. Remarkably, cellular

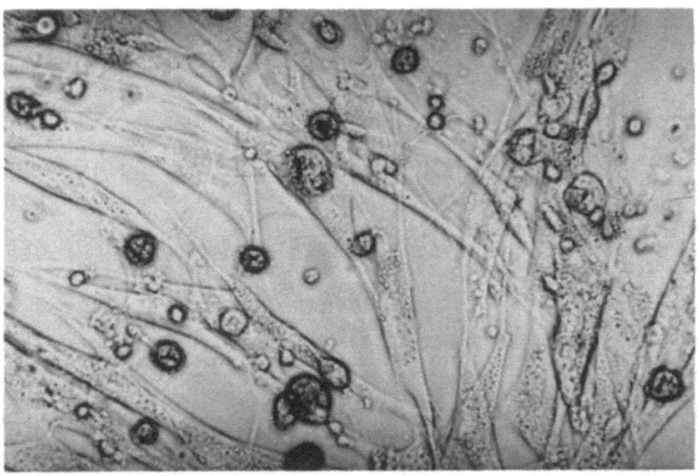

Light micrograph of a human long-term bone marrow culture after three weeks, mainly comprised of adherent elongated fibroblastoid cells, to which cells of hemopoietic origin are attached.

contacts between myeloid leukemia cells and such stromal cell layers are capable to affect the proliferation and to induce the differentiation of the tumor cells (Schölzel and Löwenberg, 1985; Ohkawa and Harigaya, 1987; Greenberg et al., 1988; Zuhrie et al., 1988; Luikart et al., 1990; Ossenkoppele et al., 1990). This proven impact of cell-to-cell contact on cellular parameters thus gives ample reason to concentrate on this model system. Concerning our experimental design to attribute observed cellular reactions to defined kinds of molecular interplay, the increasing evidence for involvement of specific protein-carbohydrate interactions in cellular recognition and biosignalling has prompted to pay attention to this aspect, especially to endogenous lectins (Liener et al., 1986; Olden and Parent, 1987; Gabius and Nagel, 1988; Bock and Harnett, 1989; Sharon and Lis, 1989; Gabius and Gabius, 1990a; Gabius, 1991). We have consequently started the investigations with experiments to verify presence of lectins. Having intentionally committed ourselves to the cell system as well as the class of supposedly relevant receptors, adequate markers need to applied for this purpose.

Lectin detection with synthetic probes

Although lectins are by definition devoid of enzymatic activity, their binding activity to carbohydrate ligands provides an appropriate means to detect their presence. Thus, chemical glycosylation has been employed to obtain effective markers for lectin detection (Stowell and Lee, 1980; Schrevel et al., 1981; Gabius and Bardosi, 1991). The attachment of various constituents of carbohydrate chains of glycoconjugates to a suitably labelled carrier enables to disclose the presence of endogenous sugar receptors in cells. Performance of this technique with a panel of carbohydrate ligands has ascertained the expression of sugar receptors for human bone marrow stromal cells (Gabius and Gabius, 1990b). The level of expression in these cells is responsive to changes in the microenvironment, caused by alterations of the medium composition (Table 1). However, the intensity of staining is evaluated semiquantitatively. The ensuing question to quantitatively disclose the amount of such activities on the cell surface of native stromal cells is addressed in binding studies that exploit a neoglycoprotein, whose association to the cell is monitored by a convenient spectrophotometric enzyme assay. In this case, carbohydrate ligands are coupled not to an inert carrier, but to an enzyme, for which a substrate is commercially available. The cell surface binding of the synthetic marker, referred to as neoglycoenzyme, can be thus evaluated non-radioactively with high sensitivity (Gabius et al., 1989; 1990a,b). On account of the assumed relevance of carbohydrate-mediated binding to cell adhesion a relatively large enzyme is

Table 1. Binding of neoglycoproteins and lectins to human bone marrow stromal cells, cultured in the presence or absence of different medium supplements

supplement / type of marker[a]	standard[e]	standard -hydrocortisone	standard -hydrocortisone -horse serum	standard +IL-3 (100 U/ml)	standard +10 % conditioned medium[f]
lactose	++	+	+++	+	+++
VAA	+++	+++	++	+++	++
RCA	+	(+)	+	(+)	++
gal-β-1,3-galNAc	(+)	(+)	(+)	-/(+)	+
PNA	+	+	+/++	+	+
melibiose	(+)	(+)	+	(+)	(+)
galNAc[b]	+	+	++	++	(+)
HPA	+	+	+	(+)	(+)
glcNAc[c]	++	+	++/+++	(+)	+/++
D-mannose	++	+	+	+	+
ConA	+++	+++	+++	+++	+++
L-fucose	++	++	+++	++	++
UEA	+	+	+	+	+
maltose	++	+/++	++/+++	+	++
cellobiose	+	(+)	+	+	+
L-rhamnose	+++	++	++/+++	+	++/+++
D-xylose	+/++	+/++	+/++	+	++/+++
sialic acid	(+)	(+)	++	++	(+)/+
glucuronic acid	++	++	+/++	(+)	+/++
D-mannose-6-P[d]	+++	+/++	++	+/++	++
D-galactose-6-P	+/++	+	++	++/+++	+
heparin	++	++	++/+++	++	++
fucoidan	(+)/++	(+)	+/++	+++	(+)/+
chondroitin-4-sulfate	(+)/+	+	+/++	+/++	++
dextran sulfate-2,3	+/++	+	+	+	++

[a]the neoglycoprotein is designated by its carbohydrate part and plant and invertebrate lectins by common abbreviations; the staining intensity with the individual probes is grouped into four categories (+), +, ++ and +++; [b]N-acetyl-D-galactosamine, exhibiting remarkable plasma membrane positivity; [c]N-acetyl-D-glucosamine; [d]phosphate; [e]the standard conditions are defined by a medium containing 12.5 % fetal calf serum, 12.5 % horse serum, 1 μM hydrocortisone, 100 μM β-mercaptoethanol, 2 mM L-glutamine, 1 % BME amino acid solution as well as 1 % MEM vitamin solution, 4 mg/100 ml inositol, 1 mg/100 ml folic acid, penicillin (100 U/ml) and streptomycin (100 μg/ml); [f]conditioned medium, containing hemopoietic growth factors, originated as supernatant from cultures of the human primary bladder carcinoma line 5637; from Gabius and Gabius, 1990b.

modified to place several ligands in spatial vicinity and, thereby, to generate high-affinity binding that is supposed to mimick cell-to-cell contact. Decrease of the size of the neoglycoenzyme should be accompanied by reduction in ligand affinity. Indeed, dissociation constants increase, when a monomeric enzyme is taken instead of the usually chosen tetrameric β-galactosidase from E. coli, with concomitant enhancement of the number of available binding sites on the tumor cells (Gabius et al., 1990a,b). Mutual glycobiological recognition can also encompass respective receptors on leukemic cells. In parallel, the neoglycoenzymes prove valuable to similarly determine the surface expression of lectins on these tumor cells. A detailed account of such a binding study is illustrated in Fig. 2.

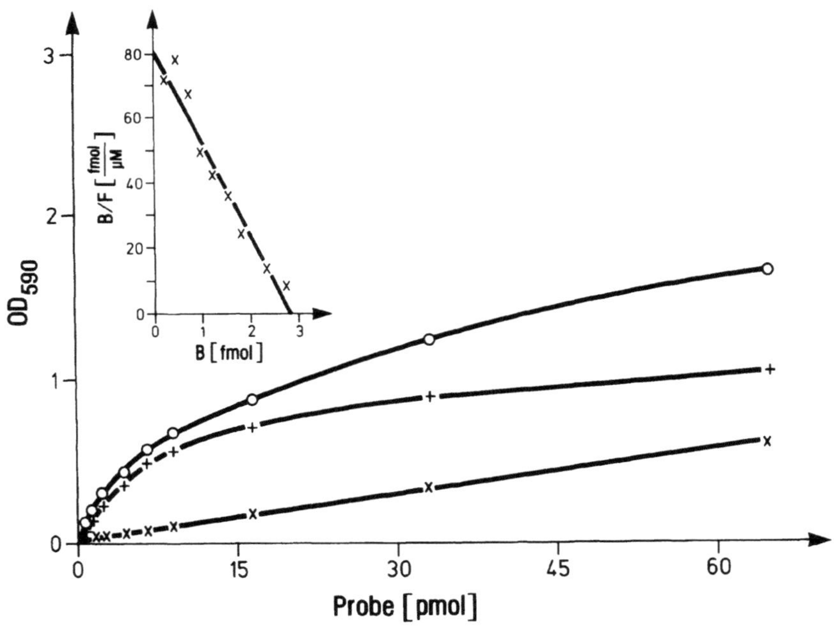

Determination of specific cell binding (+) of fucosylated E. coli β-galactosidase, obtained by subtracting the extent of unspecific binding (x) from the measureable amount of total binding (o), and Scatchard plot analysis of the binding data for the human myeloid marker-expressing cell line Croco II, established and characterized recently (Gabius et al., 1991a). Calculation yields a dissociation constant of 34 nM and 2.5 x 10^4 bound enzymes at saturation.

Due to the sensitivity of the enzymatic assay and the binding of the probes by several receptor sites that enhances the apparent affinity even rather low levels of expression of sugar receptors are amenable to detection. The hereby revealed presence of sugar receptors on human leukemic cells is by no means exceptional for blood cells. Various types of cells in the immune system exhibit this property (Gabius, 1987; Monsigny et al., 1988; Yednock and Rosen, 1989; Tavassoli and Hardy, 1990). It is now logical to devise assay systems to study their actual contribution to the establishment of adhesive contacts.

Carbohydrate-mediated adhesion to a model matrix

Carbohydrate ligands of cellular surfaces, among a variety of other types of ligands, may well serve in recognitive interactions. To unequivocally delineate this capacity of defined ligands, the inherent molecular complexity of the cellular surface inevitably needs to be reduced. Since the carbohydrate part of neoglycoproteins specifically interacts with receptor sites, its ligand properties with respect to adhesion can be examined, separated from the influence of any other type of ligand. Immobilized glycoconjugates with a defined structure offer the possibility to correlate structure of the carbohydrate sequence to the cellular response (Schnaar, 1984; Brandley and Schnaar, 1986). When neoglycoproteins are attached to a plastic surface, a model matrix is thereby constituted that exposes the chosen type of ligand. This matrix can support carbohydrate-mediated cell binding, as shown for T lymphoma lines with differences in their metastatic capacity (Table 2). The presently ongoing monitoring of human leukemic cell lines clearly raises evidence that this response can be considered to be of general importance (Gabius et al., in preparation).

Perspective

The specific interaction of tumor cells to a model matrix emphasizes to pursue two lines of experimental approach within the described conceptual framework. Firstly, any intracellular events in biosignalling after adhesion of the tumor cells can be

Table 2. Effect of carbohydrates on adhesion of cells of the weakly metastatic murine T-lymphoma line Eb and its strongly metastatic subclone ESb to plastic-immobilized neoglycoproteins, exposing the respective type of sugar moiety

	cell line	
carbohydrate[a]	Eb	ESb
lactose	-2.12 (NS)	-1.55 (NS)
melibiose	-1.02 (NS)	+3.36 (0.0165)
β-N-acetyl-D-galactosamine	-2.03 (NS)	-4.95 (0.0281)
β-N-acetyl-D-glucosamine	-1.23 (NS)	-7.70 (0.0003)
α-D-mannose	+1.00 (NS)	+5.93 (0.0042)
α-L-fucose	-4.04 (0.0350)	-1.40 (NS)
maltose	-2.16 (NS)	-5.24 (NS; 0.0567)
cellobiose	-0.84 (NS)	-4.74 (0.0226)
α-L-rhamnose	-2.35 (NS)	+6.89 (0.0043)
β-D-xylose	-5.18 (0.0016)	-4.37 (NS)
β-D-glucuronic acid	-1.61 (NS)	-2.00 (NS)

The adhesion assay was performed in neoglycoprotein-coated plastic wells for 60 min at 37°C in the absence or presence of carbohydrate-containing inhibitor solution (20 mM) with cell numbers, ranging from 4.0 - 15 x 10^4 cells per assay. Inhibition is expressed as alteration in the percentage of adhesive cells relative to the extent of adhesion within the concomitantly performed control in the absence of inhibitor (= 100 %). The Wilcoxon test was used for comparison of the data sets with < Z > assessing the probability of no difference. Any value of $Z \leq 0.05$ is considered to be significant and given; NS = not significant ($Z > 0.05$); from Gabius et al., 1990a.

studied in correlation to the type of neoglycoprotein that is accessible on the matrix. Secondly, the importance of distinct types of carbohydrate-mediated binding, measured in a model system, can be assessed in the actual situation of stromal cell-tumor cell interaction, where receptor-ligand pairs are provided by both cell types besides a variety of other mechanisms. An indication for an influence of certain carbohydrates on cell adhesion between leukemia cell lines and stromal cell layers has been reported (Gabius and Gabius, 1990b). Results in this line of experiments will help to refine the target selectivity of the tools, e.g. by purification of the endogenous ligands for cell surface lectins. They should substitute the initially used neoglycoproteins. It is notable that endogenous lectin-binding glycoproteins have even been shown to equal lectin-specific antibodies for lectin localization in glycohistochemistry (Gabius et al., 1991b; Schauer et al., this volume). Whether their application might be of similar service for the modulation of tumor cell features, can only be answered after critical assessment of the potential of glycobiological interactions in this field.

References

Bock G, Harnett S (eds) (1989) Carbohydrate recognition in cellular function. Wiley, Chichester

Brandley BK, Schnaar RL (1986) Cell-surface carbohydrates in cell recognition and response. J Leukocyte Biol 40:97-111

Coulombel L, Eaves AC, Eaves CJ (1983) Enzymatic treatment of long-term human marrow cultures reveals the preferential location of primitive hemopoietic progenitors in the adherent layer. Blood 62:291-297

Dexter TM, Allen TD, Lajtha LG (1977) Conditions controlling the proliferation of hematopoietic stem cells in vitro. J Cell Physiol 91:335-344

Gabius HJ (1987) Endogenous lectins in tumors and the immune system. Cancer Invest 5:39-46

Gabius HJ (1991) Detection and functions of mammalian lectins - with emphasis on membrane lectins. Biochim Biophys Acta, in press

Gabius HJ, Nagel GA (eds) (1988) Lectins and glycoconjugates in oncology. Springer Verlag, Heidelberg

Gabius S, Gabius HJ (1990a) Tumorlektinologie: Status und Perspektiven klinischer Anwendung. Naturwissenschaften 77:505-514

Gabius S, Gabius HJ (1990b) Sugar receptors of the stromal cell layer in human long-term bone marrow cultures: their presence, modulatory responses to changes in the microenvironment and potential role in cell adhesion. Blut 61:232-239

Gabius HJ, Bardosi A (1991) Neoglycoproteins as tools in glycohistochemistry. Progr Histochem Cytochem, in press

Gabius S, Hellmann KP, Hellmann T, Brinck U, Gabius HJ (1989) Neoglycoenzymes: a versatile tool for lectin detection in solid-phase assays and glycohistochemistry. Anal Biochem 182:447-451

Gabius S, Schirrmacher V, Franz H, Joshi SS, Gabius HJ (1990a) Analysis of cell surface sugar receptor expression by neoglycoenzyme binding and adhesion to plastic-immobilized neoglycoproteins for related weakly and strongly metastatic cell lines of murine tumor model systems. Int J Cancer 46:500-507
Gabius S, Yamazaki N, Hanewacker W, Gabius HJ (1990b) Regulation of distribution, amount and ligand affinity of sugar receptors in human colon carcinoma cells by treatment with sodium butyrate, retinoic acid and phorbol ester. Anticancer Res 10:1005-1012
Gabius S, Joshi SS, Gabius HJ, Sharp JG (1991a) Establishment, characterization and determination of cell surface sugar receptor (lectin) expression by neoglycoenzymes of a human myeloid marker-expressing B lymphoblastoid cell line. Anticancer Res, in press
Gabius HJ, Wosgien B, Hendrys M, Bardosi A (1991b) Lectin localization in human nerve by biochemically defined lectin-binding glycoproteins, neoglycoprotein and lectin-specific antibody. Histochemistry 95:269-277
Gartner S, Kaplan HS (1980) Long-term culture of human bone marrow cells. Proc Natl Acad Sci USA 77:4756-4759
Greenberg BR, Woo L, Garewal HS (1988) Effects of bone marrow fibroblastic cells and fibroblastic-conditioned medium on HL-60 and KG-1. Leukemia Res 12:61-66
Hocking WG, Golde DW (1980) Long-term human bone marrow cultures. Blood 56:118-124
Liener IE, Sharon N, Goldstein IJ (eds) (1986) The lectins: properties, functions, and applications in biology and medicine. Academic Press, New York
Luikart SD, Maniglia CA, Furcht LT, McCarthy JB, Oegema TR (1990) A heparan sulfate-containing fraction of bone marrow stroma induces maturation of HL-60 cells in vitro. Cancer Res 50:3781-3785
Monsigny M, Roche AC, Midoux P, Kieda C, Mayer R (1988) Endogenous lectins of myeloid and tumor cells: characterization and biological implications. In: Gabius HJ, Nagel GA (eds) Lectins and glycoconjugates in oncology. Springer Verlag, Heidelberg, pp 25-48
Ohkawa H, Harigaya K (1987) Effect of direct cell-to-cell interaction between the KM-102 clonal human marrow stromal cell line and the HL-60 myeloid leukemic cell line on the differentiation and proliferation of the HL-60 line. Cancer Res 47:2879-2882
Olden K, Parent JB (eds) (1987) Vertebrate lectins. Van Nostrand Reinhold Co., New York
Ossenkoppele GJ, Deukers I, Wijermans P, Huijgens PC, Nauta JJP, Beelen RJH, Langenhuijsen MMAC (1990) Differentiation induction of HL-60 cells in a long-term bone marrow culture of acute myeloid leukemia. Leukemia Res 14:611-616
Schnaar RL (1984) Immobilized glycoconjugates for cell recognition studies. Anal Biochem 143:1-13
Schölzel C, Löwenberg B (1985) Stimulation of proliferation and differentiation of acute myeloid leukemia cells on a bone marrow stroma in culture. Exp Hematol 13:664-669
Schrevel J, Gros D, Monsigny M (1981) Cytochemistry of cell glycoconjugates. Progr Histochem Cytochem 14:1-269

Sharon N, Lis H (1989) Lectins. Chapman & Hill, London
Stowell CP, Lee YC (1980) Neoglycoproteins: the preparation and application of synthetic glycoproteins. Adv Carbohydr Chem Biochem 37:225-281
Tavassoli M, Hardy CL (1990) Molecular basis of homing of intravenously transplanted stem cells to the marrow. Blood 76:1059-1070
Yednock TA, Rosen SD (1989) Lymphocyte homing. Adv Immunol 44:313-378
Zuhrie SR, Pearson JD, Wickramasinghe SN (1988) Haemoglobin synthesis in K562 erythroleukemia cells is affected by intimate contact with monolayers of various human cell types. Leukemia Res 12:567-574

Cell surface glycans influence the fate of circulating white blood cells

R. Hooghe*, M. Rampelberg, and S. Di Virgilio**

*Department Leefmilieu
Vlaamse Instelling voor Technologisch Onderzoek
B-2400 Mol
Belgium

**Algemene Biologie
Vrije Universiteit Brussel
B-1640 Sint-Genesius-Rode
Belgium

What is the role of cell surface glycans ? Thorough analysis of glycoprotein and glycolipid structure has revealed a tremendous heterogeneity. As a rule, glycans of a given glycoprotein are heterogeneous even at a given glycosylation site (reviewed in Rademacher et al., 1988). Therefore, although glycan heterogeneity potentially encodes a large amount of information, not all structural variations are of unique biological significance.

Careful analysis of glycan structure has disclosed **a relationship between the structure of cell surface carbohydrates and the level of differentiation** (Pink, 1983). Several glycoproteins and glycolipids are more sialylated on peripheral (mature) T lymphocytes than on (immature) thymocytes (Hoessli et al., 1980a ; Hoessli et al., 1980b). Similar observations were made on fetal and adult erythrocytes : the adult I antigen is more complex than the fetal i form (Watanabe et al., 1979). Such a relation between differentiation and carbohydrate complexity is seen in many cell types. Clearly, there are exceptions to the rule : the embryonic form of the N-CAM adhesion molecule is more sialylated in adult than in fetal brain and this has been correlated with changes in adhesive properties (Hoffman and Edelman, 1983).

Even before a link was proposed between glycosylation level and differentiation in normal tissues, it had been repeatedly claimed that **glycans are more complex on transformed cells than on their normal counterparts.** The differences include increased branching, increased sialylation and increased molecular weight. More recently, specific alterations in carbohydrate structure were correlated with the metastatic phenotype or with the degree of malignity (Rademacher et al., 1988).

Clearly, changes in glycosylation depend on changes in enzymatic activities : in many of the above-mentioned situations, qualitative or quantitative changes in glycosidase or glycosyltransferase activities have been documented (Di Virgilio, 1990).

The biological significance of glycosylation is still poorly understood. There is a striking parallel between the fate of desialylated soluble proteins and the fate of cells exposed to neuraminidase. Asialoproteins are trapped in the liver through binding with an endogenous lectin specific for galactose (Ashwell and Hanford, 1982 ; Breitfeld et al., 1985). Another endogenous lectin is responsible for the removal of neuraminidase-treated cells (Kolb-Bachofen et al., 1982). A large number of **endogenous lectins** have now been characterized (Aizawa and Tavassoli, 1988 ; Gabius, 1988 ; Gabius et al., 1989).

We have been interested in the homing of normal and leukemic lymphocytes (Hooghe and Pink, 1985). In the case of murine T lymphocytes, immature cells are located in the thymus whereas mature, differentiated cells have emigrated from the thymus. We wondered whether changes in cell surface glycans (as seen between immature and mature T cells, or between non-metastatic and metastatic cancer cells) may contribute to differences in homing properties. A large number of molecules involved in cell-cell or cell-matrix interactions have been characterized. Factors governing escape from the thymus, persistance in the circulation, arrest in certain tissues and specific homing ... are now better understood. Most of those factors are glycoproteins. Several cell adhesion molecules are endogenous lectins. Thus, cell surface lectins recognize glycans on the capillary endothelium (Aizawa and Tavassoli, 1987 ; Springer, 1990 ; Stoolman and Rosen, 1983 ; Watson et al., 1990).
Our own studies suggest that glycans on the surface of lymphocytes also contribute to specific homing, possibly through interaction with endogenous lectins of the endothelium.

To test this idea, cells were first treated with inhibitors of glycan processing, radioactively labeled and injected intravenously. This approach is valid if inhibitors of glycan processing affect glycanic structures only and if labeling *per se* does not alter homing properties.

Inhibitors of glycan processing

Tunicamycin prevents N-linked glycosylation of proteins and this property has been largely used to investigate the biologial function of glycans. More recently, several other compounds were identified that do not completely prevent glycosylation but rather affect particular steps in glycan processing. As a result of treatment with such agents, cell surface glycoproteins are of the high-mannose or of the hybrid type even when complex forms were expected. Such compounds have usually a low toxicity, in contrast to tunicamycin. Their biochemical and biological properties are now well documented (Saunier et al., 1982 ; Peyrieras et al., 1983 ; Elbein, 1987). They therefore represent attractive tools for studying the role of cell surface glycans in the homing process. A comprehensive investigation however would include an analysis of cell surface glycans in control and treated cells. Lectin binding usually provides useful preliminary indications in this respect.

In addition to "pure" inhibitors of glycan processing, we have also used the ionophore monensin, which interferes with many events taking place in the Golgi cisternae. Among many other pharmacological actions, monensin allows the expression of partially processed glycans at the cell surface (Tartakoff et al., 1981 ; Tartakoff, 1983).

We have also treated cells with neuraminidase. In some cases, cells were first grown in the presence of inhibitors of processing, then incubated with neuraminidase.

In vivo homing test using radioactively labeled cells

Several techniques allow to follow the fate of cells injected into a syngeneic host. Radioactive labeling combines a great sensitivity with versatility and convenience. Indeed, bulk labeling can be realized before (short) in vitro treatments. If in vitro treatments have to be long (hours), it is best to label cells after such treatments. Labeling with ^{51}Cr or ^{111}In is a very mild procedure and hardly affects viability or the availability of cell surface antigens (Issekutz et al., 1980 ; Rannie et al., 1977 ; Signore et al., 1983).

Cells are injected intravenously and their fate can be followed during several days. The half life of ^{111}In is only 2.8 days. This makes ^{111}In particularly attractive in view of greater safety but prevents long-time follow-up.

Studies with murine leukemic cells

Murine leukemic cells have been used as probes for cell-cell interactions. The information obtained may be relevant for understanding the homing of normal white blood cells, on one hand and the metastatic process on the other hand.

Our studies were done with established well characterized murine T cell lines (Hooghe and Boniver, 1985). They were grown for 24 or 48 hours in the presence of inhibitors of glycosylation, washed and immediately injected. Aliquots were stained in several cases for lectin histochemistry. Separate, large batches of control and treated cells were prepared for detailed biochemical characterization of glycans.

In our system, treatment with swainsonine (0.2 μg/ml for 24 hours) or with 1-deoxynojirimycine (10 mM) resulted in an increase of concanavaline A-binding glycoproteins labelled with [^{3}H] fucose or [^{3}H] mannose (Di Virgilio, 1990). Only marginal effects however were seen with fluorescent lectins (ConA, PNA, WGA) except when cells were stained with SBA (soybean agglutinin) : staining with SBA was increased in neuraminidase-treated cells and decreased in cells grown in the presence of swainsonine or 1-deoxynojirimycin. Clearly, the latter treatments did not simply result in the expression of, otherwise unmodified, asialoglycoproteins (Hooghe et al., 1985).

Twenty-four hours after injection of labeled cells, 50% of radioactivity is recovered, mainly in the liver. Spleen, kidney and guts are the other sites of homing. Each of all other tissues retain less than 1% of injected counts. With our cells, accumulation in the lungs was very unfrequent and accompanied by signs of lung embolism.

Leukemic cells treated with neuraminidase homed less to the spleen than untreated cells. Inhibitors of processing had similar effects, that could sometimes be increased by additional treatment with neuraminidase (Figure 1). Increased retention in the liver was not seen with any treatment (Hooghe et al., 1984 ; Hooghe et al., 1985 ; Schaaf-Lafontaine et al., 1985).

As tentative interpretations of our data, we propose that there are receptors (endogenous lectins) for complex glycans in the spleen and/or receptors for hybrid and high mannose glycans in tissues other than liver and spleen, probably scattered through the body so that accumulation escaped detection. The latter hypothesis is favored by experiments in which we loaded the mouse with soluble or particulate high molecular weight glucan or mannan (MN-Z, MN-R, lentinan or glucan P). All the compounds, given intravenously 30 minutes before labeled cells, favored retention in the spleen (Topalova et al., 1987). We thus conclude that receptors for glucose, mannose or other high-mannose or hybrid structures present in tissues other than liver, spleen, lung, gut, kidney are responsible for sequestering leukemic cells when the maturation to complex sugars has been prevented.

Thanks to the availability of in vitro tests for homing and cellular interactions, monoclonal antibodies against several homing molecules, molecular probes, purified glycans, ... our hypothesis can be further tested (Stamper and Woodruff, 1977; Watson et al., 1990). We are fully aware of the important role of proteins (regardless of the presence or the nature of their carbohydrates) and in particular of lymphocytic lectins recognizing glycans on the endothelium.
The removal of sialic acid from the surface of lymphocytes favors trapping in the liver and/or prevents homing into lymphoid organs. Less sialylated cells (aged or immature cells or blasts) are thus removed for the bloodstream. In contrast, hypersialylation (a feature of some tumor cells) could prevent removal from circulation and subsequent destruction (Schirrmacher et al., 1986). Whether cell surface

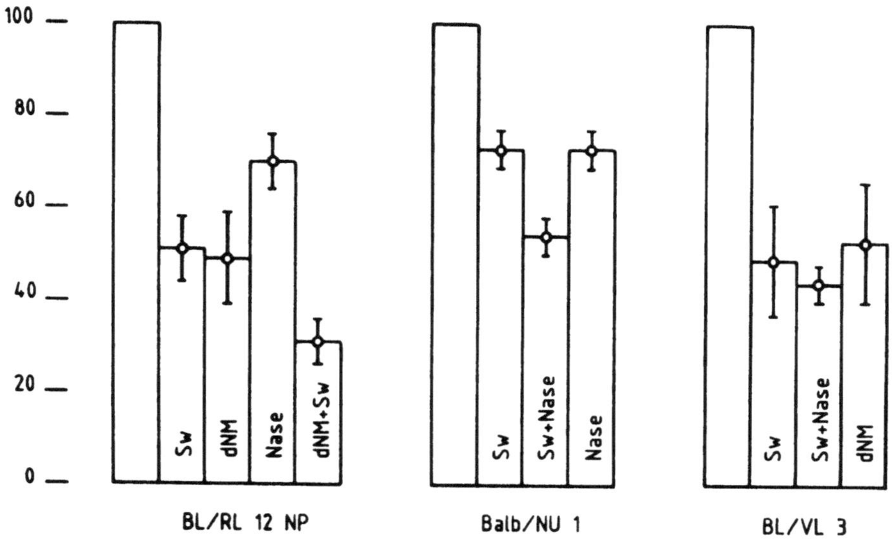

Figure 1.

Arrest in the spleen of leukemic cells grown for 48 hours RPMI 1640 with 10% fetal calf serum supplemented with swainsonine (Sw, 0,2 µg/ml) (0.2 µg/ml), 1-deoxynojirimycin (dNM, 10 mM) or both. In addition, some cells were exposed for 45 minutes to 0.05 U neuraminidase from V. cholerae. Cells were then labelled by incubation for 30 min at 37° with 100 µCi ^{51}Cr (as $Na_2{}^{51}CrO_4$) per 3 x 10^7 cells in phosphate-buffered saline (3 ml). The various drug treatments did not interfere with the uptake of chromate, nor with spontaneous release. Cells were injected i.v. into 4 syngeneic recipient mice (6 x 10^6 cells, i.e. usually 50,000 - 150,000 c.p.m. per mouse). Radioactivity in the following organs was determined after 24h : liver, spleen, kidney, lungs, thymus and lymph nodes (axillary, inguinal, peri-aortic). The results are expressed as the percent of injected c.p.m. retained by an organ. For each experimental group, the fraction of the radioactivity recovered in the spleen is compared to the same fraction in the control group (= 100). Different treatments modifying cell surface glycoconjugates also prevent arrest in the spleen. Similar data are obtained with three different cell lines.

glycans contribute specific features to the homing of normal and leukemic lymphocytes requires additional studies.

References

Aizawa S, Tavassoli M (1987) Interaction of murine granulocyte-macrophage progenitors and supporting stroma involves a recognition mechanism with galactosyl and mannosyl specificities. J Clin Inv 80:1698-1705

Aizawa S, Tavassoli M (1988) Detection of membrane lectins on the surface of hemopoietic progenitor cells and their changing pattern during differentiation. Exp Hematol 16:325-329

Ashwell G, Harford J (1982) Carbohydrate-specific receptors of the liver. Ann Rev Biochem 51:531-554

Breitfeld PP, Simmons CF jr, Strous GJAM, Geuze HJ, Schwartz AL (1985) Cell biology of the asialoglycoprotein receptor system : a model of receptor-mediated endocytosis. Int Rev Cytol 97:47-95

Di Virgilio S (1990) Glycobiologie moléculaire. PhD Thesis, Université Libre de Bruxelles

Elbein AD (1987) Inhibitors of the biosynthesis and processing of N-linked oligosaccharide chains. Ann Rev Biochem 56:497-534

Gabius HJ (1988) Mammalian lectins : their structure and their glycobiological and glycoclinical roles. ISI Atlas of Science : Biochemistry 1:210-214

Gabius S, Hellmann KP, Ciesiolka T, Nagel GA, Gabius HJ (1989) Lineage- and differentiation-dependent alterations in the expression of receptors for glycoconjugates (lectins) in different human hematopoietic cell lines and low grade lymphomas. Blut 59:165-170

Hoessli D, Bron C, Pink JRL (1980a) T-lymphocyte differentiation is accompanied by increase in sialic acid content of Thy-1 antigen. Nature 283:576-578

Hoessli DC, Vassalli P, Pink JRL (1980b) Characterization of mouse thymocyte and peripheral lymphocyte xenoantigens by two-dimensional electrophoresis. Eur J Immunol 10:814-821

Hoffman S, Edelman GM (1983) Kinetics of homophilic binding by embryonic and adult forms of the neural cell adhesion molecule. Proc Natl Acad Sci USA 80:5762-5766

Hooghe R, Boniver J (1985) Thymic lymphomas in the mouse. Immunol Today 6:240-242

Hooghe R, Pink JRL (1985) The role of carbohydrate in lymphoid cell traffic. Immunol Today 6:180-181

Hooghe R, Schaaf-Lafontaine N, Hooghe-Peters EL, Dorling P, Janowski M (1984) The carbohydrates of glycoproteins are required for in vivo homing but not for several other membrane-mediated functions. Cell Biochem Function 2:102-106

Hooghe R, Vandermeeren M, Vander Plaetse F, Greimers R (1985) Membrane carbohydrates of lymphoid cells. II. Experimental modification of N-linked sugars of membrane proteins in a

lymphoma cell line affects the binding of soybean agglutinin but not of several other lectins. Carbohydrate Res 141:172-177
Issekutz T, Chin W, Hay JB (1980) Measurement of lymphocyte traffic with indium-111. Clin exp Immunol 39:215-221
Kolb-Bachofen V, Schlepper-Schäfer J, Vogell W (1982) Electron microscopic evidence for an asialoglycoprotein receptor on kupffer cells : localization of lectin-mediated endocytosis. Cell 29:859-866
Peyrieras N, Bause E, Legler G, Vasilov R, Claesson L, Peterson P, Ploegh H (1983) Effects of the glucosidase inhibitors nojirimycin and deoxynojirimycin on the biosynthesis of membrane and secretory glycoproteins. EMBO J 2:823-832
Pink JRL (1983) Changes in T-lymphocyte glycoprotein structures associated with differentiation. Contemp Topics Molec Immunol 9:89-113
Rademacher TW, Parekh RB, Dwek RA (1988) Glycobiology. Ann Rev Biochem 57:785-838
Rannie GH, Thakur ML, Ford WL (1977) An experimental comparison of radioactive labels with potential application to lymphocyte migration studies in patients. Clin exp Immunol 29:509-514
Saunier B, Kilker RD, Tkacz JS, Quaron A, Herscovics A (1982) Inhibition of N-linked complex oligosaccharide formation by 1-deoxynojirimycin, an inhibitor of processing glucosidases. J Biol Chem 257:14155-14161
Schaaf-Lafontaine N, Hooghe R, Vander Plaetse F (1985) Membrane carbohydrates of lymphoid cells. I. Modification of blood-borne arrest properties of lymphoma cells by inhibitors of protein glycosylation suggests the existence of endogenous lectins. Carbohydrate Res. 138:315-323
Schirrmacher V, Dennis J, Waller CA, Altevogt P (1986) A role for cell surface sialic acid in liberating metastatic tumor cells from host control. In : Biochemistry and molecular genetics of cancer metastasis (eds Lapis K, Liotta LA, Rabson AS), Martinus Nijhoff Publishing, pp 251-261
Signore A, Beales P, Sensi M, Zuccarini O, Pozzilli P (1983) Labelling of lymphocytes with indium 111 oxine : effect on cell surface pheotype and antibody dependent cellular cytotoxicity. Immunol Lett 6:151-154
Springer TA (1990) Adhesion receptors of the immune system. Nature 346:425-434
Stamper HB jr, Woodruff JJ (1977) An in vitro model of lymphocyte homing. I. Characterization of the interaction between thoracic duct lymphocytes and specialized high-endothelial venules of lymph nodes. J Immunol 119:772-780
Stoolman LM, Rosen SD (1983) Possible role for cell-surface carbohydrate-biding molecules in lymphocyte recirculation. J Cell Biol 96:722-729
Tartakoff AM (1983) Perturbation of vesicular traffic with the carboxylic ionophore monensin. Cell 32:1026-1028

Tartakoff A, Hoessli D, Vassalli P (1981) Intracellular transport of lymphoid surface glycoproteins. Role of the Golgi complex. J Mol Biol 150:525-535
Watanabe K, Hakomori S, Childs RA, Feizi T (1979) Characterization of a blood group I-active ganglioside. J Biol Chem 254:3221-3228
Watson ML, Kingsmore SF, Johnston GI, Siegelman MH, Le Beau MM, Lemons RS, Bora NS, Howard TA, Weissman IL, McEver RP, Seldin MF (1990) Genomic organization of the selectin family of leukocyte adhesion molecules on human and mouse chromosome 1. J Exp Med 172:263-272

Fertilization: A model for cell-cell interaction

F. Sinowatz, E. Toepfer-Petersen, and H.-J. Gabius

Lehrstuhl für Tieranatomie II
Universität München
D-8000 München
Veterinärstraße 13

Many interactions between mammalian cells are known to involve carbohydrate moieties of plasma membrane glycoconjugates and corresponding carbohydrate-binding proteins. Examples are found in systems as diverse as embryogenesis, lymphocyte homing (Lasky, 1991), tumor invasion (Gabius and Nagel, 1988) and fertilization (Miller and Ax, 1990). Although the overall process of fertilization is unquestionably unique, it is made up of a series of steps that are not unique and are commonly used by a variety of both normal and tumor cells.

Fusion of spermatozoon and ovum involves a series of highly complex molecular interactions. Before penetration, the sperm cell must be capacitated, bound to the zona pellucida, and undergo the acrosomal reaction. In this communication we present some data on the distribution of endogenous lectins in the plasma membrane of mammalian spermatozoa and demonstrate that carbohydrate-binding proteins are implicated in several aspects of the fertilization process, such as capacitation and acrosomal reaction.

There are two possible ways in which complementary components may be distributed between the two gametes: 1) carbohydrate-binding proteins may be a constituent of the zona pellucida; in which case the carbohydrate ligand would comsequently be located on the sperm surface, or 2) the carbohydrate-binding component could be a part of the spermatozoa with the sugar receptor belonging to the zona pellucida.

Evaluation of carbohydrate-binding proteins in spermatozoa

Using histochemical methods, carbohydrate-binding proteins could be demonstrated in the sperm membranes of several species. Sinowatz et al. (1988) used 15 different neoglycoproteins and asialoglycoproteins to study the topographical distribution of carbohydrate-binding sites on the plasma membrane of bovine epididymal spermatozoa. A pronounced fluorescence was seen in the postacrosomal area of bovine spermatozoa following incubation with mannose-BSA-FTC, cellobiose-BSA-FTC, mannose-6-phosphate-BSA-FTC, lactose-BSA-FTC, maltose-BSA-FTC and the two asialoglycoproteins, desialated lactoferrin-FTC and transferrin-FTC. Weak staining of the postacrosomal area was also observed with serveral other fluorescent neoglycoproteins, whereas labelling of other portions of the bovine spermatozoa occurred only rarely. The data suggest that carbohydrate-binding proteins with specificity for D-glucose and D-mannose are likely to be present in the postacrosomal area of these cells.

The localization of fucose-binding sites in boar spermatozoa was studied by Toepfer et al. (1985) using fluoresceinyl-fucosylperoxidase, and at the ultrastructural level by Friess et al. (1987) with fucosyl-peroxidase-gold. These authors were able to demonstrate concentrated fucose-binding activity at two different areas, namely at the rostral region of the plasma membrane overlying the acrosome, and in association with the acrosomal matrix within the acrosome.

By using heat-solubilized zona glycoproteins labelled with biotin-avidin, a coincident distribution of fucose- and zona binding sites on spermatozoa has been observed. Zona-receptors are localized in the apical region of acrosome-intact sperm and in the acrosomal matrix. Various fucosylated glycoconjugates were able to block the receptor sites. Furthermore, binding of fucose-HRP to the spermatozoa was inhibited by heat-solubilized zonae pellucidae in a dose-dependent manner, thereby indicating that zona glycoproteins compete with the fucose-ligand for sperm receptor sites (Töpfer-Petersen et al. 1991).

The zona pellucida is typically composed of three sulfated glycoprotein families containing complex 0- and N-glycosidically linked carbohydrate sidechains. All asparagine-linked oligosaccharides have been shown to contain fucose in their "core" structure (Mori et al. 1991). Fucose receptor molecules on the sperm may therefore recognize specific fucose-containing structures of the zona pellucida, thus mediating gamete recognition and binding.

In order to detect carbohydrate proteins in spermatozoa, a modified enzyme-linked-lectin assay (ELLA) was also developed using cytochemical markers (Töpfer-Petersen et al. 1985). By conjugating sugar residues to peroxidase, an artificial glycoprotein was obtained whose defined sugar residues functioned as receptors for carbohydrate-binding proteins precoated to micro-ELISA plates. Binding could be identified and quantified using a suitable substrate for peroxidase. A typical binding curve was only obtained with fucose when boar spermatzoa were tested with different glycosylated peroxidases. This binding was completely inhibited by fucoidan and, to a minor extent, with the monosaccharide fucose. Lactosylperoxidase and chitobiosylperoxidase, which detect Ricinus communis agglutin-like and wheatgerm agglutinin-like activity, respectively, show no significant interaction with spermatozoa (Fig. 1).

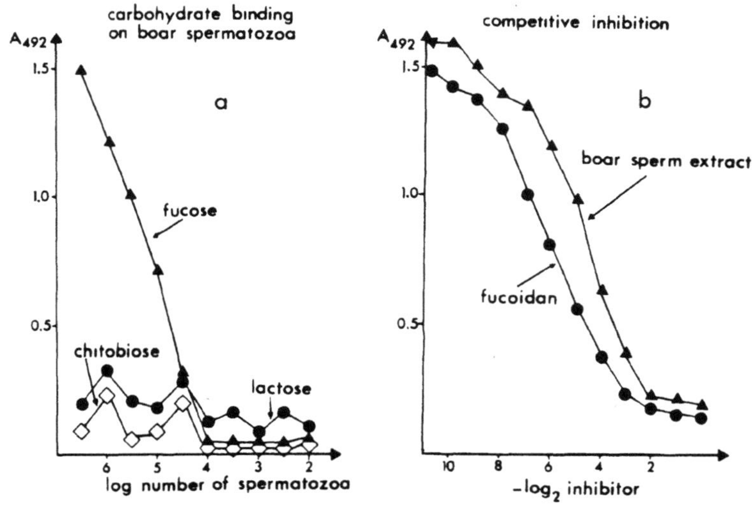

Fig. 1: Evidence of a fucose binding activity in boar spermatozoa determined by the modified ELLA. a. Titration curves of three different glycosylated peroxidases (gly-HRP, constant concentration of 10 µg/ml against boar spermatozoa as a function of sperm concentration. b. Titration of fucosyl- peroxidase (fuc-HRP) against boar spermatozoa (10^5 sperm/well) precoated to microelisa plates as a function of fuc-HRP concentration. Control experiments in the presence of 1 mg/ml fucoidan.

Fucose-binding activity is extractable from spermatozoa by mild detergent or acid treatment, and may be isolated via gel filtration and reverse-phase HPLC. Two groups of proteins, with apparent molecular weights of 53 kDa and < 20 kDa, have been shown to account for fucose-binding activity in boar spermatozoa (Töpfer-Petersen et al. 1985). The < 20 kDa and 18 kDa proteins show both fucose-binding activity and a strong affinity for the zona pellucida. On the basis of N-terminal sequences and sequence homologies, they belong to two protein families known as the AQN-familiy (AQN-1, AQN-2, AQN-3) and the AWN-family (AWN-1, AWN-2) (Jonakova et al. 1991). Both AQN-1, the dominant low molecular weight protein (Jonakova et al. 1991), and the 53 kDa fucose-binding protein agglutinate rabbit or human erythrocytes, thus meeting the classical requirement of a lectin. This strengthens the conclusion that lectin-like proteins are involved in the complex sequence of sperm-egg recognition and binding. Zona binding proteins with specificity for complex fucose structures have been found in several mammalian species (Jones 1989; Jones and Williams 1990; O'Rand et al. 1990), indicating the ubiquitous nature of the binding mechanism.

Our histochemical and biochemical studies reveal specific carbohydrate-binding proteins in the plasma membrane of mammalian spermatozoa. They appear to be concentrated in the postacrosomal (bovine) and apical (porcine) segments of the male gamete, i.e. areas of initial contact with the zona pellucida. These results lend further credence to the concept that carbohydrate-binding proteins of the sperm plasma membrane and the corresponding carbohydrates of the zona pellucida play a significant role in gamete interactions during the fertilization process.

Capacitation is stimulated by glycosaminoglycans (GAGs) such as heparin

Spermatozoa must complete a process within the female genital tract known as capacitation before further steps in the fertilization process can occur. Capacitation causes no observable morphological change in sperm. Several lines of evidence suggest that GAGs secreted from the epithelium of the female genital tract are important for capacitation. In vitro capacitation is most effectively induced by heparin. Following a 4-hour heparin preincubation, spermatozoa were immediately able to fertilize ova (Parrish et al., 1988), and also became susceptible to fusigenic agents, such as lysophosphatidylcholine or solubilized zonae pellucidae, both of which induce acrosomal reaction (Parrish et al., 1988, Florman and First, 1988). Beyond this, 4-hour heparin incubation stimulated the loss of plasma membrane antigens, which may be indicative of capacitation (Miller and Hunter, 1986). The effect of GAGs on sperm fertilizing ability is not unique to the bovine. Reyes et al. (1984) determined that porcine follicular fluid contains GAGs capable of inducing acrosome reactions in ejaculated porcine spermatozoa. In the case of ejaculated rabbit sperm, heparin, chondroitin-4- and chondroitin 6-sulfate, dermatan sulfate and hyaluronic acid all increased the percentage of acrosome reaction after 9 hours (Lenz et al., 1983). As with bovine sperm, heparin was the most potent GAG.

In view of the effects of heparin on the fertilizing ability of sperm, its binding to them may be required for the induction of capacitation. Delgado et al. (1982) found that 3H heparin bound to human spermatozoa in a manner dependent on pH and ionic strength. Further studies demonstrated (Handrow et al.,1984) heparin-sperm binding to be a typical receptor-ligand interaction and went on to characterize the heparin receptors in spermatozoa. Hurst et al. (1988) solubilized bovine epididymal spermatozoa, fractionated the material by heparin-agarose chromatography and eluted a 45-kDa protein with 3.0 M NaCl. We were recently able to localize a heparin-binding protein in the epithelium of bovine epididymis and seminal vesicles using immunocytochemistry. After secretion into the duct, it is obviously bound to the plasma membrane of spermatoza. This lends further credence to the concept of Miller and Ax (1990), that heparin-binding proteins from male accessory sex organs adhere to spermatozoa and in this way in- crease the number of heparin-binding sites on the sperm (Fig. 2).

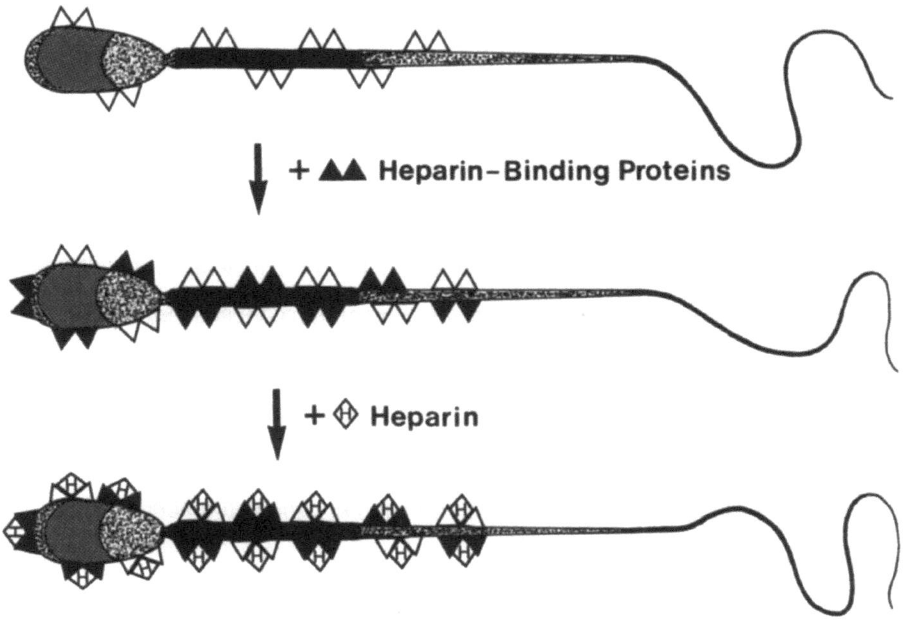

Fig. 2: A model for the regulation of heparin responsiveness by seminal plasma in bovine sperm (according to Miller and Ax, 1990). Epididymal sperm (A) have few binding sites for heparin (H) and are unable to to undergo acrosome reaction. When epididymal sperm are exposed to seminal plasma (B) and washed, some proteins are bound to sperm. Some of these are heparin binding proteins. The number of heparin binding sites on the plasma membrane of sperm is increased. Seminal plasma treated sperm are then able to respond to heparin (C) and undergo capacitation.

Spermatozoa are then able to react with heparin-like molecules or other GAGs from the female genital tract, undergo capacitation and respond to the zona pellucida with acrosome reaction.

Acrosin, a lectin-like enzyme, acts as a secondary receptor to the zona pellucida

Penetration of the zona pellucida by spermatozoa is a key event in fertilization. The mechanism by which spermatozoon overcome this barrier is not completely understood. However, the acrosome reaction, i.e. the exocytotic event allowing the release of the lytic portion of the sperm acrosome, is necessary before sperm are able to traverse the zona pellucida. Acrosin is one of the hydrolases released as a consequence of the acrosome reaction. It contributes to sperm penetration by limited and specific proteolysis of the zona glycoprotein matrix, thereby facilitating the entry of motile sperm into the zona (Urch et al. 1985, Dunbar et al. 1985).

A 53 kDa, lectin-like fucose-binding protein isolated from boar spermatozoa was recently identified as proacrosin/acrosin on the basis of immunological (Jones and Brown 1987) and sequencing data (Töpfer-Petersen and Henschen 1987). This finding is consistent with the localization of fucose-binding sites in the acrosomal matrix (Friess et al. 1987). Acrosin belongs to the familiy of serine proteases. It is a two-chain molecule, the light chain of which corresponds to the propart of a serine protease. The heavy chain displays the sequence typical of an active enzyme (Fock-Nüzel et al. 1984). Heavy and light chains are each linked by two disulfide bridges. Acrosin shows a strong affinity for the zona pellucida when tested with biotinylated zona glycoproteins. Fucose-ligand and zona pellucida appear to recognize the same carbohydrate-binding sites on the molecule (Töpfer-Petersen and Henschen 1987). The proteolytic activity can be selectively destroyed without affecting the carbohydrate binding (Töpfer-Petersen and Henschen 1988) associated with the N-terminal peptide of the acrosin heavy chain (Töpfer-Petersen et al. 1990).

Carbohydrate interaction appears to be a property unique to acrosin, as other serine proteinases lack any comparable carbohydrate affinity.

The ability of fucoidan (an algal sulfated fucan) and some other sulfated polysaccharides to block acrosin-fucose/zona binding implies that, aside from fucose, sulfate groups are critical determinants for the binding mechanism. In echinoderms, bindin, a lectin-like protein, mediates binding of acrosome-reacted sperm to the egg by recognizing fucose-containing structures within the vitelline coat (Vaquier 1986). Binding of sulfated fucans to bindin is dependent upon basic amino acids in the latter. The distinct formation of a ring-shaped bond between negatively-charged sulfate groups in the polysaccharide and the guanido moiety of certain arginine residues, suggests interactions other than simple ionic ones to be involved. Relative spatial orientation of critical arginine residues to sulfate groups on the polypeptide backbone represent a crucial parameter in the binding mechanism (DeAngelis and Glabe 1988). Some evidence suggests similar characteristics to be required for acrosin-zona binding (Töpfer-Petersen et al. 1991).

Acrosin has been recognized as the zona penetrating enzyme for several years. Carbohydrate affinity directed at the zona pellucida indicates acrosin's role as a secondary zona receptor in the maintenance of binding between acrosome-reacted sperm and zona pellucida, making it similar to bindin. The terms "primary" and "secondary" binding were introduced to describe the temporal relationship between the two events (Saling 1989). Primary binding occurs between receptors on the sperm surface, e.g. carbohydrate-binding proteins, and carbohydrates of the zona pellucida. Initiation of the acrosome reaction follows. Secondary binding occurs between acrosome-reacted sperm and the zona pellucida directly after the exocytotic event (Fig. 3).

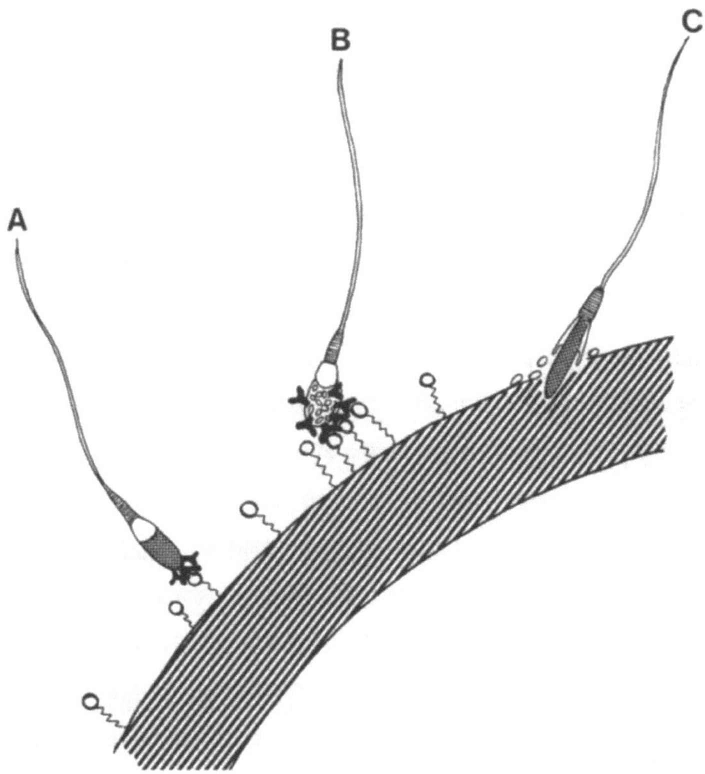

Fig. 3: A. Primary binding of a spermatozoon to receptor of the zona pellucida.
B. Secondary binding after induction of the acrosomal reaction
C. penetration of the spermatozoon through the zona pellucida

Literature references

De Angelis PL, Glabe CG (1990) Specific recognition of sulfate esters by bindin, a sperm adhesive protein from sea urchins. Biochem Biophys Acta 1037: 100-105

Delgado NM, Reyes R, Huacuja L, Merchant H, Rosado A (1982) Heparin binding sites in the human sperm plasma membrane. Arch Androl 8:87-95

Dunbar BS, Dudkiewicz AB, Bundman DS (1985) Proteolysis of specific porcine zona pellucida glycoproteins by boar acrosin. Biol Reprod 32: 619-630

Florman HM, First NL (1988) The regulation of acrosomal exocytosis. I. Sperm capacitation is required for the induction of acrosome reactions by the bovine zona pellucida in vitro. Dev Biol 128: 453-463

Fock-Nüzel R, Lottspeich F, Henschen A, Müller-Esterl CW (1984) Boar acrosin is a two chain molecule. Isolation and primary structure of the light chain:

Homology with the propart of other serine proteinases. Eur J Biochem 141: 441-446

Friess AE, Töpfer-Petersen E, Schill WB (1987) Fracture labelling of boar spermatozoa for a fucose-binding protein (FBP). Histochemistry 87: 181-183

Gabius HJ, Nagel GA (1988) Lectins and glycoconjugates in oncology, Springer, Berlin Heidelberg New York Tokyo

Handrow RR, Boehm SK, Lenz RW, Robinson JA, Ax RL (1984) Speficic binding of the glycosaminoglycan ^{3}H-heparin to bull, monkey, and rabbit spermatozoa in vitro. J Androl 5:51-63

Hurst RE, Bynum RL, Einfeldt SE, Roy J (1988) The identification of a heparin-binding protein on the surface of bovine sperm. Biochem Biophys Res Commun 153:289-293

Jonakova V, Sanzl L, Calvette JJ, Henschen A, Cechova D, Töpfer-Petersen E (1991) Isolation and characterization of a zona pellucida binding glycoprotein of boar spermatozoa. FEBS Lett 280: 183-186

Jones R (1989) Identification of carbohydrate-binding proteins in mammalian spermatozoa (human, bull, boar, ram, stallion and hamster) using ^{125} J-fucoidan and ^{125}J-neoglycoprotein probes. Human Reproduction 4 : 550-557

Jones R, Brown CR (1987) Identification of a zona-binding protein from boar spermatozoa as proacrosin. Exp Cell Res 171: 503-508

Jones R, Williams RH (1990) Identification of zona- and fucoidan-binding proteins in guinea pig spermatozoa and mechanism of recognition. Development 109: 41-50

Lasky, L.A. (1991) Lectin cell adhesion molecules (LEC-CAMs): A new family of cell adhesion proteins involved with inflammation. J. Cell. Biochem. 45: 139-146 (1991)

Lenz RW, Bellin ME, Ax RL (1983) Rabbit spermatozoa undergo an acrosome reaction in the presence of glycosaminoglycans. Gamete Res 8:11-19

Miller DJ, Hunter AG (1986) Effect of osmolality and glycosaminoglycans on motility, capacitation, acrosome reaction, and in vitro fertilizability of bovine ejaculated sperm. J Dairy Sci 69:2915-2924

Miller DJ, Ax RL (1990) Carbohydrates and fertilization in animals. Molec Reprod Devel 26: 184-198

Mori E, Takasaki S, Hedrick JL, Wardrip NJ, Mori T, Kobata A (1991) Neutral oligosaccharide structure linked to asparagines of porcine zona pellucida glycoproteins. Biochemistry 30: 2078-2087

Parrish JJ, Susko-Parrish JL, Winer MA, First NL (1988): Capacitation of bovine sperm by heparin. Biol Reprod 38:1171-1180

Reyes R, Carranco A, Hernandez O, Rosado A, Merchant H, Delgado NM (1984) Glycosaminoglycan sulfate as acrosomal reaction inducing factor of follicular fluid. Arch Androl 12:203-209

Saling P (1989) Mammalian sperm interaction with extracellular matrices of the egg. In: Milligan SR (ed) Oxford Reviews of Reproductive Biology, Oxford University Press, Oxford, p 339-388

Sinowatz F, Gabius HJ, Amselgruber W (1988) Surface sugar binding components of bovine spermatozoa as evidenced by fluorescent neoglycoproteins. Histochemistry 88: 395-399

Töpfer-Petersen E, Friess AE, Nguyen H, Schill WB (1985) Evidence for a fucose-binding protein in boar spermatozoa. Histochemistry 83: 139-145

Töpfer-Petersen, Henschen A (1987) Acrosin shows zona- and fucose binding, novel properties for a serine proteinase. FEBS Lett 226: 38-42

Töpfer-Petersen E, Henschen A (1988) Zona pellucida-binding and fucose-binding of boar sperm acrosin is not correlated with proteolytic activity. Biol Chem Hoppe Seyler 369: 69-76

Töpfer-Petersen E, Steinberger M, Ebner von Eschenbach C, Zucker A (1990) Zona pellucida binding of boar sperm acrosin is associated with the N-terminal part of the acrosin B-chain (heavy chain). FEBS Lett. 265: 51-54

Töpfer-Petersen E, Friess AE, Henschen A, Cechova P, Steinberger M (1990) Sperm acrosin and binding to the zona pellucida. In: Alexander NJ, Griffith D, Spieler JM, Waites GM (eds) Gamete Interaction, Wiley-Liss, New York, p 197-212

Töpfer-Petersen E, Weiss A, Schill WB (1991) Carbohydrate specificity of the acrosin-zona pellucida binding in the pig. Andrologia, in press

Urch UA, Wardrip NJ, Hedrick JL (1985) Limited specific proteolysis of the zona pellucida by acrosin. J Exp Zool 233: 479-483

Vacquier VD (1986) Activation of sea urchin spermatozoa during fertilization. TIBS 11: 77-81

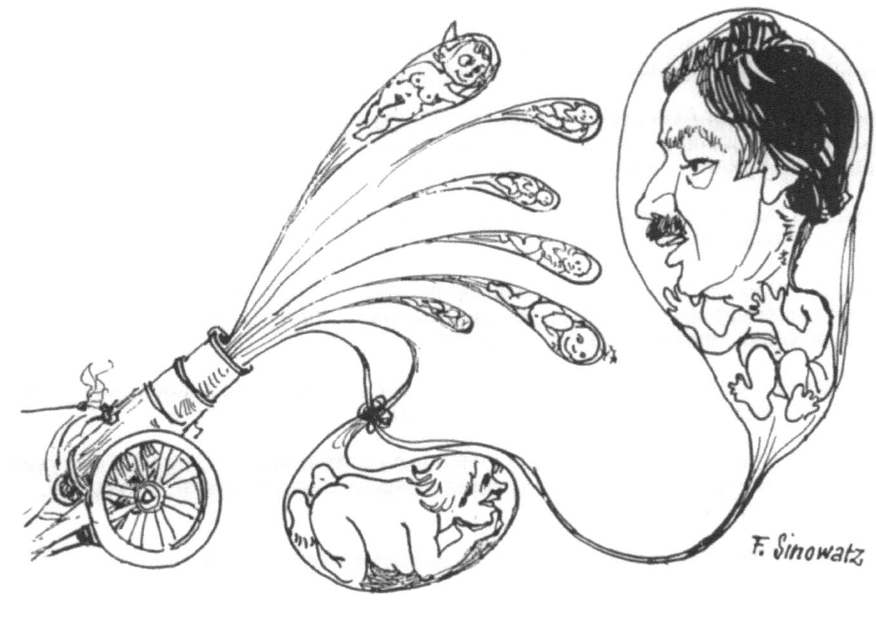

Endogenous mannose-binding lectins in brain development and function

J.-P. Zanetta, S. Kuchler, S. Lehmann, P. Marschal, M. Zaepfel, A. Meyer, A. Badache, A. Reeber, and G. Vincendon

Centre de Neurochimie du CNRS
5 rue Blaise Pascal
67084 Strasbourg cedex
France

The formation of brain tissue owes its specificity to multiple steps involving cell interactions during various stages of ontogenesis. Guided neuron migration, synaptogenesis and myelination constitute three essential steps in brain organization, where glycobiological recognition or adhesion systems play a fundamental role. The key molecules in these phenomena were postulated to belong to the family of cell adhesion molecules (CAMs; Crossin et al., 1986; Edelman, 1985 and 1986; Grumet et al., 1985; Hoffmann et al., 1986; Hoffmann and Edelman, 1987; Rieger et al., 1986; Salzer and Colman, 1989) acting by homophilic interactions. More recently two different mannose-binding endogenous lectins of similar specificity (Zanetta et al., 1985a and 1987a) seem to play important roles in relation with their endogenous glycoprotein ligands (Kuchler et al., 1988, 1989a,b and c; 1990a,b and c; Lehmann et al., 1990). It is worthy of mention that some of these glycoprotein ligands are members of the family of cell adhesion molecules (Kuchler et al., 1988, 1989a and b; 1990a and b) and share specific glycans binding to some onco-fetal antibodies. Thus, heterophilic interactions involving glycobiological system deserve attention. This paper summarizes the major observations concerned with the function in the brain of the two endogenous mannose-binding lectins, called CSL and R1 (Zanetta et al., 1985a and 1987a) and of their endogenous ligands.

STRUCTURE AND PROPERTIES OF R1 AND CSL MOLECULE.

Isolation of lectins R1 and CSL.

Lectins R1 and CSL were isolated from young rat cerebella by sequential extractions of the tissue. CSL was concentrated in a fraction specifically solubilized in the presence of 0.5 M mannose in the absence of detergent (Zanetta et al.,1987a), whereas R1 was accumulated in a fraction specifically solubilized in the presence of 0.5 M mannose and 0.5% Triton X-

100 (Zanetta et al., 1985a and 1987b). The rationale for this purification is that during homogeneization, the lectins CSL and R1 bind major endogenous ligands which are highly insoluble glycoproteins. The purified protein bands eluted from preparative gel electrophoresis (Zanetta et al., 1985a and 1987a), were initially used to produce antibodies in rabbits. Antibodies, in return, were employed to isolate the antigens by immunoaffinity chromatography (Zanetta et al., 1987a; Marschal et al., 1989) from the material solubilized in the presence of mannose as indicated above.

Using several immobilized antibodies, active lectin R1 was obtained as a major 45 kDa band and a minor 65 kDa band. The 45 kDa band derived from a unique proteolytic cleavage during the purification of the initial lectin of Mr 65 kDa. *In vivo*, lectin R1 behaves as a dimer of Mr 130 kDa. Active CSL lectin (possessing agglutinating activity) was obtained as a major doublet band of Mr 33 and 31.5 kDa and having a minor component of Mr 45 kDa. Antidodies raised in rabbits against individual protein bands reacted with all the three components as revealed by immunoaffinity chromatography. This demonstrated a clear immunochemical relationship between the three components. Amino-acid composition (Zanetta et al., 1987a) showed an homology between 33 and 31.5 kDa components since they differed only by a short polypeptide of about 15-20 aminoacids (Zanetta et al., 1987a). Recent data obtained from cloning of CSL molecules indicated that two mRNA may coexist. Active CSL molecules are found as macromolecular complexes formed of about 40 different identical monomers (Zanetta et al., 1987a) consisting either of the 33 kDa protein or of the 31.5 kDa protein.

Based on different experimental approaches, a mannose-binding protein has been detected in white matter (Bardosi et al., 1988; Gabius et al., 1988) which is probably identical to CSL. Similarly, a partially purified protein fraction solubilized from axons by heparin (DeCoster and DeVries, 1989) contains probably a high proportion of CSL.

Carbohydrate specificity of R1 and CSL.

Lectins R1 and CSL are totally unrelated proteins but they display similar, although not identical, carbohydrate specificity (Marschal et al.,1989). Mannose is a relatively poor inhibitor and shows inhibition at a concentration of 150 mM, whereas the other monosaccharides did not show any inhibition. Glycosaminoglycans were poorly inhibitory, the best being hyaluronic acid for R1 and heparin for CSL (acting at the concentration of 185 μg/ml; the concentration range at which heparin-binding proteins are inhibited). Horseradish peroxidase isoenzyme 6 and above all isoenzyme 8 (11 μg/ml) are good inhibitors of both lectins. In contrast, ovalbumin a good inhibitor of R1 requires much higher concentration (2.5 mg/ml) for inhibition of agglutination induced by CSL. Endogenous glycoproteins B1 and B2 transiently expressed in the cerebellum (Zanetta et al., 1978; Reeber et al., 1981) are very good inhibitors (2–3 μg/ml). The glycopeptides isolated from these molecules inhibit at a concentration of 1 μg/ml and correspond to the best inhibitors yet identified. Studies with glycopeptides of known structures (Marschal et al., 1989) suggest that the first N-acetyl-glucosamine residue of N-glycans is part of the binding site of the lectins. In the cerebellum (Marschal et al., 1989) and in other parts of the nervous tissue (Kuchler et al., 1988; 1989a,b and c; 1990 a, b and c; Lehmann et al., 1990), R1 and CSL display specificity for a smaller number of glycoprotein subunits than that revealed using the plant lectin Concanavalin A (ConA). This indicates that, in most non transformed cells, the endogenous lectins R1 and CSL have a few number of endogenous ligands. This may be the bases for specific cell recognition and adhesion. There is two important identified ligands of CSL: i) myelin-associated glycoprotein, MAG and ii) the major glycoprotein of the peripheral nervous system, P0 (Kuchler et al., 1989b). Only a few of the polypeptide chains of these molecules bind to CSL, although they are all ConA-binding (Kuchler et al., 1989b; Badache et al., submitted). Several of these chains have glycans recognized by HNK-1 or Elec-39 antibodies.

ROLE OF LECTIN R1 IN THE NERVOUS TISSUE

In the nervous tissue, R1 is present in endothelial cells, macrophages and neurons. Its localisation in the first two types of cells is not surprising since R1 is probably identical to a receptor for circulating mannosyl-glycoproteins found in liver sinusoidal cells (endothelial and Kuppfer cells; Zanetta et al., 1987b). In neurons, lectin R1 plays a role in neuronal recognition during synaptogenesis. The lectin is transiently externalized at the surface of the dendrites of Purkinje cells at the period of synaptogenesis (Zanetta et al., 1985a; Dontenwill et al., 1985). R1 binds to specific glycoproteins at the surface of incoming axons (the parallel fibres). The glycans of these glycoproteins bind to ConA (Zanetta et al., 1978; Reeber et al., 1981), CSL, R1 (Kuchler et al., 1989a; Marschal et al., 1989) and to Elec-39 monoclonal antibody, an antibody of the HNK-1 family revealing onco-foetal glycan antigen (Bon et al., 1987, Musset et al., 1987, Mailly et al., 1989; Kuchler et al., 1990c). The complex formed between lectin R1 and glycoproteins is internalized with part of the two membranes into Purkinje cells through coated pits and double walled coated vesicles (Dontenwill et al., 1985) in ageement with earlier reports (Eckenhoff and Pysh, 1979; Palacios-Prü et al., 1981). The axonal glycoproteins are degraded in the target cells since blocking lysosomal degradation with chloroquine induces a dramatic accumulation of glycoproteins of parallel fibres in Purkinje cells (Dontenwill et al., 1983).

This recognition step constitutes probably the first stage of synapse formation and a step necessary for the formation of synapses. If it does not occur (due to transient action of tunicamycin during synthesis of the axonal glycoprotein, this recognition mechanism fails and synapses cannot be formed (Zanetta et al., 1987c). A decrease in the level of the glycoproteins leads (under the effect of thyroid hormone defficiency; Zanetta et al., 1985b) to reduction in the number of synapses (Nicholson and Altman, 1972a and b).

In the adult, R1 is located exclusively intracellularly and the axonal glycoproteins with their special type of glycans are no more present on the axonal surface. But, both partners of the glycobiological recognition system can reappear (Lehmann et al., submitted) at the surface of dendrites (for R1) and on axons (for the glycoproteins) during a neosynaptogenesis in the adult induced by production of lesions of these axons (Chen and Hillmann, 1982). Similar mechanism of endocytosis of glycoprotein components of the axonal membrane into the target cell was revealed morphologically (Lehmann et al., submitted). We have proposed (Zanetta et al., 1987c) that a diffusive factor is released after lesions as a signal for both partner cells to resynthesize recognition molecules. This mode of signalling for repair is in agreement with previous observations (Henderson et al., 1983) on the peripheral nervous system and probably may operate in other systems where repair takes place (Holder and Clarke, 1988). Independent of the nature of signal, this observation constitutes first demonstration that reexpression of recognition molecules may be a basis for repair mechanism, analogous to the one occurring during normal CNS formation.

ROLE OF ENDOGENOUS LECTIN CSL IN THE NERVOUS TISSUE

CSL and contact guidance of neuron migration

The concept of contact guidance of migration emerged primarily from the morphological studies of normal or mutant cerebella deficient in cell migration (Ramon y Cajal, 1911; Rakic, 1971; Rakic and Sidman, 1973; Sidman and Rakic, 1973; Sotelo and Changeux, 1974a and b; Altman 1972a, b and c; Palay and Chan-Palay, 1974). As shown in Fig. 1, in the cerebellum, granule cells proliferate in external zone in the proximity of pia mater, the external germinative layer (egl). Subsequently cells are ready to migrate in the premigratory zone situated at the transition of molecular layer (ml) and egl. Then cells cross the ml and the Purkinje cell layer (Pcl), finally reaching internal granular layer (igl). Such a migration occurs along pre-existing radial glial processes (Bergmann fibers). The cultures of cerebellar explants provided meaningful tools to understand molecular basis of contact guidance (Hemmendinger and Caviness, 1988; Moonen et al., 1982; Lindner et al., 1983 and 1986). It has been observed that protease inhibitors (Moonen et al., 1982; Lindner et al., 1983 and 1986) or Fab fragments derived from antibodies specific of some glycoprotein components of the family of Cell Adhesion Molecules (namely L1/Ng-CAM molecules (Hemmendinger and Caviness, 1988; Lindner et al., 1983 and 1986) inhibited neuron migration in cultured cerebellar explants. But, Fab fragments obtained from the antibodies raised against CSL, are much powerful inhibitors (20–200 folds) of neuron migration in cerebellar explants than the others (Lehmann et al., 1990). A similar powerful inhibition was obtained using antibodies against the 31 kDa neuronal glycoprotein (Lehmann et al., 1990), a ligand of CSL which is concentrated at the surface of migrating neurons and localized as CSL (Kuchler et al., 1989d) in the areas of contact between migrating neuroblasts and astrocyte processes (Kuchler et al., 1989a). The proposed mechanism (Lehmann et al., 1990) is that CSL is synthesized and externalized by astrocytes. The polyvalent lectin allows formation of bridges between glycans of glycoproteins of the neuroblast (Nbl) surface (Kuchler at al;, 1989a) and similar glycans shared by different glycoproteins of the astrocyte membrane (Kuchler et al., 1989c). Thus, CSL has a role of adhesion molecule ensuring the tight contact between the migrating cell and its guide. The property of 31 kDa glycoprotein of being anchored in the membrane through a phosphatidylinositol glycan anchor could be of great interest. It could allow a free relative mobility of the surfaces in the plane of the membrane, whereas binding of the glycan through CSL to the astrocyte surface would ensure a strong and specific transversal interaction. In this system, proteases (Moonen et al., 1982; Lindner et al., 1983 and 1986) and predictably phospholipases, glycanases, endo-N-acetyl-glucosaminidases or mannosidases may inhibit this migration as efficiently as the antibodies to CSL and 31 kDa glycoprotein.

Endogenous lectin CSL and myelination

Myelin is a specialized compact wrapping around axons made of the processes of the oligodendrocytes in the CNS and myelinating Schwann cells in the PNS. The studies of the mechanisms involved in myelination and myelin compaction constitute a broad field of neurobiology since myelin plays an important role in conductance of nerve impulse both in central and peripheral nervous system. The dramatic consequences of demyelination is illustrated by multiple sclerosis. Several hypothesis have been presented for explaining myelin compaction. From the studies of *mld* dysmyelinating mutant (Doolittle and Schweikart, 1977), it has been proposed (Matthieu, 1982) that myelin basic protein (MBP) is involved in the mechanism of myelin compaction at the cytoplasmic surface of the oligodendrocyte membrane. The adhesion could be due to interaction of the positively charged MBP with negatively charged sulfo-galactocerebrosides. But it could be due to an interaction of MBP with other galactolipids since a ß-galactoside-binding activity associated with it has been reported (Ikeda and Yamamoto, 1985). Similarly, the proteolipid protein (PLP) has been suggested to be involved in myelin compaction at the level of extracellular face of the oligodendrocyte (Dautigny et al., 1986; Duncan et al., 1987). This assumption was based on the observation that myelin of the jimpy mutant, having a specific anomaly of PLP protein (Nave et al., 1986), is not compacted. Furthermore, PLP molecules aggregate very easily. But, so far as the interactions between the extracellular surface of myelinating cells are concerned, several hypothesis assumed that myelin glycoproteins were involved in myelin compaction and particularly the MAG (Matthieu, 1981). But this hypothesis was not sustained because MAG is not a constituant of compact myelin. In contrast it could participate at junctions between axons and myelinating cells. In the PNS tissue, a major glycoprotein constituent of myelin is glycoprotein P0 (Kitamura et al., 1976; 1981; Lemke and Axel, 1985; Sakamoto et al., 1987). P0 is a 29 kDa glycoprotein of the family of CAMs (Salzer and Colman, 1989) found in compact myelin (Trapp et al,. 1981) and possessing cell adhesion properties. The gene transfection experiments for this protein carried out in CHO cells shows that these cells increase their adhesion properties (Filbin et al., 1990). P0 glycoprotein has been involved in myelin compaction through two different kinds of interactions: hydrophobic interactions between external surfaces of Schwann cells (Lemke et al.,1985); yet to be documented a mechanism involving the glycan (Lemke et al., 1985); yet to be documented a mechanism involving the adhesion specificity of CAMs (Salzer and Colman, 1989) and glycan lectin interactions (Kuchler et al., 1989b; Zanetta et al., 1990a; Badache et al., 1991). A number of molecules in the family of the Cell Adhesion Molecules have been found in myelin of the central or peripheral nervous system (Martini and Schachner, 1986; Poltorak et al.,1987). Most of the myelin glycoproteins so far identified (MAG and P0) share a common L2/HNK-1 epitope (McGarry et al., 1983; Inuzuka et al., 1984; O'Shannessy et al., 1985), which were also detected on a glycolipid containing sulfated glucuronic acid residue (Inuzuka et al., 1984).

Presence of CSL in myelin.

In the CNS, CSL is present in oligodendrocytes and in myelin of the central nervous tissue *in vivo* or in cultures (Kuchler et al., 1987 and 1988) as observed by immunocytochemical techniques at the level of light and electron microscopy. In cultures of oligodendrocytes (Espinosa de Los Monteros et al., 1986; McCarthy and Devellis, 1980), CSL lectin is present intracellularly, on the external face of the oligodendrocyte membrane, in the areas of contact between oligodendrocytes and in compact myelin formed by the cultured oligodendrocytes (Kuchler et al., 1988). CSL is directly involved in myelin compaction since CSL glycoprotein ligands are present at the oligodendrocyte surface and since very small amounts ($4 \mu g/ml$) of anti-CSL Fab fragments are able to dissociate compact myelin (Kuchler et al., 1988). The role of CSL in myelin compaction was further documented by studying dysmyelinating mutants (Kuchler et al., 1990 a and b): CSL displays a non homogeneous distribution and is present only in the areas where myelin is normally compacted and is totally absent from the areas where myelin is fully disorganized (Kuchler et al., 1990a and b).

In the PNS, the immunocytochemical detection at the ultrastructural level reveals the presence of CSL in compact myelin of the adult rat sciatic nerve (Kuchler et al., 1989b). The lectin is very concentrated intracellularly in myelinating Schwann cells and can be detected in low amounts in compact myelin. A mannose-binding protein constituted of two polypeptide chains of Mr 31.5 kDa and 33 kDa has been detected recently in adult pig sciatic nerve (Gabius et al., 1988). It corresponds probably to CSL. CSL immunoreactivity is detected very early in rat sciatic nerve (postnatal day 1). It is present in oblong cells which become more and more elongated then produce processes surrounding axons at the period of myelination (Kuchler et al., 1989b). Recent studies on Schwann cell cultures demonstrated that CSL is present in Schwann cells in cultures, although its levels seem to be less than what found in oligodendrocytes.

Ligands of CSL in myelinating cells.

Two major ConA-binding ligands of CSL were found in cultured oligodendrocytes. One with a Mr in the range of 100 kDa is probably MAG whereas the second with a Mr of 16 kDa, remains unknown (Kuchler et al., 1988). These ligands are in part at the surface (Fressinaud et al., 1988) since oligodendrocytes adhere rapidly on CSL coat with an immediate and significant increase in proliferation.

Studies performed on purified myelin fractions from normal animals indicate that the population of CSL glycoprotein ligands found in myelin preparations was much more complex than in oligodendrocyte cultures (Kuchler et al., 1990 a and b). A doublet glycoprotein band in the region of 100 kDa was seen along with several glycoprotein bands with Mr of 50, 31, 29, 24, 19 and 16 kDa (Fig.1e). Glycoproteins with similar Mr have been previously detected in purified myelin fractions (Linington et al., 1984; Matthieu et al., 1974; Neskovic et al., 1986; Poduslo, 1981; Poduslo et al., 1977 and 1980; Quarles et al., 1979; Schluesener et al., 1987; Zanetta et al., 1977a and 1977b). The 16 kDa protein corresponds to that found in oligodendrocyte cultures. In *mld*, quaking and jimpy dysmyelinating mutants, the bands at 24 and 19 kDa were absent (Kuchler et al., 1990 a and b). In contrast, the 16 kDa constituent was unchanged, suggesting that it does not constitute a structural component of compact myelin. The 31 kDa glycoprotein corresponds to the 31 kDa axonal glycoprotein ligand of CSL (Kuchler et al., 1989a and b).

The study of CSL glycoprotein ligands during sciatic nerve development (Kuchler et al., 1989b) indicates considerable quantitative variations. Four major components are present at day 1 with Mr of 100, 50, 31 and 16 kDa. The 16 kDa glycoprotein decreases slowly between day 1 and 15, disappearing almost completely in the adult. It corresponds to the same molecule which is found in oligodendrocyte cultures and is considerably increased in myelin of *mld*, jimpy and quaking mutants. The 31 kDa decreases progressively until adult stage where it is present in very small amount. It corresponds to the 31 kDa axonal glycoprotein previously identified in CNS axons (Kuchler et al., 1989a and b). The 50 kDa component decreases progressively with age and does not seem to be specific for the PNS since it corresponds to a similar component in the CNS. The molecule of high Mr decreases considerably with maturation, but is still present in low amounts in the adult sciatic nerve: it is probably MAG. Major modifications were observed for a protein having Mr at 29 kDa which has all the characteristics of P0 glycoprotein. The 31 kDa glycoprotein, is definitely different from the P0, and this corresponds to the 31 kDa axonal glycoprotein (Kuchler et al., 1989a and b).

All the components binding to CSL are also ConA-binding. However, the quantity of CSL binding material is not proportionnal to that of having ConA-binding properties. In other word, it is evident that, for one given molecular glycoprotein entity, not all the glycans are CSL binding. In contrast, it is probable that changes in proportion of polypeptide chains binding to CSL occur during development, differently from that changes of ConA-binding polypeptide chains.

Adhesive and signalling properties of CSL during myelination

So far as myelin compaction is concerned, bridging of the external surface of the myelinating cell membrane by interactions of the polyvalent lectin CSL with glycans of surface glycoproteins seems a possibility. The basis for this is the localization of CSL and the effect of anti-CSL Fab fragments. The ligands involved in this compaction *in vitro*, could be only MAG and the 16 kDa glycoprotein in oligodendrocyte cultures, other low Mr glycoproteins in rat CNS myelin and P0 in PNS myelin. The localization of both CSL and 31 kDa axonal glycoprotein in the areas of contact provide support for the hypothesis that they are involved in the initial contact between axon and myelinating cells during initial period of myelination and the maintenance of this contact. The actual glycoproteins of the myelinating cell membrane are still not identified and could correspond to MAG, the 50 kDa or the 16 kDa glycoprotein both in CNS and PNS. The observation that CSL is absent from non myelinating Schwann cells whereas the axons are still immunoreactive for 31 kDa is perfectly in agreement with the postulated role of CSL. Under certain circonstances CSL could serve for inducing signals for proliferation of myelinating cells or for differentiation of neurons. It can be assumed that initial contact between neuroblast and oligodendrocyte surface *in vivo* could be the signal for the symbiotic development of neurons and oligodendrocytes. This could be the salient property of CSL as a Cell Adhesion Recognition Lectin (CARL).

CONCLUSIONS.

The binding of CSL (and at a lesser extent of R1) to its endogenous ligands occurs through special types of glycans. Particularly, HNK-1 or Elec-39 reactive glycan epitopes seem to play important roles in this phenomenon. Thus, it is tempting to hypothesize that CSL is actually a Cell Adhesion Lectin having specific binding for carbohydrate epitopes, implicated in cell adhesion. In conclusion, we may say that CSL (and at a lesser extent R1) is possibly one of the "missing link" in the mechanism of cell adhesion. Furthermore, it has been demonstrated that CSL shows adhesive properties in hepatocytes (Zanetta et al., 1990c), kidney, intestine etc. (unpublished results). In normal cells, the quantity and number of CSL glycoprotein ligands is very small. In contrast, in transformed cells, the quantity and number of CSL glycoprotein ligands is increased about fifty folds (Zanetta et al., 1990b). This increase in CSL ligands correlates with the observed increase of level of onco-foetal antigens detected by antibodies of the HNK-1 family in transformed cells (Abo and Balch, 1981). These observations tempted us to hypothesize that cell transformation results from the generation of polysemic signal in cells which are maintained in contact through CSL and its numerous membrane-bound glycoprotein ligands. The resultant of which is a non-sense signal for cell i.e. transformation (Zanetta et al., 1990b).

The authors thank Prof A. N. Malviya for reviewing the manuscript. These works were supported by grants of the "Association Française contre les Myopathies, AFM" and of the "National Multiple sclerosis Society".

REFERENCES

ABO T, BALCH C M (1981) A differentiation antigen of human NK and K cells identified by a monoclonal antibody (HNK-1). J Immunol 127:1024-1029

ALTMAN J (1972a) Postnatal development of the cerebellar cortex in the rat. I The external germinal layer and the transitional molecular layer. J Comp Neurol 145:353-398

ALTMAN J (1972b) Postnatal development of the cerebellar cortex in the rat. II Phases in the maturation of Purkinje cells and of the molecular layer. J Comp Neurol 145:399-464

ALTMAN J (1972c) Postnatal development of the cerebellar cortex in the rat. III Maturation of the components of the granular layer. J Comp Neurol 145:465-514

BADACHE A, BURGER D, KUCHLER S, VINCENDON G, ZANETTA J-P, STECK A (1991) Heterogeneity of the carbohydrate moieties of myelin glycoproteins. J Neurochem submitted

BARDOSI A, DIMITRI T, GABIUS H-J (1988) Endogenous carbohydrate binding proteins in oligodendrogliomas. A histochemical study. Acta Neuropathol 446: 1-7

BOLLENSEN E, SCHACHNER M (1987) The peripheral myelin glycoprotein P0 expresses the L2/HNK-1 and L3 carbohydrate structures shared by neural adhesion molecules. Neurosci Lett 82:77-82

BON S, MEFLAH K, MUSSET F, GRASSI J, MASSOULIE J (1987) An immunoglobulin M monoclonal antibody, recognizing a subset of acetyl-cholinesterase molecules from electric organs of *Electrophorus electricus* and *Torpedo*, belongs to the HNK-1 anticarbohydrate family. J Neurochem 49:1720-1731

CHEN S, HILLMAN D E (1982) Plasticity of the parallel fiber-Purkinje cell synapse by spine takeover and new synapse formation in the adult rat. Brain Res 240:205-220

CROSSIN KL, HOFFMAN S, GRUMET M, THIERY J-P, EDELMAN GM (1986) Site restricted expression of cytotactin during development of the chicken embryo. J Cell Biol 102:1917-1930

DAUTIGNY A, MATTEI M-G, MORELLO D, ALLIEL PM, PHAM-DINH D, AMAR L, ARNAUD D, SIMON D, MATTEI J-F, GUENET J-L, JOLLES P, AVNER P (1986) The structural gene coding for myelin-associated proteolipid protein is mutated in jimpy mice. Nature 321:867-869

DOOLITTLE DP, SCHWEIKART KM (1977) Myelin deficient, a new neurological mutant in the mouse. J Hered 68:331-332

DECOSTER M A, DEVRIES GH (1989) Evidence that the axolemmal mitogen for cultured Schwann cells is a positively charged, heparan sulfate proteoglycan-bound, heparin displaceable molecule. J Neurosci Res 22:283-289

DONTENWILL M, DEVILLIERS G, LANGLEY O K, ROUSSEL G, HUBERT P, REEBER A, VINCENDON G, ZANETTA J-P (1983) Arguments in favour of endocytosis of glycoprotein components of the membranes of parallel fibers by Purkinje cells during the development of the rat cerebellum. Dev Brain Res 10:287-299.

DONTENWILL M, ROUSSEL G, ZANETTA J-P (1985) Immunohistochemical localization of a lectin like molecule "R1" during the postnatal development of the rat cerebellum. Dev Brain Res 17:245-252

DUNCAN ID, HAAMMANG JP, TRAPP BD (1987) Abnormal compact myelin in the myelin-deficient rat: absence of proteolipid protein correlates with a defect in the intraperiod line. Proc Natl Acad Sci USA 84:6287-6291

ECKENHOFF MF, PYSH JJ (1979) Double walled coated vesicle formation: evidence for massive and transient conjugate internalization of plasma membranes during cerebellar development. J Neurocytol 8:623-638

EDELMAN GM (1985) Cell adhesion and the molecular processes of morphogenesis. Ann Rev Biochem 54:135-169

EDELMAN GM (1986) Cell adhesion molecules in the regulation of animal form and tissue pattern. Ann Rev Cell Biol 2:81-116

ESPINOSA DE LOS MONTEROS A, ROUSSEL G, NUSSBAUM J-L (1986). A procedure for long-term culture of oligodendrocytes. Dev Brain Res 24:117-125

FILBIN M T, WALSH F S, TRAPP B D, PIZZEY J A, TENNEKOON X (1990) Role of myelin P0 protein as a homophilic adhesion molecule. Nature 344:871-872.

FRESSINAUD C, KUCHLER S, SARLIEVE LL, VINCENDON G, ZANETTA J-P (1988) Adhesion on a matrix made of the nervous lectin CSL induces increased proliferation of oligodendrocytes. C R Acad Sci Paris 307:863-868

GABIUS H-J, KOHNKE B, HELLMANN T, DIMITRI T, BARDOSI A (1988) Comparative histochemical and biochemical analysis of endogenous receptors for glycoproteins in human and pig peripheral nerve. J Neurochem 51:756-763

GRUMET M, HOFFMAN S, CROSSIN KL, EDELMAN GM (1985) Cytotactin, an extracellular matrix protein of neural and non neural tissues that mediates glia neuron interaction. Proc Natl Acad Sci USA 82:8075-8079

HEMMENDINGER L M, CAVINESS Jr V S (1988) Cellular migration in developing cerebral wall explants in vitro. Dev Brain Res 38:290-295

HENDERSON CE, HUCHET M, CHANGEUX J-P (1983) Denervation increases a neurite-promoting activity in extracts of skeletal muscle. Nature 302:609-611

HOFFMAN S, EDELMAN GM (1987) A proteoglycan with HNK-1 antigenic determinants is a neuron-associated ligand for cytotactin. Proc Natl Acad Sci 84:2523-2527

HOFFMAN S, FRIEDLANDER DR, CHUONG C-M, GRUMET M, EDELMAN GM (1986) Differential contributions of Ng-CAM and N-CAM to cell adhesion in different neural regions. J Cell Biol 103:145-158

HOLDER N, CLARKE DW (1988) Is there a correlation between continuous neurogenesis and directed axon regeneration in the vertebrate nervous system? TINS 11:94-99

IKEDA K, YAMAMOTO T (1985) Myelin basic protein has lectin like properties. Brain Res 329:105-108

INUZUKA T, QUARLES RH, NORONHA AB, DOBERSEN MJ, BRADY RO (1984) A human lymphocyte antigen is shared with a group of glycoproteins in peripheral nerve. Neurosci Lett 51:105-111

KITAMURA K, SUZUKI M, UYEMURA K (1976) Purification and partial characterization of the two glycoproteins in bovine peripheral nerve myelin membrane. Biochim Biophys Acta 455:806-816

KITAMURA K, SAKAMOTO Y, SUZUKI M, UYEMURA K (1981) In: Yamatawa T, Osawa T, Handa S (eds) Glycoconjugates. Japan Scientific Societies Press, Tokyo, p 273-274

KUCHLER S, VINCENDON G, ZANETTA J-P (1987) Immunocytochemical localization of an endogenous cerebellar lectin during development of rat cerebellum. C R Acad Sci Paris, Série III, 305:317-320

KUCHLER S, FRESSINAUD C, SARLIEVE LL, VINCENDON G, ZANETTA J-P (1988) Cerebellar soluble lectin is responsible for cell adhesion and participates in myelin compaction in cultured rat oligodendrocytes. Dev Neurosci 10:199-212

KUCHLER S, ROUGON G, MARSCHAL P, LEHMANN S, REEBER A, VINCENDON G, ZANETTA J-P (1989a) Localization of a transiently expressed glycoprotein in developing cerebellum delineating its possible ontogenetic roles. Neuroscience 33:111-124.

KUCHLER S, HERBEIN G, SARLIEVE LL, VINCENDON G, ZANETTA J-P (1989b) An endogenous lectin CSL interacts with major glycoprotein components in peripheral nervous system myelin. Cell Molec Biol 35:581-596

KUCHLER S, PERRAUD F, SENSENBRENNER M, VINCENDON G, ZANETTA J-P (1989c) An endogenous lectin found in rat astrocyte culture has a role in cell adhesion but not in cell proliferation. Glia 2:437-445

KUCHLER S, ZANETTA J-P, ZAEPFEL M, BADACHE A, SARKIEVE L L,GUMPEL M, BAUMANN N, VINCENDON G (1990a) Endogenous lectin CSL and its ligands in central nervous system of myelin of Quaking and Jimpy mutant mice. Dev Neurosci 12:382-397

KUCHLER S, ZANETTA J-P, ZAEPFEL M, BADACHE A, SARKIEVE L L, VINCENDON G, MATTHIEU J-M (1990b) Endogenous lectin CSL and its ligands in central nervous system of myelin deficient (mld) mutant mice. J Neurochem 56:436-445

KUCHLER S, ZANETTA J-P, BON S, ZAEPFEL M, MASSOULIE J, VINCENDON G (1991) Expression and localization in the developing cerebellum of the carbohydrate epitopes revealed by Elec-39, an IgM monoclonal antibody related to HNK-1. Neuroscience in press

LEHMANN S, KUCHLER S, THEVENIAU M, VINCENDON G, ZANETTA J-P (1990) An endogenous lectin and one of its neuronal glycoprotein ligands are involved in contact guidance of neuron migration. Proc Natl Acad Sci USA 87:6455-6459

LEHMANN S, KUCHLER S, GOBAILLE S, MARSCHAL P, BADACHE A, REEBER A, VINCENDON G, ZANETTA J-P (1991) Lesion-induced reexpression of recognition molecules in adult rat cerebellum. Proc Natl Acad Sci USA submitted

LEMKE G, AXEL R (1985) Isolation and sequence of a cDNA encoding the major structural protein of peripheral myelin. Cell 40:501-508

LINDNER J, RATHJEN F G, SCHACHNER M (1983) L1 mono- and polyclonal antibodies modify cell migration in early postnatal mouse cerebellum. Nature 305:427-430

LINDNER J, GUENTHER J, NICK H, ZINSER G, ANTONICEK H, SCHACHNER M, MONARD D (1986) Modulation of cell migration by a glia derived protein. Proc. Natl. Acad. Sci. USA 83:4568-4571

LININGTON C, WEBB M et WOODHAMS PL (1984) A novel myelin associated glycoprotein defined by a mouse monoclonal antibody. J Neuroimmunol 6:387-396

MAILLY P, BEN YOUNES-CHENNOUFI A, BON S (1989) The monoclonal antibodies Elec-39, HNK-1 and NC-1 recognize common structures in the nervous system and muscles of vertebrates. Neurochem Int 15:517-530

MARSCHAL P, REEBER A, NEESER J-R, VINCENDON G, ZANETTA J-P (1989) Carbohydrate and glycoprotein specificity of two endogenous cerebellar lectins. Biochimie 71:645-653

MARTINI R, SCHACHNER M (1986) Immunoelectron microscopic localization of neural cell adhesion molecules (L1, N-CAM, and MAG) and their shared carbohydrate epitope and myelin basic protein in developing sciatic nerve. J Cell Biol 103:2439-2448

MATTHIEU J-M (1981) Glycoproteins associated with myelin in the central nervous system. Neurochem Int 3:355-363

MATTHIEU J-M (1982) Myelin basic protein and the stability of the multilamellar myelin structure. Bull Schweiz Acad Med Wiss 101:108

MATTHIEU J-M, DANIEL A, QUARLES R H, BRADY R O (1974) Interaction of Concanavalin A and other lectins with CNS myelin. Brain Res 81:348-353

MC CARTHY KD, DE VELLIS J (1980) Preparation of separate astroglial and oligodendroglial cell cultures from rat cerebral tissue. J Cell Biol 85:890-902

MC GARRY RC, HELFAND SL, QUARLES RH, RODER JC (1983) Recognition of myelin associated glycoprotein by the monoclonal antibody HNK-1. Nature 306:376-378

MUSSET F, FROBERT Y, GRASSI J, VIGNY M, BOULLA G, BON S, MASSOULIE J (1987) Monoclonal antibodies against acetylcholinesterase from electric organs of *Electrophorus* and *Torpedo*. Biochimie 69:147-156

MOONEN G, GRAU-WAGEMANS M P, SELAK I (1982) Plasminogen activator-plasmin system and neuronal migration. Nature 298:753-755

NAVE K-A, LAI C, BLOOM FE, MILNER RJ (1986) Jimpy mutant mouse : a 74-base deletion in the mRNA for myelin proteolipid protein and evidence for a primary defect in RNA splicing. Proc Natl Acad Sci USA 83:9264-9268

NESKOVIC NM, ROUSSEL G, NUSSBAUM J-L (1986) UDP-Galactose: ceramide galactosyltranferase of rat brain: a new method of purification and production of specific antibodies. J Neurochem 47:1412-1418

NICHOLSON J L, ALTMAN J (1972a) The effect of early hypothyroidism on the development of rat cerebellar cortex. I Cell proliferation and differentiation. Brain Res 44:13-24

NICHOLSON J L, ALTMAN J (1972b) The effect of early hypothyroidism on the development of rat cerebellar cortex. II Synaptogenesis in the molecular layer. Brain Res 44:25-32

O'SHANNESSY DJ, WILLISON HJ, INUZUKA T, QUARLES RH (1985) The species distribution of nervous system antigens that react with anti myelin associated glycoprotein antibodies. J Neuroimmunol 9:255-268

PALACIOS-PRU E L, PALACIOS L, MENDOZA R V (1981) Synaptogenic mechanism during chicken cerebellar cortex development. J Submicrosc Cytol 13:145-167

PALAY S L, CHAN-PALAY V (1974) in "Cerebellar cortex" PALAY SL and CHAN-PALAY V Eds, Springer Verlag, Berlin, Heidelberg

PODUSLO JF (1981) Developmental regulation of the carbohydrate composition of glycoproteins associated with central nervous system myelin. J Neurochem 36:1924-1931

PODUSLO JF, EVERLY JL, QUARLES RH (1977) A low molecular weight glycoprotein associated with isolated myelin: distinction from myelin proteolipid protein. J Neurochem 28:977-986

PODUSLO JF, HARMAN JL, MC FARLIN DE (1980) Lectin receptors in central nervous system myelin. J Neurochem 34:1733-1744

POLTORAK M, SADOUL R, KEILHAUER G, LANDA C, FAHRIG T, SCHACHNER M (1987) Myelin-associated glycoprotein, a member of the L2 / HNK-1 family of neural cell adhesion molecules, is involved in neuron- oligodendrocyte and oligodendrocyte-oligodendrocyte interaction. J Cell Biol 105:1893-1899

QUARLES RH, MCINTYRE LJ, PASNAK CF (1979) Lectin binding proteins in central nervous system myelin. Biochem J 183:213-221

RAKIC P (1971) Neuron glia relationship during granule cell migration in developing cerebellar cortex. A golgi and electron microscopy study. J Comp. Neurol. 141:283-312

RAKIC P, SIDMAN R L (1973) Sequence of developmental abnormalities leading to granule cell deficit in cerebellar cortex of weaver mutant mice. J Comp Neurol 152:103-132

RAMON y CAJAL S (1911) Histologie du système nerveux de l'homme et des vertébrés. Vol. I and II, Maloine, Paris .

REEBER A, VINCENDON G, ZANETTA J-P (1981) Isolation and immunohistochemical localization of a Purkinje cell specific glycoprotein subunit from rat cerebellum. Brain Res 229:53-65

RIEGER F, DANILOFF JK, PINCON-RAYMOND M, CROSSIN KL, GRUMET M, EDELMAN M (1986) Neuronal cell adhesion molecules and cytotactin are colocalized at the node of Ranvier. J Cell Biol 103:379-391

SAKAMOTO Y, KITAMURA K, YOSHIMURA K, NISHIJIMA T, UYEMURA K (1987) Complete amino acid sequence of PO protein in bovine peripheral nerve myelin. J Biol Chem 262:4208-4214

SALZER J L, COLMAN D R (1989) Mechanisms of cell adhesion in the nervous system. Role of the immunoglobulin gene superfamily. Dev Neurosci 11:377-390

SCHLUESENER HJ, SOBEL RA, LININGTON C, WEINER HL (1987) A monoclonal antibody against a myelin oligodendrocyte glycoprotein induces relapses and demyelination in central nervous system autoimmune disease. J Neuroimmunol 12:4016-4021

SIDMAN R L, RAKIC P (1973) Neuron migration with special reference to developing human brain: a review. Brain Res 62:1-35

SOTELO C, CHANGEUX J-P (1974a) Bergmann fibers and granular cell migration in the cerebellum of homozygous weaver mutant mouse. J Cell Biol 53:271-289

SOTELO C, CHANGEUX J-P (1974b) Trans-synaptic degeneration en cascade in the cerebellar cortex of staggerer mutant mouse. Brain Res 77:484-491

TRAPP BD, QUARLES RH, SUZUKI K (1984b) Immunocytochemical studies of quaking mice support a role for the myelin-associated glycoprotein in forming and maintaining the periaxonal space and periaxonal cytoplasmic collar of myelinating Schwann cells. J Cell Biol 99:594-606

ZANETTA J-P, SARLIEVE LL, MANDEL P, VINCENDON G, GOMBOS G (1977a) Fractionation of glycoproteins associated to adult rat brain myelin fractions. J Neurochem 29:827-838

ZANETTA J-P, SARLIEVE LL, REEBER A, VINCENDON G, GOMBOS G (1977b) A protein fraction enriched in all myelin associated glycoproteins from adult rat central nervous system. J Neurochem 29:355-357

ZANETTA J-P, ROUSSEL G, GHANDOUR MS, VINCENDON G, GOMBOS G (1978) Postnatal development of rat cerebellum : massive and transient accumulation of Concanavalin A binding glycoproteins in parallel fiber axolemma. Brain Res 142:301-319

ZANETTA J-P, DONTENWILL M, MEYER A, ROUSSEL G (1985a) Isolation and immunohistochemical localization of a lectin like molecule from the rat cerebellum. Dev Brain Res 17:233-243

ZANETTA J-P, DONTENWILL M, REEBER A, VINCENDON G, LEGRAND CH, CLOS J, LEGRAND J (1985b) ConA-binding glycoproteins in the developing cerebellum of control hypothyroid rats. Develop Brain Res 21:1-6

ZANETTA J-P, MEYER A, KUCHLER S, VINCENDON G (1987a) Isolation and immunochemical study of a soluble cerebellar lectin delineating its structure and function. J Neurochem 49:1250-1257

ZANETTA J-P, BINGEN A, DONTENWILL-KIEFFER M, REEBER A, MEYER A, VINCENDON G (1987b) Cerebellar lectin R1 is related to the receptor of circulating mannosyl glycoproteins of liver sinusoidal cells. Cell Molec Biol 33:423-434

ZANETTA J-P, DONTENWILL M, REEBER A, VINCENDON G (1987c) Expression of recognition molecules in the cerebellum of young and adult rats. NATO ASI Serie H 2:92-104

ZANETTA J-P, KUCHLER S, MARSCHAL P, ZAEPFEL M, MEYER A, BADACHE A, REEBER A, LEHMANN S, VINCENDON G (1990) Role of an endogenous mannosyl lectin in myelination and stabilization of myelin structure. NATO ASI Series H 43:433-450

ZANETTA J-P, STAEDEL C, KUCHLER S, ZAEPFEL M, MEYER A, VINCENDON G (1991) Malignant transformation in hepatocytes is associated xwith the general increase of glycoprotein ligands specifically binding to the endogenous lectin CSL. Carbohyd Res 211 in press

Subject index

MIX
Papier aus verantwortungsvollen Quellen
Paper from responsible sources
FSC® C105338

If you have any concerns about our products,
you can contact us on
ProductSafety@springernature.com

In case Publisher is established outside the EU,
the EU authorized representative is:
Springer Nature Customer Service Center GmbH
Europaplatz 3, 69115 Heidelberg, Germany

Printed by Libri Plureos GmbH
in Hamburg, Germany